SETS AND STABILITY DOMAINS

The book establishes a unified stability theory for general time-varying nonlinear systems, based on the consistent stability methodology. This theoretical work broadens and generalizes stability theory beyond the existing Lyapunov stability framework, permitting the dependence of stability properties on the initial instant of the system. The book focuses on the stability domains of positive invariant sets, which can include equilibrium states, limit cycles, or any positive invariant set of the system.

1. Contributes new concepts, definitions of sets and system features
2. Aims to open new directions for research in the stability theory as an essential introduction to the tracking theory and enables new various engineering applications.
3. Includes time-varying and nonlinear nature of the systems which ensures the generality of the concepts, definitions, conditions and criteria

This book is aimed towards professors, researchers, engineers and advanced students in mathematics, physics, dynamical systems science, control science and engineering.

SETS AND STABILITY DOMAINS

Time-Varying Nonlinear Systems

Second Edition

Lyubomir T. Gruyitch

CRC Press
Taylor & Francis Group
Boca Raton London New York

CRC Press is an imprint of the
Taylor & Francis Group, an **informa** business

Designed cover image: Shutterstock

Second edition published 2026
by CRC Press
2385 NW Executive Center Drive, Suite 320, Boca Raton FL 33431

and by CRC Press
4 Park Square, Milton Park, Abingdon, Oxon, OX14 4RN

CRC Press is an imprint of Taylor & Francis Group, LLC

First edition published by CRC Press 2004

ISBN: 978-1-041-25144-6 (hbk)
ISBN: 978-1-041-24984-9 (pbk)
ISBN: 978-1-003-74452-8 (ebk)

DOI: 10.1201/9781003744528

Typeset in CMR10 font
by KnowledgeWorks Global Ltd.

Publisher's note: This book has been prepared from camera-ready copy provided by the authors.

Contents

Part I

Dedication

Dedicated to

ANTHONY N. MICHEL

November 17, 1935 – February 1, 2020

The Matthew H. McCloskey Dean Emeritus of the College of Engineering
The Frank M. Freimann Professor Emeritus at the University of Notre Dame
Life fellow of the Institute of Electrical and Electronics Engineers
The world leader in the qualitative theory of dynamical systems
with emphasis and fundamental contributions to stability theory
and an international pioneer in the stability study of time-varying sets.
For
his great human soul,
invaluable permanent support of my scientific work,
the honorable invitation to be Visiting Professor of the
E E Department, University of Notre Dame, 1988/1989
and for opening to me the area of neural networks and
fruitful scientific cooperation on their stability.

Part II

Preface

Systems, theory, and computers

0.1 Dynamical systems

Is there a real (dynamical) physical system that is not nonlinear? Probably not. The boundedness of energy and matter sources available for system operation cause the limitation of nonlinearity. There are often geometric and kinematic limitations. The path and its length of every displacement are limited. The maximal speed and acceleration of every technical object are bounded. Some physical processes are inherently nonlinear (e.g., friction). Sophisticated control algorithms are mainly nonlinear. Linear control algorithms cannot ensure a finite reachability *time*. The study of nonlinear dynamical systems has been attracting more and more research efforts. Their study is challenging owing to their complexity and for their theoretical and practical importance.

Task 1 ***The unifying stability treatment of the various classes of nonlinear dynamical systems***

The task of the book is to present a unifying treatment of both the known and the following classes of the newly discovered time-varying nonlinear dynamical systems [232]:

- ***The Input-Output (IO) systems*** *described by an m-th order time-varying nonlinear vector differential equation in the output vector* $\mathbf{Y}$*, which can contain derivatives of the input vector, not only the input vector.*
- ***The Input-State-Output (ISO) systems*** *(known as* ***state-space systems****) described by the first order time-varying nonlinear vector differential equation in the state vector* $\mathbf{S} = \mathbf{X}$*, and described by the algebraic nonlinear time-varying output vector equation that defines the dependence of the output vector* $\mathbf{Y}$ *on the state vector* $\mathbf{S} = \mathbf{X}$ *and on the input vector* $\mathbf{I}$*; none of the equations contains derivatives of the input vector.*
- ***The Higher order Input-State-Output (HISO) systems*** *described by the first order time-varying nonlinear vector differential equation in the state vector* $\mathbf{S} = \mathbf{X}^{m-1}$*, which can contain derivatives of the input vector, not only the input vector, and described by the algebraic nonlinear time-varying output vector equation that defines the dependence of the output vector* $\mathbf{Y}$ *on the state vector* $\mathbf{S} = \mathbf{X}^{m-1}$ *and on the input vector* $\mathbf{I}$*, but not on the derivatives of the input vector.*
- ***The Input-Internal and Output (IIO) dynamical systems*** *described by the first order time-varying nonlinear vector differential equation in the state vector* $\mathbf{S} = \mathbf{X}^{m-1}$*, which can contain derivatives of the input vector, and by the* $\nu - th$ *order time-varying nonlinear vector differential output equation in the output vector* $\mathbf{Y}$*, which determines the dependence of the output vector* $\mathbf{Y}$ *on the state vector* $\mathbf{S} = \mathbf{X}^{m-1}$ *and on the input vector* $\mathbf{I}$*.*

- ***The Extended Input-State-Output (EISO) systems*** *described by the first order time-varying nonlinear vector differential equation in the vector* $\mathbf{S}$*, which expresses the influence of the extended input vector* $\mathbf{I}^{\xi}$*, i.e., of the input vector* $\mathbf{I}$ *and of its derivatives up to the order* ξ*, and described by the algebraic nonlinear time-varying output vector equation that defines the dependence of the output vector* $\mathbf{Y}$ *on the state vector* $\mathbf{S}$ *and on the input vector* $\mathbf{I}$.

Note 2 *The EISO system represents the unifying state form for all five system families. The proof is given herein. We call them simply* **the systems** *in the sequel. They represent the general nonlinear systems in the state form.*

0.2 Dynamical systems and computers

Should the research on stability of dynamical and control systems stop due to the remarkable advancement of computer science and engineering, of AI development?

Problems that need the computer's big memory, fast operations, and small computer space are now solvable by computer engineers, computer designers, and by software specialists. The computers enable the development of the artificial intelligence. The digital computer now succeeds in carrying out simulations of dynamical processes practically equally well as the analog computer.

The nonlinear systems can abruptly change the character of their dynamical behavior due to a very small, sometimes infinitesimal, variation of initial conditions and/or of an external action. The qualitative properties (e.g., controllability, observability, optimality, stability and trackability [232], [243], [244]) of some systems can be valid only for initial conditions belonging to a connected bounded infinite set (i.e., can be only local). This largest set is ***the domain*** $\mathcal{D}_{(.)}$ *of the corresponding qualitative dynamical property*. It can be open, or closed, but need not be either. The problem of how to determine the shape, the size/volume, the location and the boundary of the property domain can be extremely difficult and complex. The computer simulations are very illustrative, but they cannot resolve the problem of the stability domain determination [247]. In order to cope effectively, the corresponding theory should be well developed. It serves as the basis for computer simulations, for engineering design, for its applications, for the tracking and the trackability theories [232] and for further research.

0.3 Dynamical systems and artificial intelligence

Artificial intelligence has been making new scientific, engineering and societal inroads, including political challenges and changes. Can artificial intelligence eliminate any need for scientific theories, e.g. stability theory?

Until artificial intelligenc can effectively resolve stability problems by using only stability definitions, without using analytical stability conditions and/or criteria, it remains a wonderful computational tool only. It then just emphasizes the necessity for a well-established stability theory.

Dynamical systems, stability and the book

Stability is one of the basic qualitative dynamical properties of dynamical systems. It is basic due to its *necessity* for a satisfactory dynamical behavior of the system under arbitrary actions on it. The fact that the whole history of a dynamical system until the initial moment t_0 is condensed in the system initial state $\mathbf{S}_0$ at the initial moment t_0 and that the initial state $\mathbf{S}_0$ is most often unpredictable and untouchable impose the need to assure satisfactory system behavior for any initial state $\mathbf{S}_0$ belonging to an appropriate set $\mathfrak{S}_0$. The number of different initial states in the set $\mathfrak{S}_0$ is infinite. It is impossible to simulate system behavior for every initial state $\mathbf{S}_0 \in \mathfrak{S}_0$. This fact justified the establishment and the development of the theory aimed to test the system behavior for any $\mathbf{S}_0 \in \mathfrak{S}_0$. The most developed and used in engineering applications is the stability theory founded by Alexander Mihailovich Lyapunov in his famous doctoral dissertation [335].

Lyapunov treated stability of *time*-invariant and *time* varying linear and nonlinear systems under two crucial conditions that restrict the theory.

The following condition is implicit, tacit in the Lyapunov stability theory:

Assumption 3 ***Lyapunov assumption*** *The external actions are nominal, i.e., the input vector is nominal, in terms of the total values.*

Equivalently, *the deviations of the external actions are equal to zero, i.e., the input vector deviation is zero vector all the time since the initial moment* t_0.

This is an inherent condition and characteristic of the Lyapunov stability theory. It also characterizes this book. The tracking theory [232] is free of that restrictive condition. It deals with arbitrary and unknown external actions, i.e., with an arbitrary and unknown input vector. The stability theory is unavoidable for the tracking theory.

The following condition Lyapunov stated explicitly:

Condition 4 ***Lyapunov condition*** *"We shall be exclusively interested in those cases in which the solution of the considered question does not depend on a choice of the instant* t_0*, when perturbations start acting on the system" [335, page 13].*

This condition restricts the Lyapunov stability theory to the stability phenomena and properties that are independent of the initial moment t_0. It characterizes all *time*-invariant systems. Such stability phenomena and properties are uniform in the initial moment t_0.

The Lyapunov method avoids the need for solving the system mathematical model. It does not require knowledge of the system motions. The semidefinite and definite functions together with their derivatives form its basis.

For *time*-invariant linear systems, Lyapunov established the methodology that guarantees the complete stability solution. The methodology permits an arbitrary choice of a negative definite function in the differential equation of an unknown function differentiated along system motions. Since the motions are not known, the differentiation is to be carried out by using the mathematical model of the system. In order to avoid higher derivatives of the system state, the mathematical model should be in the state-space form. For the asymptotic stability it is necessary and sufficient that the solution function of its differential equation is positive definite. This methodology is **the Lyapunov direct methodology**.

However, for the nonlinear systems, Lyapunov established **the inverse methodology**. It starts with a free choice of a positive definite function. Then the sign of its total *time* derivative along system motions determined by applying the system mathematical model is tested. If the function derivative is negative definite, then the system possesses the asymptotic stability property. *Otherwise, the whole procedure should be repeated and can be endless!* The inverse Lyapunov methodology does not enable us to solve the following crucial problems in the framework of nonlinear systems:

- *What are both necessary and sufficient stability conditions?*
- *What is the system Lyapunov function or how to find, to determine, to construct, the system Lyapunov function?*
- *How to determine the exact asymptotic stability domain of the system?*

A new methodology in the Lyapunov sense, established in [134], [175], and further developed in [128], [140], [127]-[129], [139], [135], [136], [138], [142], [146], [154]-[160], [176], [190], [219]-[224], [247] as **the consistent (Lyapunov) stability methodology.** It permits us to resolve all three fundamental stability problems in the framework of the state-space nonlinear systems in the Cauchy, i.e., normal, form. It is further developed in this book to all five classes of both *the time*-invariant and the *time*-varying nonlinear systems. Their linear counterpart is presented in the framework of stability [230], observability and controllability [233], trackability and tracking [243].

On the book

0.4 Goal of the book

For the references on the Lyapunov stability theory see Appendix. This book is complementary to them, and generalizes and broadens the stability theory beyond the existing Lyapunov stability theory.

Goal 5 ***The stability domain theory for time-varying nonlinear systems.***
The goal of the book is to establish the unified stability theory for the general time-varying nonlinear systems by permitting the dependence of the stability properties on the initial instant t_0 *of the system.*

This permits us to enrich essentially the corresponding university courses, to open new directions for research in stability theory as an essential introduction to tracking theory [232], and enables new various engineering applications.

There is not in this book an available space for more examples, for simulations, or for applications. They are left open to everybody for future works. The book is theoretical.

0.5 The book structure and composition

There are in the sequel nine parts, four of which compose the main body of the book. The parts mainly contain chapters that incorporate sections. Some sections contain subsections. The systems defined by any order nonlinear vector differential equation that describes the system state behavior and by the nonlinear algebraic or differential vector equation that determines the system output dependence on the system internal dynamics and on the input action are **the input-internal dynamics-output** ($IIDO$) **systems**, for short, **the input-internal and output dynamics** (IIO) **systems**, and **the extended input-state-output** ($EISO$) **systems**. Their special classes are **the known input-state-output** (ISO) **systems**, and the **input-output** (IO) **systems**. The first-order nonlinear vector differential equation defines the internal dynamics, i.e., the state dynamics, and the nonlinear algebraic vector equation defines the output behavior of the ISO systems. The mathematical model of the IO systems is the m-th-order nonlinear vector differential equation that shows directly how the system output behavior depends on the external (control and disturbance) actions on the system. Their extensions are **the higher order input-state-output** ($HISO$) **systems** and **the input, internal and output dynamics** (IIO) systems. The book treats, in detail, the $EISO$, systems so that all the results are directly applicable to IO, ISO, $HISO$ and IIO systems. For the sake of simplicity the $EISO$ systems are simply called ***the systems*** in the sequel.

The *time*-varying and nonlinear nature of the systems ensures the generality of the concepts, definitions, conditions and criteria.

The main scientific body of the book begins with the essential topics for dynamical systems. The following parts start with new concepts that incorporate definitions of new systems and their properties. It ends with the stability criteria or conditions on the system.

The fact that the book contributes new concepts, definitions of sets and system features, the author suggests that the text be read by the reader without omitting any part, chapter, section, or subsection.

In order to describe the systems, to define their various dynamical properties, and to cope effectively with the dynamical problems caused by the systems complexity, the book uses the new simple vector and matrix notation analogous to scalar notation, [217], [232], [244]. It is the basis of the novel **compact calculus** (Subsections: **1.1.3** and **1.1.4**).

0.6 In gratitude

The author is grateful to Mr. George PEARSON with MacKichan Company for his unprecented, continuous, and very kind and effective assistance through the decades to resolve various problems related to the fast effective Scientific Work Place application caused by the author's lack of the high software competence.

The author expresses his gratitude to Ms Nora KONOPKA, Editorial director of Engineering, for effectively and leading the whole publication process.

The author is thankful to Mr. Paul BREWER for his very careful lectoring the book.

*Lyubomir T. Gruyitch**, the author .

* The correct transcription from the original written in Serb language into English reads: *Lyubomir T. Gruyitch* and into Croatian: *Ljubomir T. Grujić* that is wrongly simplified to: *Ljubomir T. Grujic*.

Part III

SYSTEMS FUNDAMENTALS

Chapter 1

Introduction

1.1 Notational preliminary

Let $\alpha, \eta, \mu, \nu, \rho, i, j, k$, K, l, L, m, M, n and N be natural numbers.

The logical symbol $\wedge$ means "*and*"; e.g., $\alpha \wedge \beta$ means "α *and* β".

1.1.1 Sets, spaces, and systems

Lower-case ordinary letters denote scalars, bold (lower-case and capital, Greek and Roman) block letters signify vectors, capital block and italic letters stand for matrices and matrix functions, and we use capital $\mathfrak{Fraktur}$ letters for families of functions, sets and spaces. System motions (solutions) or special sets are denoted by capital $\mathcal{CALLIGRAPHIC}$ letters.

Let us explain the meaning of the basic notation and symbols.

In what follows:

$\mathfrak{C}^{ki} = \mathfrak{C}^k(\mathfrak{R}^i)$ is *the family of all functions defined and k-times continuously differentiable on* $\mathfrak{R}^i$.

$\mathfrak{C}^k(\mathfrak{T}_0)$ is *the family of all functions defined, continuous, and k-times continuously differentiable on* $\mathfrak{T}_0$, $\mathfrak{C}^k = \mathfrak{C}^k(\mathfrak{T})$ and $\mathfrak{C}^0 = \mathfrak{C}$.

$\mathfrak{C}^{k-}(\mathfrak{R}^N)$ is *the family of all functions defined and continuous everywhere on* $\mathfrak{R}^N$, $(k-1)$ *times continuously differentiable on* $\mathfrak{R}^N$, *and k-times continuously differentiable on* $\mathfrak{R}^N \backslash \{\mathbf{0}_N\}$, which have defined the left and the right k-th-order derivative at the origin $\mathbf{0}_N$.

$\mathfrak{C}^{-}(\mathfrak{R}^N)$ is *the family of all functions defined everywhere on* $\mathfrak{R}^N$.

$\mathcal{J}^i$ is a given, or to be determined, family of all bounded i-times continuously differentiable on $\mathfrak{T}_0$ permitted input vector functions $\mathbf{I}(.)$,

$$\mathcal{J}^i \subset \mathfrak{C}^i(\mathfrak{T}). \tag{1.1}$$

$\mathfrak{L} \subseteq \mathfrak{T}$ is *the largest temporal set of the initial moments* t_0 *for which the system has a solution*, $\mathfrak{L} = (l_m, l_M)$.

$\mathfrak{R}$ is *the set of all real numbers*. The division of two real numbers x and y, which are different from zero, is well-defined in $\mathfrak{R}$:

$$x \in \mathfrak{R}, \quad y \in \mathfrak{R}, \quad xy \neq 0 \Longrightarrow \exists \frac{x}{y} \in \mathfrak{R} \ \ and \ \ \exists \frac{y}{x} \in \mathfrak{R}.$$

$\mathfrak{R}_+$ is the set of all nonnegative real numbers.

DOI: 10.1201/9781003744528-4

$\mathfrak{R}^+$ is the set of all positive real numbers.

$\mathfrak{R}^k$ is the k dimensional real vector space, the elements of which are $k-$dimensional real-valued vectors. Notice that $\mathfrak{R}^1 \neq \mathfrak{R}$. The division of two vectors $\mathbf{x}$ and $\mathbf{y}$ from $\mathfrak{R}^n$ was not defined in $\mathfrak{R}^n$ perhaps until 2016 [232, p. 11] despite the fact that the divisions xy^{-1} and yx^{-1} have been well-defined in $\mathfrak{R}$ for $xy \neq 0$. The division of two vectors $\mathbf{x}$ and $\mathbf{y}$ from $\mathfrak{R}^n$ is further extended in the sequel not only for the sake of pure mathematical development, but also due to its usefulness for sophisticated control algorithms [232, Part VIII, pp. 263 - 360].

$\mathfrak{R}^{M\times N}$ is $M\times N$ matrix space, the zero matrix in which is $O_{M\times N}$. If M=N then, for short $O_{N\times N}=O_N$ and I_N is the identity matrix in $\mathfrak{R}^{N\times N}$.

$\mathfrak{R}^N\backslash\{\mathbf{0}_N\}$ *is the set difference between* $\mathfrak{R}^N$ *and the singleton* $\{\mathbf{0}_N\}$, *i.e., the complement of* $\{\mathbf{0}_N\}$ *to* $\mathfrak{R}^N$.

$\mathfrak{Y}\subseteq\mathfrak{R}^N$ is *the total output vector space of the system.*

$\mathfrak{Y}_d^k$ is a given, or to be determined, family of all bounded and $(k+1)$-*times* continuously differentiable realizable desired output vector functions $\mathbf{Y}_d(.)\in\mathfrak{C}^{k+1}$ such that they and their first k+1 derivatives obey $TCUP$ (Principle 27), i.e. $\mathfrak{Y}_d^k$ is a given, or to be determined, family of all bounded $(k+1)-$ continuously differentiable realizable desired output vector functions $\mathbf{Y}_d(.)\in\mathfrak{C}^{k+1}$,

$$\mathfrak{Y}_d^k\subset\mathfrak{C}^{k+1}(\mathfrak{R}^N). \tag{1.2}$$

$\mathfrak{Y}_d^0=\mathfrak{Y}_d$ is the family of all bounded continuous, continuously differentiable realizable desired output vector functions $\mathbf{Y}_d(.)\in\mathfrak{Y}_d$ such that they and their first derivatives obey $TCUP$,

$$\mathfrak{Y}_d\subset\mathfrak{C}^1(\mathfrak{R}^N). \tag{1.3}$$

1.1.2 Systems, vectors and vector functions

$\mathbf{e}\in\mathfrak{R}^N$ is the system output error vector,

$$\mathbf{e}=\mathbf{Y_d}-\mathbf{Y}, \tag{1.4}$$

$\mathbf{I}$ is the total input vector

$$\mathbf{I}=[I_1\ I_2\ ...\ I_M]^T\in\mathfrak{R}^M, \tag{1.5}$$

where the values I_i, $i=1,2,..,M$, are the total values of the system input variables; the total value of a variable is its value relative to its total zero value,

$\mathbf{I}(.)$ *is the system input vector function.*

$\mathbf{S}\in\mathfrak{R}^n$ is *the system state vector*, for short: *state of the system,*

$$\mathbf{S}=[S_1\ S_2\ ...\ S_n]^T\in\mathfrak{R}^n, \tag{1.6}$$

where the values S_j, $j=1,2,..,n$, are the total values of the state variables.

$\mathcal{S}(.;t_0;\mathbf{S}_0)$ *is the system motion (solution) passing through (starting from) the initial state* $\mathbf{S}_0$ *at the initial moment* t_0 if the input vector does not act on the system. Its vector value at any moment t is $\mathbf{S}(t;t_0;\mathbf{S}_0)$, for short $\mathbf{S}(t)$ for a fixed initial pair $(t_0,\mathbf{S}_0)$,

$$\mathbf{S}(t;t_0;\mathbf{S}_0)=\mathbf{S}(t)\in\mathfrak{R}^n\ \textit{for a fixed pair}\ (t_0,\mathbf{S}_0), \tag{1.7}$$

$\mathbf{Y}\in\mathfrak{Y}$ is *the output vector of the system,*

$$\mathbf{Y}=[Y_1\ Y_2\ ...\ Y_N]^T\in\mathfrak{R}^N, \tag{1.8}$$

where the values Y_k, $k=1,2,..,N$, are the total values of the output variables.

1.1.3 Extended vectors and matrices

In order to define and use the system properties effectively, we need new, simple, and elegant notation proposed in [126] and advantageously applied in [55], [217], [230], [232], [233], [243].

Definition 6 ***Extended vectors***

The vector $\mathbf{I}^{\mu}$ *is* ***the extended input vector*** *composed of the input vector* $\mathbf{I}$ *itself and its first* μ *derivatives* $\mathbf{I}^{(1)}, \mathbf{I}^{(2)}, \ldots, \mathbf{I}^{(\mu)}$:

$$\mathbf{I}^{\mu} = \begin{bmatrix} \mathbf{I} \\ \mathbf{I}^{(1)} \\ \cdots \\ \mathbf{I}^{(\mu)} \end{bmatrix} = \begin{bmatrix} \mathbf{I}^{(0)} \\ \mathbf{I}^{(1)} \\ \cdots \\ \mathbf{I}^{(\mu)} \end{bmatrix} = \left[\mathbf{I}^{T} \vdots \mathbf{I}^{(1)T} \vdots \ldots \vdots \mathbf{I}^{(\mu)T}\right]^{T} \in \mathfrak{R}^{(\mu+1)N}, \quad \mathbf{I}^{0} = \mathbf{I}. \tag{1.9}$$

The superscript μ is not in the parentheses in $\mathbf{I}^{\mu}$ because $\mathbf{I}^{(\mu)}(t)$ denotes the $\mu - th$ derivative $d^{\mu}\mathbf{I}(t)/dt^{\mu}$ of $\mathbf{I}(t)$,

$$\mathbf{I}^{\mu}(t) = \left[\mathbf{I}^{T}(t) \vdots \mathbf{I}^{(1)T}(t) \vdots \ldots \vdots \mathbf{I}^{(\mu)T}(t)\right]^{T} \neq \mathbf{I}^{(\mu)}(t) = \frac{d^{\mu}\mathbf{I}(t)}{dt^{\mu}}. \tag{1.10}$$

Definition 7 ***Extended matrices***

Let $A_k \in \mathfrak{R}^{N\times N}$, $\forall k = 0, 1, 2, \ldots, \nu$. *The matrices* A_k *form* ***the extended matrix*** $A^{(\nu)}$ *defined by*

$$A^{(\nu)} = \begin{bmatrix} A_0 & A_1 & A_2 & \ldots & A_{\nu} \end{bmatrix} \in \mathfrak{R}^{N\times(\nu+1)N}, \tag{1.11}$$

which is different from the ν*-th power* A^{ν} *of the matrix* A:

$$A^{(\nu)} = \begin{bmatrix} A_0 & A_1 & A_2 & \cdots & A_{\nu} \end{bmatrix} \neq A^{\nu} = \underbrace{AA\cdots A}_{\nu-times}. \tag{1.12}$$

Note 8 *The higher the system order and/or the higher the system dimension, the more advantageous the new notation.*

It is useful to apply the symbolic matrix and vector notation and operations in the elementwise sense as follows:

Definition 9 ***Matrices induced by vectors***

The (N*-dimensional) zero vector* $\mathbf{0}_N$ *and the* (N*-dimensional) unit vector* $\mathbf{1}_N$:

$$\mathbf{0}_N = [0 \ \ 0 \ \cdots \ 0]^T \in \mathfrak{R}^N, \quad \mathbf{1}_N = [1 \ \ 1 \ \cdots \ 1]^T \in \mathfrak{R}^N. \tag{1.13}$$

The diagonal matrix V *is associated elementwise with a vector* $\mathbf{v}$:

$$\mathbf{v} = [v_1 \ \ v_2 \ \cdots \ v_N]^T \in \mathfrak{R}^N \Longrightarrow V = diag\{v_1 \ \ v_2 \cdots \ v_N\} \in \mathfrak{R}^{N\times N}. \tag{1.14}$$

The vector and matrix absolute values,

$$|\mathbf{v}| = [|v_1| \ \ |v_2| \cdots \ |v_N|]^T \in \mathfrak{R}_+^N, \quad |V| = diag\{|v_1| \ \ |v_2| \cdots \ |v_N|\} \in \mathfrak{R}_+^{N\times N}. \tag{1.15}$$

The elementwise vector equality:

$$\mathbf{w} = \mathbf{v} \Longleftrightarrow w_i = v_i, \quad \forall i = 1, 2, \cdots, N,$$

The sign function $sign(.) : \mathfrak{R} \to \{-1, 0, 1\}$ *is the scalar signum function,*

$$sign(v) = |v|^{-1} v \ if, \ and \ only \ if \ v \neq 0, \ and \ sign(0) = 0. \tag{1.16}$$

Note 10 *$\nexists \mathbf{Y}^{-1}$ signifies that $\mathbf{Y}^{-1}$ does not exist in general. However, if all entries Y_i of $\mathbf{Y}$ are nonzero real numbers, $Y_i \neq 0$, $\forall i = 1, 2, ..., N$, then we can define $\mathbf{Y}^{-1}$ as follows:*

Definition 11 ***The vector inversion***
The vector $\mathbf{Y}$,

$$\mathbf{Y} = [Y_1 \ \ Y_2 \cdots \ Y_N]^T \in \mathfrak{R}^{N \times 1} = \mathfrak{R}^N, \ Y_i \in \mathfrak{R}, \ \forall i = 1, 2, \cdots, N, \tag{1.17}$$

has its inverse vector $\mathbf{Y}^{-1}$ if and only if all its entries Y_i are nonzero numbers,

$$\exists \mathbf{Y}^{-1} = \left[Y_1^{-1} \ Y_2^{-1} \cdots \ Y_N^{-1}\right] \in \mathfrak{R}^{1\text{x}N} \iff Y_i \neq 0, \ \ \forall i = 1, 2, \cdots, N. \tag{1.18}$$

Then the product $\mathbf{Y}^{-1}\mathbf{Y}$ equals N:

$$\mathbf{Y}^{-1}\mathbf{Y} = N, \ \ i.e., \ \frac{1}{N}\mathbf{Y}^{-1}\mathbf{Y} = 1, \tag{1.19}$$

For

$$\begin{gathered} diag\mathbf{Y} = diag\left\{Y_1 \ \ Y_2 \cdots \ Y_N\right\} \in \mathfrak{R}^{N \times N} \Longrightarrow \\ \exists diag\mathbf{Y}^{-1} = diag\left\{Y_1^{-1} \ \ Y_2^{-1} \cdots \ Y_N^{-1}\right\} \in \mathfrak{R}^{N \times N} \iff Y_i \neq 0, \ \ \forall i = 1, 2, \cdots, N., \end{gathered} \tag{1.20}$$

the following relations hold:

$$\begin{gathered} diag\left(\mathbf{Y}\mathbf{Y}^{-1}\right) = diag\mathbf{Y} diag\mathbf{Y}^{-1} = \begin{bmatrix} 1 & 0 & ... & 0 \\ 0 & 1 & ... & 0 \\ ... & ... & ... & ... \\ 0 & 0 & ... & 1 \end{bmatrix} = I_N \in \mathfrak{R}^{N \times N}, \\ \left(diag\mathbf{Y}\right)\left(\mathbf{Y}^{-1}\right)^T = \left(diag\mathbf{Y}^{-1}\right)\mathbf{Y} = \mathbf{1}_N = [1 \ \ 1 \cdots \ 1]^T \in \mathfrak{R}^{N\text{x}1}, \\ \mathbf{Y}^T\left(diag\mathbf{Y}^{-1}\right) = \mathbf{Y}^{-1} \ diag\mathbf{Y} = [1 \ \ 1 \cdots \ 1] = \mathbf{1}_N^T \in \mathfrak{R}^{1\text{x}N}. \end{gathered} \tag{1.21}$$

Notice that $\mathbf{1}_N \neq I_N$.

Definition 12 ***The vector division***

The vector $\mathbf{b} \in \mathfrak{R}^{L \times 1}$,

$$\mathbf{b} = [b_1 \ \ b_2 \cdots \ b_L]^T \in \mathfrak{R}^{L \times 1}, \ b_i \in \mathfrak{R}, \ b_i \neq 0 \ \forall i = 1, 2, \cdots, L, \tag{1.22}$$

is devisible by the vector $\mathbf{c} \in \mathfrak{R}^{M \times 1}$,

$$\mathbf{c} = [c_1 \ \ c_2 \cdots \ c_M]^T \in \mathfrak{R}^{M \times 1}, \ c_i \in \mathfrak{R}, \ \forall i = 1, 2, \cdots, M, \tag{1.23}$$

if and only if:
a) $L = M$,
b) $c_i \neq 0$, $\forall i = 1, 2, ..., L = M$.
Then:

$$\left(diag\mathbf{c}^{-1}\right)\mathbf{c} = \begin{bmatrix} 1 \\ 1 \\ ... \\ 1 \end{bmatrix} = \mathbf{1}_L, \tag{1.24}$$

$$\mathbf{c}^{-1}\mathbf{b} = \sum_{i=1}^{i=L=M} b_i c_i^{-1} \in \mathfrak{R}, \implies \mathbf{b}^{-1}\mathbf{b} = L = M \in \mathfrak{R}, \tag{1.25}$$

$$\mathbf{bc}^{-1} = \left\{ \begin{array}{ccccc} b_1c_1^{-1} & b_1c_2^{-1} & \cdots & b_1c_{L-1}^{-1} & b_1c_L^{-1} \\ b_2c_1^{-1} & b_2c_2^{-1} & \cdots & b_2c_{L-1}^{-1} & b_2c_L^{-1} \\ \cdots & \cdots & \cdots & \cdots & \cdots \\ b_{L-1}c_1^{-1} & b_{L-1}c_2^{-1} & \cdots & b_{L-1}c_{L-1}^{-1} & b_{L-1}c_L^{-1} \\ b_Lc_1^{-1} & b_Lc_2^{-1} & \cdots & b_Lc_{L-1}^{-1} & b_Lc_L^{-1} \end{array} \right\} \in \mathfrak{R}^{L\times L}. \tag{1.26}$$

Other new notation is defined at its first use in the text and in Appendix A.

1.1.4 Compact calculus

Equations (1.9)-(1.11) permit us to set the vector differential equation:

$$\sum_{k=0}^{k=\nu} A_k \mathbf{Y}^{(k)}(t) = \sum_{m=0}^{m=\mu} B_m \mathbf{I}^{(m)}(t),\ A_k \in \mathfrak{R}^{N\times N},\ \forall k = 0,1...,\nu,\ B_m \in \mathfrak{R}^{N\times M}, \forall m = 0,1...,\mu, \tag{1.27}$$

into the compact form:

$$A^{(\nu)}\mathbf{Y}^{\nu}(t) = B^{(\mu)}\mathbf{I}^{\mu}(t). \tag{1.28}$$

This is the basis of the simple and elegant ***compact calculus*** that is effectively applicable not only to linear continuous-*time* systems [217], [230], [233], [243], but also to linear discrete-*time* systems [55] and to nonlinear dynamical systems *[232, Definition 49, p. 21]* as used throughout this book.

If the matrix A_ν has the full rank N then the matrix $A^{(\nu)}$ (1.11) has also the full rank N and the NxN matrix $(A^{(\nu)}A^{(\nu)T})$ is nonsingular (regular), which leads to the solution of (1.28) in the following form:

$$\mathbf{Y}^{\nu}(t;t_0;\mathbf{I}_0^{\mu-1}) = A^{(\nu)T}(A^{(\nu)}A^{(\nu)T})^{-1}B^{(\mu)}\mathbf{I}^{\mu}(t),\ \forall (t,t_0) \in \mathfrak{T}_0 \times \mathfrak{T}_i. \tag{1.29}$$

The solution $\mathbf{Y}^{\nu}(t;t_0;\mathbf{I}_0^{\mu-1})$ is the particular solution of the equation (1.28) determined at every moment $t \in \mathfrak{T}_0$ for the given initial data $(t_0;\mathbf{I}_0^{\mu-1})$. It is obtained algebraically rather than solving linear differential equation (1.28).

Note 13 *Throughout the book the matrix $A^{(\nu)}$ (1.11) is nonsingular, i.e. it has the full rank N.*

Chapter 2

Time

2.1 *Time* definition and properties

All processes, motions, and movements, all behaviors of dynamical systems and their responses, as well as all external actions on the systems, occur and propagate in *time*. All stability processes take place in *time*.

A variation of the value of every physical variable is in *time*.

It is natural from the physical point of view to study the system behavior and its dynamical properties directly in the temporal domain. Stability properties of *time*-varying systems crucially depend on *time*. These facts require us to be clear in how we understand what *time* is and what its properties are, which we explain as follows (for the complete exact analysis of the *time* phenomenon and its relativity see: [225], [228], [235]-[237]).

Definition 14 ***Time*** *[232, Definition 12, p. 3], [233, Definition 1, p. 3], [235, Definition 46, pp. 66, 67], [243, Definition 2, p. 1], [245], [242].*

***Time** (i.e., **the temporal variable**), denoted by t or by τ, is an independent scalar physical informational, massless and energyless variable such that:*

*- Its value called **instant** or **moment** determines uniquely **when** the existence of somebody or something started or was interrupted.*

*- Its values determine uniquely **since when and until when** somebody or something existed/exists or will exist.*

*- Its values determine uniquely **how long** somebody or something existed/exists or will exist.*

*- Its values uniquely determine whether an event E_1 occurs **before** another event E_2 has happened, or the event E_1 took/takes/will take place just the event E_2 was/is/will be happening, or the event E_1 occurs **after** the event E_2 has happened.*

*- Its value **occupies (covers, encloses, imbues, impregnates, is over and in, penetrates) equally** everybody and everything (i.e., beings, objects, energy, matter, and space) **everywhere and always.***

*- Its value has been, is, and will be **permanently changing smoothly, strictly monotonously continuously, equally** in all spatial directions and their senses, in and around everybody and everything, **independently** of everybody and everything (i.e., independently of beings, objects, energy, matter, and space), **independently** of all other variables, **independently** of all happenings, movements, and processes.*

Comment 15 *Continuity, permanent monotonous motion of the instants, the invariance and the independence are among the essential properties of time.*

DOI: 10.1201/9781003744528-5

Comment 16 *Time is a basic constituent of the existence of everybody and of everything. All human trials for millennia have failed to explain, to express, the nature, the phenomenon, of time in terms of other well-defined notions [235]-[237]. The nature of time, its physical content, cannot be explained in terms of other basic constituents of the existence (in terms of energy, matter, space) or in terms of other physical phenomena or variables. Time has its own, original, self-contained, nature since it is massless and energyless, which we can only call the nature of time, i.e., the temporal nature or the time nature [228], [235]-[237]. We mostly refer to time as an informational variable.*

Comment 17 ***The value of time*** *, i.e.,* ***instant*** *or* ***moment****, is denoted by t (or by τ). It is an* ***instantaneous (momentous)*** *and* ***elementary time value****. It is the* ***temporal value****. It does not contain either energy or matter or space. It can happen exactly once and then it is the same everywhere for, and in, everybody and everything (i.e., for, and in, beings, energy, matter, objects, and space), for all other variables, for all happenings, for all movements, for all processes, and for all biological, physical, and social systems. It is not repeatable. Nobody and nothing can influence the flow of instants [228], [235]-[237].*

Comment 18 *The physical dimension of time is denoted by $[T]$, where T stands for time dimension. It cannot be expressed in terms of the physical dimension of another variable. Its physical dimension is one of the basic physical dimensions. It is used to express the physical dimensions of most of the physical variables. A selected unit 1_t of time can be arbitrarily chosen and then fixed. If it is seconds, s, then $1_t = s$, which we denote by $t\langle 1_t \rangle = t\langle s \rangle$ [228], [235]-[237].*

There can be assigned *exactly one* (which is denoted by $\exists!$) real number to every moment (instant) and vice versa. The numerical value *num t* of the moment *t* is a real number, *num* $t \in \mathfrak{R}$, where $\mathfrak{R}$ is the set of all real numbers [228], [235]-[237].

2.2 *Time* set

Time set $\mathfrak{T}$ is the set of all moments. It is an open, unbounded, and connected set. It is in the biunivoque (one-to-one) correspondence with the set $\mathfrak{R}$ of all real numbers,

$$\begin{gathered}\mathfrak{T} = \{t : num\ t \in \mathfrak{R},\ dt > 0,\ \ v_t = \frac{dt}{dt} = t^{(1)} \equiv 1\left[TT^{-1}\right]\},\\ \forall t \in \mathfrak{T},\ \exists! x \in \mathfrak{R} \Longrightarrow x = num\ t,\ and\\ \forall x \in \mathfrak{R},\ \exists! t \in \mathfrak{T} \Longrightarrow x = num\ t,\\ num\ inf\ \mathfrak{T}\ = num\ t_{\text{inf}} = -\infty\ and\ num\ sup\ \mathfrak{T} = num\ t_{\text{sup}} = \infty.\end{gathered} \tag{2.1}$$

The rule of the correspondence determines an **accepted relative zero numerical *time* value** t_{zero}, a ***time* scale,** and a ***time* unit** denoted by 1_t (or by 1_τ). The *time* unit can be ... , millisecond, second, minute, hour, day, ..., as well noted by *Sir Isaac Newton* [403, p. 8: Scholium]. He clearly explained therein the sense and the meaning of *relative time* in the same sentence in which he explained the sense and meaning of *absolute time.*

Newton's *absolute time* is *the temporal variable*: *time*. Newton's *relative time* is *the value of the temporal variable*: *the value of time* measured in a chosen *time* scale T_t with the corresponding *time* unit 1_t [228], [235]-[237]. *Why has modern physics been ignoring these facts?*

Although the *time* set $\mathfrak{T}$ is in biunivoque (one-to-one) correspondence with *the set* $\mathfrak{R}$ *of the real numbers*, they are inherently different. The crucial difference between them is the dynamic property of the *time* set $\mathfrak{T}$ while *the set* $\mathfrak{R}$ is static. The permanent, monotonous, continuous flow of the instants (which are the elements of the *time* set $\mathfrak{T}$) expresses the dynamics of the *time* set $\mathfrak{T}$. The real numbers are immobile [228], [235]-[237].

We accept in this book, as in [217], [228], [230], [232], [233] [235]-[237], [244], the *relative zero moment* t_{zero} to have the zero numerical value, $num\ t_{zero} = 0$. However, we need not adopt it to be also *the initial moment* t_0, $t_0 \neq t_{zero}$, $num\ t_0 \neq 0$ is permissible, $num\ t_0 > -\infty$, in view of the *time*-varying nature of the systems to be studied. This determines the subset $\mathfrak{T}_0$ of $\mathfrak{T}$,

$$\mathfrak{T}_0 = \{t : t \in \mathfrak{T},\ num\ t \in [num\ t_0, \infty[\},\ inf\mathfrak{T}_0 > -\infty,\ sup\mathfrak{T}_0 = \infty. \tag{2.2}$$

If the system properties do not depend on the choice of the initial instant t_0, which means that the properties are uniform (relative to t_0), then we can accept $t_0 = t_{zero}$, $num\ t_0 = 0$. This holds for all *time*-invariant systems.

Various properties of *time*-varying systems can vary in *time*. For example, stability properties can depend essentially on the initial moment t_0 [128], [139], [223], [224]. *The set of the initial moments* t_0 of our interest is $\mathfrak{T}_i$. It is a subset of $\mathfrak{T}$. In general:

$$\begin{gathered}\mathfrak{T}_i = \{t : t[T]\langle s\rangle,\ t \in \mathfrak{T}\} \subseteq \mathfrak{T},\ inf\ \mathfrak{T}_i \geq -\infty,\ sup\mathfrak{T}_i \leq \infty,\\ (inf\ \mathfrak{T}_i = -\infty\ and\ sup\mathfrak{T}_i = \infty) \Longleftrightarrow \mathfrak{T}_i = \mathfrak{T}.\end{gathered} \tag{2.3}$$

For example, if the initial moment t_0 is fixed and unique, then $\mathfrak{T}_i$ is singleton, $\mathfrak{T}_i = \{t_0\}$. Then $inf\ \mathfrak{T}_i = sup\mathfrak{T}_i = t_0$, $(num\ t_0) \in \mathfrak{R}$.

In the framework of the Lyapunov stability properties study, the set $\mathfrak{T}_i$ represents the set of initial moments t_0 defined by:

$$\begin{gathered}\mathfrak{T}_i = \{t : t[T]\langle s\rangle,\ t \in \mathfrak{T}\} \subseteq \mathfrak{T},\ inf\ \mathfrak{T}_i \geq -\infty,\ sup\mathfrak{T}_i = sup\mathfrak{T} = \infty,\\ inf\ \mathfrak{T}_i = -\infty \Longleftrightarrow \mathfrak{T}_i = \mathfrak{T}.\end{gathered} \tag{2.4}$$

From the above explanations and from (2.2)-(2.4) follows that $\mathfrak{T}_0$ is a subset of $\mathfrak{T}_i$:

$$\mathfrak{T}_0 \subseteq \mathfrak{T}_i. \tag{2.5}$$

Note 19 *[236], [237] It has been the common custom to use the letters t and τ to denote not only time itself, and an arbitrary instant, but also the numerical value of the arbitrary instant relative to the chosen zero instant, e.g., $t = 0$ is used in the sense $num\ t = 0$. From the physical point of view, this is incorrect. The numerical value $num\ t$ of the instant t is a real number without a physical dimension, whereas the instant t is a temporal value that has a physical dimension: the temporal dimension T of time. We overcome this by using the normalized, dimensionless, mathematical temporal variable, denoted by $\bar{t}$ and defined by*

$$\bar{t} = \frac{t}{1_t}\ [-],$$

so that the time set $\mathfrak{T}$ is to be replaced by

$$\overline{\mathfrak{T}} = \{\bar{t}\ [-] : \bar{t} = num\ \bar{t} = num\ t \in \mathfrak{R},\ d\bar{t} > 0,\ \bar{t}^{(1)} \equiv 1\}.$$

With this in mind we use in the sequel the letter t also for $\bar{t}$, and $\mathfrak{T}$ also for $\overline{\mathfrak{T}}$ in the mathematical sense. Hence,

$$t\ [-] = num\ t\ [-].$$

Between any two different instants $t_1 \in \mathfrak{T}$ and $t_2 \in \mathfrak{T}$ there is a third instant $t_3 \in \mathfrak{T}$, either $t_1 < t_3 < t_2$ or $t_2 < t_3 < t_1$. The *time* set $\mathfrak{T}$ is a **continuum**. It is also called **the continuous-*time* set**.

Let $\tau \in \mathfrak{T}$ or $\tau = \infty$, $\tau = t_0$ is permitted. The instant τ determines the *time* sets $\mathfrak{T}_\tau^-$ and $\mathfrak{T}_\tau$,

$$\mathfrak{T}_\tau^- = \{t : \ t \in \mathfrak{T}, \ t \leq \tau\} \subseteq \mathfrak{T} \tag{2.6}$$

$$\mathfrak{T}_\tau = \{t : \ t \in \mathfrak{T}, \ t \geq \tau\} \subseteq \mathfrak{T}, \ \mathfrak{T}_\tau^- \cup \mathfrak{T}_\tau = \mathfrak{T}, \tag{2.7}$$

$$t_0 \in \mathfrak{T}_\tau \Longrightarrow \mathfrak{T}_0 \subseteq \mathfrak{T}_\tau, \ \tau = t_0 \Longleftrightarrow \mathfrak{T}_\tau = \mathfrak{T}_0. \tag{2.8}$$

Being a physical variable, *time* possesses some general properties valid for all physical variables. We summarize them in the form of the principles, [228], [235]-[237].

2.3 *Time* speed law

Theorem 20 ***Universal time speed law*** *[228], [235]-[237], [245], [242].*
Time is the unique physical variable such that the speed v_t (denoted also with v_τ) of the evolution of its value (of the flow of its values), and of its numerical value (of their numerical values):

a) is invariant with respect to a choice of a relative zero moment t_{zero}, of an initial moment t_0, of a time scale and of a time unit 1_t, i.e., invariant relative to a choice of a time axis, invariant relative to a selection of spatial coordinates, and invariant relative to everybody and everything.

b) Its value (its numerical value) is invariant relative to everybody and everything, and equals one arbitrary time unit per the same time unit (equals one), respectively,

$$v_t = \frac{dt}{dt} = 1[\mathrm{TT}^{-1}]\left\langle 1_t 1_t^{-1}\right\rangle = 1[TT^{-1}]\left\langle 1_\tau 1_\tau^{-1}\right\rangle = \frac{d\tau}{d\tau} = v_\tau,$$
$$num \ v_t = num \ v_\tau = 1, \tag{2.9}$$

relative to arbitrary time axes T and T_τ; i.e., its numerical value equals 1 (one) with respect to all time axes (with respect to any accepted relative zero instant t_{zero}, any chosen initial instant t_0, any time scale and any selected time unit 1_t), with respect to all spatial coordinate systems, and with respect to all beings and all objects.

Comment 21 *The derivative of every variable x over itself equals identically one:*

$$\frac{dx}{dx} \equiv 1,$$

but such derivative is the speed of the variation of the value of the variable x if and only if the variable x is time t (or τ). Otherwise, dx/dx is not the speed of the variable x value variation if it is not time, i.e., if $x \neq t$. Its speed is dx/dt (or $dx/d\tau$). This explains why time is the unique variable with such speed properties. This is so obvious. Physics ignores this time speed property so much that it has become able to accept since 1905 the meaningless claim on the existence of different times with different time speeds (for more on this see [228], [235]-[237], [245], [242]).

2.4 Physical variables, *time* and principles

Common qualitative properties of the physical variables are important for the study of dynamical (in general) and control systems (in particular). We summarize some of them in the following principles by referring to the books [228], [235]-[237].:

Principle 22 ***Physical Continuity and Uniqueness Principle (*****PCUP*****): scalar form***

A physical variable can change its value from one value to another one only by passing through every intermediate value, and it possesses a unique local instantaneous real value in any place (in any being or in any object) at any moment.

Comment 23 *Time obeys the Physical Continuity and Uniqueness Principle.*

Principle 24 ***Physical Continuity and Uniqueness Principle (*****PCUP*****): matrix and vector form***

A vector physical variable or a matrix (vector) of physical variables can change, respectively, its vector or matrix (vector) value from one vector or matrix (vector) value to another one only by passing elementwise through every intermediate vector or matrix (vector) value, and it possesses a unique local instantaneous real vector or matrix (vector) value in any place (i.e., in any being or in any object) at any moment, respectively.

Principle 25 ***Physical Continuity and Uniqueness Principle (*****PCUP*****): system form***

The system physical variables (including those of their derivatives and integrals that appear in the system, as well as system solutions) can change, respectively, their (scalar or vector or matrix) values from one (scalar or vector or matrix) value to another one only by passing elementwise through every intermediate (scalar or vector or matrix) value, and they possess unique local instantaneous real (scalar or vector or matrix) values in any place at any moment.

The *PCUP* is required for accurate modeling of physical systems.

Corollary 26 ***Mathematical model of a physical variable, mathematical model of a physical system and PCUP***

a) For a mathematical (scalar or vector) variable to be, respectively, an adequate description of a physical (scalar or vector) variable, it is necessary that it obeys the physical continuity and uniqueness principle.

b) For a mathematical model of a physical system to be an adequate description of the physical system it is necessary that all its system variables obey the physical continuity and uniqueness principle; i.e., that the mathematical model obeys the physical continuity and uniqueness principle.

The synthesis of the properties of *time* and of the common properties of the physical variables expressed by *PCUP* (Principle 22 through Principle 25) results in the following:

Principle 27 ***Time Continuity and Uniqueness Principle (TCUP)***

Any (scalar or vector) physical variable and any vector / matrix of physical variables can change, respectively, its scalar / vector / matrix value from one scalar / vector / matrix value to another one only continuously in time by passing (elementwise) through every intermediate scalar / vector / matrix value, respectively, and it possesses a unique local instantaneous real scalar / vector / matrix value, respectively, in any place (in any being or in any object) at any moment.

Definition 28 ***The system form of the*** $TCUP$

The system form of the $TCUP$ *means that all system variables (including their derivatives and integrals that appear in the system, as well as system solutions) satisfy the* $TCUP$.

The effective application of the $TCUP$ to the stability study of dynamical systems and to their control synthesis is in [219], [220], [229], [234], [238]-[240].

Corollary 29 *[217], [228], [236], [237]* ***Mathematical representation of a physical variable, mathematical model of a physical system and TCUP***

a) *For a mathematical (scalar or vector) variable to be, respectively, an adequate description of a physical (scalar or vector) variable, it is necessary that it obeys the time continuity and uniqueness principle* $TCUP$.

b) *For a mathematical model of a physical system to be an adequate description of the physical system it is necessary that all its system variables obey the time continuity and uniqueness principle* $TCUP$*; or equivalently, that the mathematical model obeys the time continuity and uniqueness principle* $TCUP$.

c) *For a mathematical model of a physical system to be an adequate description of the physical system, it is necessary that its solutions are unique and continuous in time.*

The books [225], [228], [235]-[237] contain the complete study of *time*, the establishment of the novel, consistent, physical, and mathematical relativity theory, as well as its relationship to systems.

Chapter 3

Sets

This chapter largely broadens [232, Chapter 2: SETS, pp. 13-20].

3.1 Set basis: *time*-invariant sets

The meaning and the notion of the term **set** is in the following sense:

Definition 30 ***Set***

A ***Set****, also called a* ***constant set*** *or* ***time-invariant set****, which is denoted by $\mathfrak{S}$, is* ***separateness*** *(at any moment $t \in \mathfrak{T}$, i.e., independent of time) of some beings, objects, symbols, numbers, letters, ..., which are its* ***members,*** *i.e., its* ***elements****. The nature of its members can be any (abstract, biological, chemical, economical, financial, mathematical, physical, social, ..., fictional, ...).*

A Set is mathematically denoted by the brackets $\{..\}$ between which are the symbols in biunivoque correspondence with the set members.

If and only if x is a member of the set $\mathfrak{S}$, then it is denoted by $x \in \mathfrak{S}$ meaning "x belongs to $\mathfrak{S}$", i.e., "x is an element of $\mathfrak{S}$", i.e., "x is a member of $\mathfrak{S}$".

If and only if y is not a member of $\mathfrak{S}$, then it is denoted by $y \notin \mathfrak{S}$ meaning "y does not belong to $\mathfrak{S}$"; "y is not element of $\mathfrak{S}$", "y is not a member of $\mathfrak{S}$".

Comment 31 *The biunivoque correspondence among the set members and their symbols means that one symbol is associated only with one set member and one set member is denoted only by one symbol.*

For the basic set properties, operations and relations, see, for example, [329] and [378].

A set can be *time*-invariant, (*time*-independent), i.e., *constant.* It is the classical and general meaning of sets, Definition 30. The set $\mathfrak{R}$ of the real numbers is *time*-invariant. The *time* set $\mathfrak{T}$ is also a *time*-invariant set.

The singleton is the set that has exactly one element. If its unique element is the origin $\mathbf{0}_n$ of the real vector space $\mathfrak{R}^n$, then the singleton is denoted by $\mathfrak{O}$ for any natural number n,

$$\mathfrak{O} = \{\mathbf{0}_n\} \subset \mathfrak{R}^n, \quad n \in \{1, 2, ...\}. \tag{3.1}$$

A set that does not contain any element is an **empty set** denoted by ϕ. It is a constant set. The number of its elements is zero. It is without any element,

$$\phi = \{\}. \tag{3.2}$$

DOI: 10.1201/9781003744528-6

Note 32 *Throughout the book the set $\mathfrak{S}$ is a time-invariant, i.e., constant, connected set,* $\mathfrak{S} \subseteq \mathfrak{R}^n$.

Definition 33 *A set $\mathfrak{S}$ is* ***finite*** *if and only if the number of its elements is finite.*

Definition 34 *A set $\mathfrak{S}$ is* ***infinite*** *if and only if the number of its elements is infinite. This holds throughout the book for the set $\mathfrak{S}$.*

Definition 35 *The* ***union*** $\mathfrak{S}_1 \cup \mathfrak{S}_2$ *of sets $\mathfrak{S}_1$ and $\mathfrak{S}_2$ is the set $\mathfrak{S}$ that contains all elements of both $\mathfrak{S}_1$ and $\mathfrak{S}_2$,* $\mathfrak{S}_1 \cup \mathfrak{S}_2 = \{\mathbf{S} : \mathbf{S} \in \mathfrak{S}_1 \vee \mathbf{S} \in \mathfrak{S}_2\}$.

Definition 36 *The* ***intersection*** $\mathfrak{S}_1 \cap \mathfrak{S}_2$ *of sets $\mathfrak{S}_1$ and $\mathfrak{S}_2$ is the set $\mathfrak{S}$ that contains all common elements of $\mathfrak{S}_1$ and $\mathfrak{S}_2$,*

$$\mathfrak{S}_1 \cap \mathfrak{S}_2 = \{\mathbf{S} : \mathbf{S} \in \mathfrak{S}_1 \wedge \mathbf{S} \in \mathfrak{S}_2\}. \tag{3.3}$$

Definition 37 *A set $\mathfrak{A}$ is a* ***subset*** *of a set $\mathfrak{S}$ if and only if it contains only some or all elements of the set $\mathfrak{S}$, which is denoted by $\mathfrak{A} \subseteq \mathfrak{S}$, or equivalently $\mathfrak{S} \supseteq \mathfrak{A}$. The empty set ϕ is a subset of every nonempty set. Notice that the set $\{0\}$ is not the empty set. It contains the number zero.*

A set $\mathfrak{A}$ is a ***proper subset*** *of a set $\mathfrak{S}$ if and only if it contains only some, but not all, elements of the set $\mathfrak{S}$, which is denoted by $\mathfrak{A} \subset \mathfrak{S}$, or equivalently $\mathfrak{S} \supset \mathfrak{A}$. The empty set ϕ is a proper subset of every nonempty set.*

Let $\mathcal{C}$ be the set of all complex numbers.

Definition 38 *A scalar function denoted by $\|.\|$, $\|.\| : \mathfrak{R}^n \to \mathfrak{R}_+$, is* ***norm on*** $\mathfrak{R}^n$ *if and only if it fulfills the following five conditions:*

1) It is defined and nonnegative on $\mathfrak{R}^n : \|\mathbf{S}\| \in \mathfrak{R}_+, \forall \mathbf{S} \in \mathfrak{R}^n$.

2) It is continuous on $\mathfrak{R}^n : \|\mathbf{S}\| \in \mathfrak{C}(\mathfrak{R}^n)$.

3) It vanishes only at the origin: $\|\mathbf{S}\| = 0$ *if and only if* $\mathbf{S} = \mathbf{0}_n$.

4) $\|\alpha\mathbf{S}\| = |\alpha| \, \|\mathbf{S}\|, \forall (\alpha, \mathbf{S}) \in \mathcal{C} \times \mathfrak{R}^n$.

5) $\|\mathbf{S}_1 + \mathbf{S}_2\| \leq \|\mathbf{S}_1\| + \|\mathbf{S}_2\|, \forall \mathbf{S}_k \in \mathfrak{R}^n, k = 1, 2$.

The norm function $\|.\|$ induces ***the distance function*** $d(.)$ ***on*** $\mathfrak{R}^n$, $d(.) : \mathfrak{R}^n \to \mathfrak{R}_+$, $d(\mathbf{S}_1, \mathbf{S}_2) = \|\mathbf{S}_1 - \mathbf{S}_2\| \equiv \|\mathbf{S}_2 - \mathbf{S}_1\|$ *is the* ***distance between vectors*** *(points, states)* $\mathbf{S}_1$ *and* $\mathbf{S}_2$.

Let $\mathbf{1}$ be **the unity vector,** $\mathbf{1} = \begin{bmatrix} 1 & 1 & ... & 1 \end{bmatrix}^T \in \mathfrak{R}^n$.

The norm $\|.\|_1$ called **norm 1** is defined for a vector $\mathbf{S} = \begin{bmatrix} S_1 & S_2 & ... & S_n \end{bmatrix}^T \in \mathfrak{R}^n$, for which $|\mathbf{S}| = \begin{bmatrix} |S_1| & |S_2| & ... & |S_n| \end{bmatrix}^T$, by

$$\|\mathbf{S}\|_1 = \mathbf{1}^T |\mathbf{S}| = \sum_{i=1}^{n} |S_i| . \tag{3.4}$$

The Euclidean norm $\|.\|$, called **the norm 2**, denoted also by $\|.\|_2$, is defined for the vector $\mathbf{S} = \begin{bmatrix} S_1 & S_2 & ... & S_n \end{bmatrix}^T \in \mathfrak{R}^n$ by

$$\|\mathbf{S}\| = \|\mathbf{S}\|_2 = \left(\mathbf{S}^T\mathbf{S}\right)^{1/2} = \sqrt{S_1^2 + S_2^2 + ... + S_n^2}. \tag{3.5}$$

The **norm** $\|.\|_\infty$, called **the max-norm**, is determined by:

$$\|\mathbf{S}\|_\infty = max \{|S_i| : \ i = 1, 2, ..., n\} . \tag{3.6}$$

Note 39 *If not otherwise specified, the norm $\|.\|$ can be any norm in the sequel.*

Definition 40 *Any accepted scalar norm $\|.\| : \mathfrak{R}^n \rightarrow \mathfrak{R}_+$ on $\mathfrak{R}^n$ induces* ***the distance function*** *$d(.) : \mathfrak{R}^n \rightarrow \mathfrak{R}_+$* ***on*** *$\mathfrak{R}^n$, which determines* ***the distance*** *$d(\mathbf{a}_1, \mathbf{a}_2)$* ***between vectors*** *(or,* ***points****) $\mathbf{a}_1 \in \mathfrak{R}^n$* ***and*** *$\mathbf{a}_2 \in \mathfrak{R}^n$ defined by*

$$d(\mathbf{a}_1, \mathbf{a}_2) = \|\mathbf{a}_1 - \mathbf{a}_2\| = \|\mathbf{a}_2 - \mathbf{a}_1\| = d(\mathbf{a}_2, \mathbf{a}_1). \tag{3.7}$$

Definition 41 *The set containing all the subsets of a set $\mathfrak{S}$ and the set $\mathfrak{S}$ itself is* ***the Power set*** *$\mathfrak{P}(\mathfrak{S})$ denoted also as $2^{\mathfrak{S}}$. The Power set $\mathfrak{P}(\mathfrak{R}^n)$ of $\mathfrak{R}^n$ is $2^{\mathfrak{R}^n}$.*

Definition 42 *Let $d(.) : \mathfrak{R}^n \times 2^{\mathfrak{R}^n} \rightarrow \mathfrak{R}_+$ be* ***the distance function on*** *$\mathfrak{R}^n \times 2^{\mathfrak{R}^n}$. Its value $d(\mathbf{S}, \mathfrak{S})$ is* ***the distance of a vector*** *$\mathbf{S}$* ***from a time-invariant set*** *$\mathfrak{S} \subseteq \mathfrak{R}$ defined by*

$$d(\mathbf{S}, \mathfrak{S}) = inf\left(\|\mathbf{S} - \mathbf{W}\| : \mathbf{W} \in \mathfrak{S}\right). \tag{3.8}$$

Definition 43 ***The distance between two constant sets*** *$\mathfrak{S}_k \subseteq \mathfrak{R}^n$, $k = 1, 2$, is denoted by $d(\mathfrak{S}_1, \mathfrak{S}_2)$, $d(\mathfrak{S}_1, \mathfrak{S}_2) = inf\left[\|\mathbf{S} - \mathbf{W}\| : \mathbf{S} \in \mathfrak{S}_1,\ \mathbf{W} \in \mathfrak{S}_2\right]$.*

Definition 44 *$\mathfrak{B}_\mu(\mathbf{Z})$ is* ***the open hyperball centered at*** *$\mathbf{Z}$* ***with the radius*** *μ,*

$$\mathfrak{B}_\mu(\mathbf{Z}) = \{\mathbf{S} : \|\mathbf{S} - \mathbf{Z}\| < \mu\}. \tag{3.9}$$

Definition 45 ***The interior*** *$In\mathfrak{S}$* ***of the set*** *$\mathfrak{S}$ is the set of all points $\mathbf{S} \in \mathfrak{S}$ for which there is a positive real number $\mu \in \mathfrak{R}^+$, determined by $\mathbf{S}$ and $\mathfrak{S}$, such that the open hyperball $\mathfrak{B}_\mu(\mathbf{S})$ centered at $\mathbf{S}$ with the radius μ is a subset of the set $\mathfrak{S}$, $\mathfrak{B}_\mu(\mathbf{S}) \subseteq \mathfrak{S}$,*

$$In\mathfrak{S} = \left\{\mathbf{S} : \mathbf{S} \in \mathfrak{S} \Longrightarrow \exists \mu \in \mathfrak{R}^+ \Longrightarrow \mathfrak{B}_\mu(\mathbf{S}) \subseteq \mathfrak{S}\right\}. \tag{3.10}$$

The point $\mathbf{S}$ is ***an interior point of the set*** *$\mathfrak{S}$.*

Definition 46 *A set $\mathfrak{S}$ is an* ***open set*** *if and only if it is its own interior $In\mathfrak{S}$, $\mathfrak{S} = In\mathfrak{S}$.*

Definition 47 *$\partial\mathfrak{S}$ is* ***the boundary of the set*** *$\mathfrak{S}$, which is the set of all points $\mathbf{S}$ such that in every open hyperball $\mathfrak{B}_\eta(\mathbf{S})$ centered at $\mathbf{S}$ there is a point $\mathbf{Y} \in \mathfrak{B}_\eta(\mathbf{S})$ belonging to the set $\mathfrak{S}$, $\mathbf{Y} \in \mathfrak{B}_\eta(\mathbf{S}) \cap \mathfrak{S}$, and a point $\mathbf{W} \in \mathfrak{B}_\eta(\mathbf{S})$ that is not in the set $\mathfrak{S}$, $\mathbf{W} \in \mathfrak{B}_\eta(\mathbf{S}) \backslash \mathfrak{S}$.*

The point $\mathbf{S}$ is ***a boundary point*** *of the set $\mathfrak{S}$.*

Definition 48 *$Cl\mathfrak{S}$ is* ***the closure of the set*** *$\mathfrak{S}$, which is the set of all interior and boundary points of the set $\mathfrak{S}$,*

$$Cl\mathfrak{S} = \partial\mathfrak{S} \cup In\mathfrak{S}. \tag{3.11}$$

Definition 49 *A set $\mathfrak{S}$ is a* ***closed set*** *if and only if it is its own closure $Cl\mathfrak{S}$, $\mathfrak{S} = Cl\mathfrak{S}$,*

$$\mathfrak{S} = Cl\mathfrak{S}. \tag{3.12}$$

Definition 50 *A set $\mathfrak{S}$ is a* ***compact set*** *if and only if it is both a closed set and a bounded set.*

Definition 51 ***The power set*** *$\mathfrak{P}(\mathfrak{S})$ of the set $\mathfrak{S}$ is the set containing all subsets of a set $\mathfrak{S}$ including the set $\mathfrak{S}$ itself and the empty set ϕ. It is denoted also by $2^{\mathfrak{S}}$ or by $\mathfrak{P}(\mathfrak{S})$, $\mathfrak{P}(\mathfrak{S}) = 2^{\mathfrak{S}}$.*

$$\mathfrak{P}(\mathfrak{S}) = 2^{\mathfrak{S}}. \tag{3.13}$$

The power set of $\mathfrak{R}^n$ is denoted by $\mathfrak{P}(\mathfrak{R}^n)$ or by $2^{\mathfrak{R}^n}$,

$$\mathfrak{P}(\mathfrak{R}^n) = 2^{\mathfrak{R}^n}. \tag{3.14}$$

For example, the power set of the empty set ϕ is the nonempty set $\mathfrak{P}(\phi) = 2^\phi$, $\mathfrak{P}(\phi) = 2^\phi = \{\phi\} \neq \phi$. It is a singleton. The power set $\mathfrak{P}(\mathfrak{S})$ of any singleton $\mathfrak{S} = \{\mathbf{S}\}$ is the two member set, $\mathfrak{S} = \{\mathbf{S}\} \Longrightarrow \mathfrak{P}(\mathfrak{S}) = \{\phi, \mathfrak{S}\}$.

3.1.1 Set diameter

Definition 52 ***Set diameter***

a) A nonnegative real number $\mathfrak{d} \in \mathfrak{R}_{+}$, $\mathfrak{d} = \mathfrak{d}(\mathfrak{S})$, *or* $\mathfrak{d} = \mathfrak{d}(\mathfrak{S}) = \infty$, *is:*

a-1. the ***inner diameter of a time-invariant set*** $\mathfrak{S}$, *which is denoted by* $\mathfrak{d}_m = \mathfrak{d}_m(\mathfrak{S})$, *if and only if it is the diameter of the closure* $Cl\mathfrak{B}_{\mathfrak{d}_m/2}$ *of the largest hyperball* $\mathfrak{B}_{\mathfrak{d}_m/2}$ *such that* $Cl\mathfrak{B}_{\mathfrak{d}_m/2} \subseteq Cl\mathfrak{S}$

a-2. the ***outer diameter of a time-invariant set*** $\mathfrak{S}$, *which is denoted by* $\mathfrak{d}_M = \mathfrak{d}_M(\mathfrak{S})$, *if and only if it is the diameter of the closure* $Cl\mathfrak{B}_{\mathfrak{d}_M/2}$ *of the smallest hyperball* $\mathfrak{B}_{\mathfrak{d}_M/2}$*such that* $Cl\mathfrak{B}_{\mathfrak{d}_M/2} \supseteq Cl\mathfrak{S}$

b) A nonnegative real number $\mathfrak{d}$, $\mathfrak{d} = \mathfrak{d}[\mathfrak{S}(t)] \equiv \mathfrak{d}(t;\mathfrak{S}) \in \mathfrak{R}_{+}$, *or* $\mathfrak{d}(t;\mathfrak{S}) = \infty$, *is:*

b-1. ***the inner diameter of a time-varying set*** $\mathfrak{S}(t)$ ***at a moment*** $t \in \mathfrak{T}$, *which is denoted by* $\mathfrak{d}_m[\mathfrak{S}(t)] = \mathfrak{d}_m(t;\mathfrak{S})$, *if and only if it is the diameter of the closure* $Cl\mathfrak{B}_{\mathfrak{d}_m/2}(t)$ *of the largest hyperball* $\mathfrak{B}_{\mathfrak{d}_m/2}(t)$ *at the moment* t *such that* $Cl\mathfrak{B}_{\mathfrak{d}_m/2}(t) \subseteq Cl\mathfrak{S}(t)$

b-2. ***the outer diameter of a time-varying set*** $\mathfrak{S}(t)$ ***at a moment*** $t \in \mathfrak{T}$, *which is denoted by* $\mathfrak{d}_M[\mathfrak{S}(t)] = \mathfrak{d}_M(t;\mathfrak{S})$, *if and only if it is the diameter of the closure* $Cl\mathfrak{B}_{\mathfrak{d}_M/2}(t)$ *of the smallest hyperball* $\mathfrak{B}_{\mathfrak{d}_M/2}(t)$ *at the moment* t *such that* $Cl\mathfrak{B}_{\mathfrak{d}_M/2}(t) \supseteq Cl\mathfrak{S}(t)$

c) A nonnegative real number $\mathfrak{d}$, $\mathfrak{d} = \mathfrak{d}(\mathfrak{T}_0;\mathfrak{S}) \in \mathfrak{R}_{+}$, *or* $\mathfrak{d}(\mathfrak{T}_0;\mathfrak{S}) = \infty$, *is:*

c-1. ***the inner diameter of a time-varying set*** $\mathfrak{S}(t)$ ***on*** $\mathfrak{T}_0$, *which is denoted by* $\mathfrak{d}_{m\mathfrak{T}_0} = \mathfrak{d}_m(\mathfrak{T}_0;\mathfrak{S})$, *if and only if*

$$\mathfrak{d}_m(\mathfrak{T}_0;\mathfrak{S}) = inf\,[\mathfrak{d}_m(t;\mathfrak{S}) : t \in \mathfrak{T}_0] = \mathfrak{d}_{m\mathfrak{T}_0}(\mathfrak{S}) = \mathfrak{d}_m(\mathfrak{S}_{m\mathfrak{T}_0}),$$
$$\mathfrak{S}_{m\mathfrak{T}_0} = \cap\,[\mathfrak{S}(t) : t \in \mathfrak{T}_0].$$

c-2. ***the outer diameter of a time-varying set*** $\mathfrak{S}(t)$ ***on*** $\mathfrak{T}_0$, *which is denoted by* $\mathfrak{d}_M = \mathfrak{d}_M(\mathfrak{T}_0;\mathfrak{S})$, *if and only if*

$$\mathfrak{d}_M(\mathfrak{T}_0;\mathfrak{S}) = sup\,[\mathfrak{d}_M(t;\mathfrak{S}) : t \in \mathfrak{T}_0] = \mathfrak{d}_{M\mathfrak{T}_0}(\mathfrak{S}) = \mathfrak{d}_M(\mathfrak{S}_{M\mathfrak{T}_0}),$$
$$\mathfrak{S}_{M\mathfrak{T}_0} = \cup\,[\mathfrak{S}(t) : t \in \mathfrak{T}_0].$$

*The expression "****on*** $\mathfrak{T}_0$*" and subscript "*$\mathfrak{T}_0$*" are to be omitted if and only if* $\mathfrak{T}_0 = \mathfrak{T}$. *Then, and only then,* $\mathfrak{d}_{(.)}(\mathfrak{T}_0;\mathfrak{S}) = \mathfrak{d}_{(.)}(\mathfrak{S})$, $(.) \in \{m, M\}$.

Using the above equations, we can now define the boundedness of a time-varying set in terms of its outer diameter as follows

Definition 53 *A set* $\mathfrak{S}(t)$ *is a bounded set at a moment* $t \in \mathfrak{T}_0$ *if and only if its outer diameter is a positive real number at the moment* t, $\mathfrak{d}_M[\mathfrak{S}(t)] \in \mathfrak{R}^{+}$. *It is bounded on* $\mathfrak{T}_0$ *if and only if its outer diameter on* $\mathfrak{T}_0$ *is a positive real number,* $\mathfrak{d}_M(\mathfrak{T}_0;\mathfrak{S}(t)) \in \mathfrak{R}^{+}$.

The same holds for a time-varying set $\mathfrak{S}(t;t_0;\mathfrak{S}_0)$.

3.1.2 Neighborhood

Fraktur letter $\mathfrak{N}$ will denote a neighborhood.

Let a time set $\mathfrak{T}_\omega$ be any subset of $\mathfrak{T}_i$ or of $\mathfrak{T}_0$ or of $\mathfrak{T}$:

$$\mathfrak{T}_\omega \subset \mathfrak{T}_i \veebar \mathfrak{T}_0 \veebar \mathfrak{T}. \tag{3.15}$$

In what follows, $\mu \in \mathfrak{R}^{+}$, or $\mu = \infty$.

Definition 54 ***Neighborhood of a time-invariant nonempty set*** $\mathfrak{S}$

A set $\mathcal{N}$, $\mathcal{N} \subseteq \mathfrak{R}^n$, *is a* ***neighborhood of a time-invariant nonempty set*** $\mathfrak{S}$ *denoted by* $\mathcal{N}(\mathfrak{S})$ *if and only if there exists* $\mu \in \mathfrak{R}^+$, *or* $\mu = \infty$, *such that the set* $\{\mathbf{S} : d(\mathbf{S},\mathfrak{S}) < \mu\}$ *is a proper subset of the set* $\mathcal{N}$,

$$\mathcal{N}{=}\mathcal{N}(\mathfrak{S}) \Longleftrightarrow \exists \mu \in \mathfrak{R}^+ \cup \{\infty\} \Longrightarrow \{\mathbf{S} \notin \mathfrak{S} : d(\mathbf{S},\mathfrak{S}) < \mu\} \subseteq \mathcal{N} \Longrightarrow \mathfrak{S} \cap \mathcal{N}(\mathfrak{S}) = \phi. \tag{3.16}$$

Every neighborhood $\mathcal{N}(\mathfrak{S})$ *of a nonempty set* $\mathfrak{S}$ *is also a nonempty set. If the set* $\mathfrak{S}$ *is additionally a connected set, then its neighborhood* $\mathcal{N}(\mathfrak{S})$ *is also connected.*

Every neighborhood $\mathcal{N}(\phi)$ *of the empty set* ϕ *is also the empty set* ϕ,

$$\mathcal{N}(\phi) = \phi. \tag{3.17}$$

Definition 55 ***The*** μ***-neighborhood of a time-invariant nonempty set*** $\mathfrak{S}$

A time-invariant set $\mathcal{N}_\mu(\mathfrak{S})$ *is* **the** $\mu-$**neighborhood of a time-invariant nonempty set** $\mathfrak{S}$ *if and only if* μ *is a fixed positive real number or* $\mu = \infty$ *and* $\mathcal{N}_\mu(\mathfrak{S})$ *contains all points* $\mathbf{S} \in \mathfrak{R}^n$, *the distance of which from the set* $\mathfrak{S}$ *is smaller than* μ,

$$\mu \in \mathfrak{R}^+ \cup \{\infty\} \Longrightarrow \mathcal{N}_\mu(\mathfrak{S}) = \{\mathbf{S} : \mathbf{S} \in \mathfrak{R}^n,\ d(\mathbf{S},\mathfrak{S}) < \mu\}. \tag{3.18}$$

Comment 56 ***Structure of the set neighborhood***

A neighborhood $\mathcal{N}(\mathfrak{S})$ *of a nonempty set* $\mathfrak{S}$ *can be an open set:* $\mathcal{N}(\mathfrak{S}) = In\mathcal{N}(\mathfrak{S})$, *or a closed set:* $\mathcal{N}(\mathfrak{S}) = Cl\mathcal{N}(\mathfrak{S})$, *or none of them. It can be a disconnected set. If it is a disconnected set, then it is a union of disjoint sets* $\mathcal{N}_1, \mathcal{N}_2, \ldots, \mathcal{N}_k$, $k \geq 2$. *One of them can be both a connected set and a neighborhood of the set* $\mathfrak{S}$. *It is a connected neighborhood of the set* $\mathfrak{S}$. *It need not be a unique connected neighborhood of the set* $\mathfrak{S}$.

The $\mu-$*neighborhood* $\mathcal{N}_\mu(\mathfrak{S})$ *of the time-invariant set* $\mathfrak{S}$ *is an open set,* $\mathcal{N}_\mu(\mathfrak{S}) = In\mathcal{N}_\mu(\mathfrak{S})$.

If the set $\mathfrak{S}$ *is bounded and* $\mu \in \mathfrak{R}^+$, *then its* $\mu-$*neighborhood* $\mathcal{N}_\mu(\mathfrak{S})$ *is also bounded and has the boundary* $\partial\mathcal{N}_\mu(\mathfrak{S}) = \{\mathbf{S} : \mathbf{S} \in \mathfrak{R}^n,\ d(\mathbf{S},\mathfrak{S}) = \mu\}$.

Comment 57 *The set* $\mathfrak{S}$ *can be the zero equilibrium singleton set* $\mathfrak{S}_{E0} = \{\mathbf{0}_n\}$.*A set* $\mathcal{N}$, $\mathcal{N} \subseteq \mathfrak{R}^n$, *is a* **neighborhood** $\mathcal{N}(\mathfrak{S}_{E0})$ **of the origin** $\mathbf{S} = \mathbf{0}_n$ *if and only if there is an open hyperball* $\mathfrak{B}_\mu$ *centered at the origin* $\mathbf{S} = \mathbf{0}_n$ *with the radius* $\mu \in \mathfrak{R}^+$ *such that it is a subset of the set* $\mathcal{N}$,

$$\exists \mu \in \mathfrak{R}^+,\ \mathfrak{B}_\mu \subseteq \mathcal{N} \Longrightarrow \mathfrak{S}_{E0} = \{\mathbf{0}_n\} \subset In\mathcal{N}\{\mathfrak{S}_{E0}\}. \tag{3.19}$$

Note 58 *We use neighborhoods of the origin* $\mathbf{S} = \mathbf{0}_n$ *denoted, for example, by* $\mathcal{N}_\delta$, $\mathfrak{N}\varepsilon$, *or* $\mathcal{N}_\Delta$. *In general, they are different from the hyperballs* $\mathfrak{B}_\delta$, $\mathfrak{B}\varepsilon$, *and* $\mathfrak{B}_\Delta$, *respectively,*

$$\mathcal{N}_\delta \neq \mathfrak{B}_\delta,\ \mathcal{N}_\varepsilon \neq \mathfrak{B}_\varepsilon,\ \text{ and }\ \mathcal{N}_\Delta \neq \mathfrak{B}_\Delta \text{ in general,}$$
$$In\mathcal{N}_\delta \supseteq \mathfrak{B}_\delta,\ In\mathcal{N}_\varepsilon \supseteq \mathfrak{B}_\varepsilon,\ \text{ and }\ In\mathcal{N}_\Delta \supseteq \mathfrak{B}_\Delta,$$

in the sense that they can be of any shape, of any form. The subscripts δ, ε, *and* Δ *of* $\mathcal{N}_\delta$, $\mathfrak{N}\varepsilon$, *and* $\mathcal{N}_\Delta$ *are the inner radii of the neighborhoods* $\mathcal{N}_\delta$, $\mathfrak{N}\varepsilon$, *and* $\mathcal{N}_\Delta$, *respectively. The neighborhoods* $\mathcal{N}_\delta$, $\mathfrak{N}\varepsilon$, *and* $\mathcal{N}_\Delta$ *can be equal to the hyperballs* $\mathfrak{B}_\delta$, $\mathfrak{B}\varepsilon$, *and* $\mathfrak{B}_\Delta$, *respectively, in special cases.*

3.1.3 Uniquely bounded sets

What follow presents a new set family of ***uniquely bounded sets [180], [180], [181],*** **[160]:**

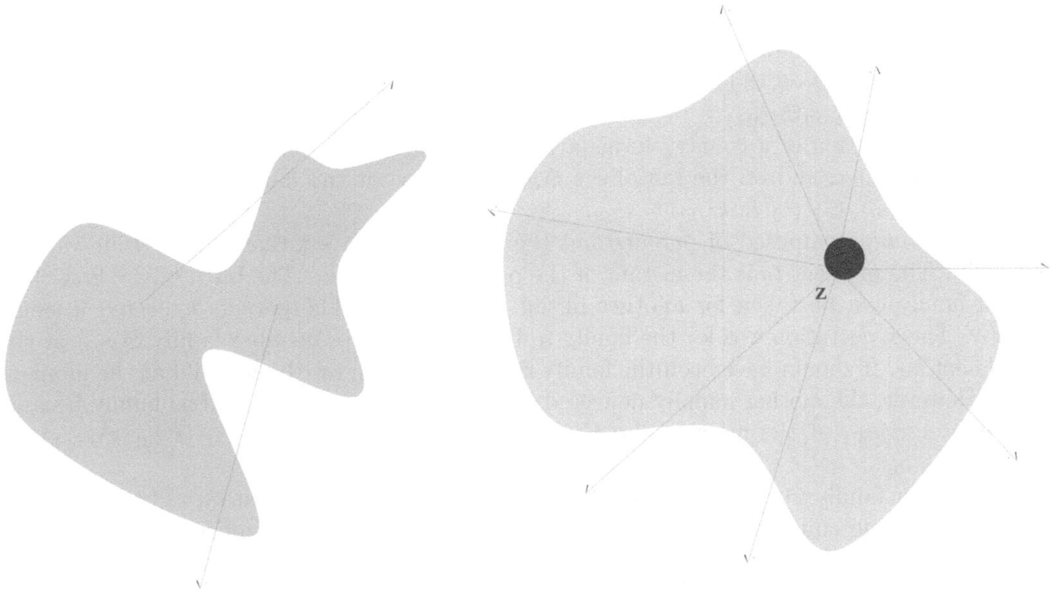

Figure 3.1: Not uniquely bounded set and uniquely bounded set.

Definition 59 *A set $\mathfrak{S}$ is* ***uniquely bounded*** *if and only if it satisfies the following conditions:*

1. it is bounded,

2. there is a point denoted by Z such that any ray from it in any direction intersects the set boundary exactly once; see Figure 3.1.

Definition 60 *A uniquely bounded set is an* ***O-uniquely bounded set*** *if and only if point Z is the Origin of the coordinate system.*

3.2 Set advances: *time*-varying sets

For the general theories of time-varying sets see the books [13], [280], [334], [351].

What follows presents the theory of time-varying sets directly applicable to the stability domains study.

A set can depend on *time*, i.e., it can be *time-varying*, i.e., a *variable set.* It is denoted by $\mathfrak{S}(t)$ if it is independent of the initial moment t_0, or if the initial moment can be any in $\mathfrak{T}$, or it is a short notation for a *time*-varying set $\mathfrak{S}(t;t_0;\mathfrak{S}_0)$ that depends on $(t,t_0,\mathfrak{S}_0) \in \mathfrak{T}_0 \times \mathfrak{T}_i \times \mathfrak{P}(\mathfrak{R}^n)$ so that $\mathfrak{S}(t_0;t_0;\mathfrak{S}_0) \equiv \mathfrak{S}(t_0) \equiv \mathfrak{S}_0$.

The set of all matter (substance) at a moment $t \in \mathfrak{T}$ is variable: $\mathfrak{S}_M(t)$. The set of all energy at the moment $t \in \mathfrak{T}$ is also variable: $\mathfrak{S}_E(t)$. Their union $\mathfrak{S}_{EM}(t) = \mathfrak{S}_E(t) \cup \mathfrak{S}_M(t)$ at the moment $t \in \mathfrak{T}$ is *time*-invariant, i.e., a constant set: $\mathfrak{S}_{EM}(t) = \mathfrak{S}_{EM}$,

$$\mathfrak{S}_{EM}(t) = \mathfrak{S}_E(t) \cup \mathfrak{S}_M(t) = \mathfrak{S}_{EM}, \ \forall t \in \mathfrak{T}. \tag{3.20}$$

The mass $m(.)$ of a space vehicle is an example of a *time*-varying set $\mathfrak{S}(t;t_0;\mathfrak{S}_0)$, $m(t;t_0;m_0) = \mathfrak{S}(t;t_0;\mathfrak{S}_0)$.

The same holds for a fleet of moving ships or flying aircraft. The space occupied by the fleet is a *time*-varying set $\mathfrak{S}_{Fl}(t;t_0)$. Its position, shape, members and their number can change. It depends on the initial moment t_0 and on the initial fleet $\mathfrak{S}_{Fl0}$, $\mathfrak{S}_{Fl}(t;t_0;\mathfrak{S}_{Fl0})$, $\mathfrak{S}_{Fl}(t_0;t_0;\mathfrak{S}_{Fl0}) = \mathfrak{S}_{Fl0}$.

The population of a country forms a *time*-varying set $\mathfrak{S}_P(t)$. It depends on both the current instant t, and on the initial set $\mathfrak{S}_{P0} = \mathfrak{S}_P(t_0)$ at the initial instant $t_0, \mathfrak{S}_P(t) = \mathfrak{S}_P(t;t_0;\mathfrak{S}_{P0})$, $\mathfrak{S}_P(t_0;t_0;\mathfrak{S}_{P0}) = \mathfrak{S}_{P0}$.

The *family* composed of parents and their children is a set that can be *time*-varying $\mathfrak{S}_F(t)$. The moment t_0 is the instant of the parents' marriage. The birth of the first child in a family can be taken for another initial moment t_{01}. The parents expected a single baby. The expectation was for the family to become the three member family $\mathfrak{S}_{F301}$ at the moment t_{01}. It should have been the family $\mathfrak{S}_F(t;t_{01};\mathfrak{S}_{F301})$ with one child at the moment t_{01}. However, the mother happily delivered a triplet at the moment t_{01}. The family became a quintet $\mathfrak{S}_{F501}$ at the moment t_{01}. Evidently, $\mathfrak{S}_F(t;t_{01};\mathfrak{S}_{F301}) \neq \mathfrak{S}_F(t;t_{01};\mathfrak{S}_{F501})$ for every $t \geq t_{01}$.

Another family $\mathfrak{S}_F(t)$ also expected the first child at the moment t_{01}. The mother did deliver a single child at the moment t_{01}. They became a three-member family $\mathfrak{S}_{F301}$ at the moment t_{01}. Since then they have constituted the family $\mathfrak{S}_F(t;t_{01};\mathfrak{S}_{F301})$. The birth of their second child happened at another initial moment t_{02}. They became a four-member family $\mathfrak{S}_{F402}$ at the moment t_{02}. The family $\mathfrak{S}_F(t;t_0;\mathfrak{S}_{F20})$ considered on the time interval $[t_0, t_{01}[$ is different from itself, $\mathfrak{S}_F(t;t_{01};\mathfrak{S}_{F301})$, at the moment t_{01}. At the initial moment t_{02}, it is the family $\mathfrak{S}_F(t;t_{01};\mathfrak{S}_{F402})$ that is not the same as itself $\mathfrak{S}_F(t;t_{01};\mathfrak{S}_{F301})$ since the initial moment t_{01} up to t_{02}. In this example, $\mathfrak{S}_F(t;t_{01};\mathfrak{S}_{F301})$ for $t \in [t_{01}, t_{02}[$ is different from $\mathfrak{S}_F(t;t_{02};\mathfrak{S}_{F402})$ for $t \geq t_{02}$. However,

$$\mathfrak{S}_F(t;t_{02};\mathfrak{S}_{F402}) = \mathfrak{S}_F(t;t_{01};\mathfrak{S}_{F301}) = \mathfrak{S}_F(t;t_0), \ \forall t \geq t_{02}.$$

This means that the set obeys the *Set uniqueness property:*

Property 61 *63 The following formula defines the Set uniqueness property:*

$$\begin{aligned}\mathfrak{S}_F(t;t_{02};\mathfrak{S}_{F402}) &= \mathfrak{S}_F[t;t_{02};\mathfrak{S}_F(t_{02};t_{01};\mathfrak{S}_{F301})] = \mathfrak{S}_F(t;t_{01};\mathfrak{S}_{F301}) = \\ &= \mathfrak{S}_F[t;t_{01};\mathfrak{S}_F(t_{01};t_0;\mathfrak{S}_{F0})] = \mathfrak{S}_F(t;t_0;\mathfrak{S}_{F20}). \end{aligned} \tag{3.21}$$

Besides, in this example, the initial set value $\mathfrak{S}_{F0} = \mathfrak{S}_F(t_0)$ depends on the initial moment t_0 :

$$\mathfrak{S}_{F20} = \mathfrak{S}_{F2}(t_0) \neq \mathfrak{S}_{F301} = \mathfrak{S}_{F3}(t_{01}) \neq \mathfrak{S}_{F402} = \mathfrak{S}_{F4}(t_{02}). \tag{3.22}$$

Sets of viruses in a human body, in a family, in a city, in a country are also variable, i.e., *time*-varying.

The set of the money in a bank account is variable.

A biological cell is a *time*-varying set. It can divide itself at some moment.

In conclusion, we can say that *time*-varying sets exist in mathematics, physics, chemistry, engineering, social, economical, financial and biological systems.

We recall Definition 14 of *time* in Section 2.1. *Time* and its properties are crucial for *time*-varying sets.

Definition 62 ***Set function and time-varying set***

A set function is a set mapping either $\mathfrak{S}(.) : \mathfrak{T} \longrightarrow \mathfrak{P}(\mathfrak{R}^n)$ ***or*** $\mathfrak{S}(.) : \mathfrak{T}_0 \times \mathfrak{T}_i \times \mathfrak{P}(\mathfrak{R}^n) \longrightarrow \mathfrak{P}(\mathfrak{R}^n)$. *Its* ***instantaneous set value at a moment*** *t is the set $\mathfrak{S}(t)$ or the set $\mathfrak{S}(t;t_0;\mathfrak{S}_0)$, which is* ***the instantaneous separateness*** *of some beings, objects* *of any nature, symbols, numbers, letters, ... that are its* ***instantaneous members,*** *i.e.,* ***instantaneous elements, at the moment*** $t \in \mathfrak{T}$ ***or*** $t \in \mathfrak{T}_0$, ***if it is independent of the initial data*** $(t_0, \mathfrak{S}_0) \in \mathfrak{T} \times \mathfrak{P}(\mathfrak{R}^n)$ ***or if it depends on the initial moment*** $t_0 \in \mathfrak{T}_i$ *and on the initial set value* $\mathfrak{S}_0$ *at* t_0, *respectively.*

The nature of its members can be any *(abstract, biological, chemical, economical, fictitious, financial, human, mathematical, physical, social, ...) at the moment* $t \in \mathfrak{T}$ *or* $t \in \mathfrak{T}_0$, *and can vary with the variation of the moment* t *and/or with the variation of both* t *and* t_0.

The initial set value $\mathfrak{S}_0$ *of the time-varying set* $\mathfrak{S}(t)$ *at any initial moment* $t_0 \in \mathfrak{T}$, *or of the time-varying* $\mathfrak{S}(t; t_0; \mathfrak{S}_0)$ *at the initial moment* $t_0 \in \mathfrak{T}_i$, *is determined by:*

$$\mathfrak{S}_0 = \mathfrak{S}(t_0),\ t_0 \in \mathfrak{T},\ \text{or}\ \mathfrak{S}_0 = \mathfrak{S}(t_0; t_0; \mathfrak{S}_0),\ t_0 \in \mathfrak{T}_i, \tag{3.23}$$

respectively.

If an initial set $\mathfrak{S}_{01} = \mathfrak{S}(t_{01}; t_{01}; \mathfrak{S}_{01})$ at an instant t_{01} is a result of the temporal set evolution from the previous initial set $\mathfrak{S}_0$ at a previous initial moment t_0, $t_0 < t_{01}$, then the set $\mathfrak{S}(t; t_0; \mathfrak{S}_0)$ can possess the following uniqueness property:

Property 63 ***Set uniqueness property***

For every $(t, t_0, \tau, \mathfrak{S}_0) \in \mathfrak{T}_0 \times \mathfrak{T}_i \times \mathfrak{T}_0 \times \mathfrak{P}(\mathfrak{R}^n)$ *every instantaneous set value of the set function* $\mathfrak{S}(.) : \mathfrak{T}_0 \times \mathfrak{T}_i \times \mathfrak{P}(\mathfrak{R}^n) \longrightarrow \mathfrak{P}(\mathfrak{R}^n)$ *satisfies the following* ***set uniqueness property:***

$$\begin{gathered}\mathfrak{S}[t; t_0 + \tau; \mathfrak{S}(t_0 + \tau; t_0; \mathfrak{S}_0)] = \mathfrak{S}(t; t_0; \mathfrak{S}_0), \\ \forall (t, t_0, \tau, \mathfrak{S}_0) \in \mathfrak{T}_0 \times \mathfrak{T}_i \times \mathfrak{T}_0 \times \mathfrak{P}(\mathfrak{R}^n).\end{gathered} \tag{3.24}$$

Note 64 ***Sets*** $\mathfrak{S}(t)$ ***and*** $\mathfrak{S}(t; t_0; \mathfrak{S}_0)$ ***can be, but need not be, independent of some system***

Throughout the book, the sets $\mathfrak{S}(t)$ *and* $\mathfrak{S}(t; t_0; \mathfrak{S}_0)$ *can be, but need not be, independent of some system, hence of the system motions.*

Note 65 ***Simplified set notation***

Sometimes for the sake of simplicity we can use the notation $\mathfrak{S}(t; t_0)$ *as the simplified notation of* $\mathfrak{S}(t; t_0; \mathfrak{S}_0)$:

$$\begin{gathered}\mathfrak{S}(t; t_0; \mathfrak{S}_0) = \mathfrak{S}(t; t_0),\ \forall (t, t_0, \mathfrak{S}_0) \in \mathfrak{T}_0 \times \mathfrak{T}_i \times \mathfrak{P}(\mathfrak{R}^n), \\ \mathfrak{S}(t_0; t_0; \mathfrak{S}_0) = \mathfrak{S}(t_0; t_0) = \mathfrak{S}_0,\ \forall (t_0, \mathfrak{S}_0) \in \mathfrak{T}_i \times \mathfrak{P}(\mathfrak{R}^n).\end{gathered} \tag{3.25}$$

We remind ourselves that the constant open hyperball $\mathfrak{B}_\mu(\mathbf{Z})$ centered at the constant vector $\mathbf{Z} \in \mathfrak{R}^n$ and with the constant radius μ is defined by

$$\mathfrak{B}_\mu(\mathbf{Z}) = \{\mathbf{X} : \mathbf{X} \in \mathfrak{R}^n,\ \|\mathbf{X} - \mathbf{Z}\| < \mu\}. \tag{3.26}$$

Analogously, if the radius of the hyperball is *time*-varying $\mu(.) : \mathfrak{T} \longrightarrow \mathfrak{R}^+$, and/or its center is *time*-varying $\mathbf{Z}(.) : \mathfrak{T} \longrightarrow \mathfrak{R}^n$, then the hyperball becomes **time-varying hyperball** $\mathfrak{B}_{\mu(t)}[\mathbf{Z}(t)]$, or for short, $\mathfrak{B}_\mu(t; \mathbf{Z})$,

$$\mathfrak{B}_\mu(t; \mathbf{Z}) = \mathfrak{B}_{\mu(t)}[\mathbf{Z}(t)] = \{\mathbf{X} : \mathbf{X} \in \mathfrak{R}^n,\ \|\mathbf{X} - \mathbf{Z}(t)\| < \mu(t)\}. \tag{3.27}$$

Definition 66 *A time-varying set* $\mathfrak{S}(t)$ *is* ***bounded at*** $t \in \mathfrak{T}$ *if and only if there is a constant bounded nonempty set* $\mathfrak{A}$ *such that* $\mathfrak{S}(t) \subseteq \mathfrak{A}$, $t \in \mathfrak{T}$.

A time-varying set $\mathfrak{S}(t)$ *is* ***bounded on*** $\mathfrak{T}$ *if and only if there is a constant bounded nonempty set* $\mathfrak{A}$ *such that* $\mathfrak{S}(t) \subseteq \mathfrak{A}, \ \forall t \in \mathfrak{T}$.

A time-varying set $\mathfrak{S}(t;t_0;\mathfrak{S}_0)$ *is* ***bounded at*** $t \in \mathfrak{T}_0$ *for* $t_0 \in \mathfrak{T}_i$ *if and only if there is a constant bounded nonempty set* $\mathfrak{A}$ *such that* $\mathfrak{S}(t;t_0;\mathfrak{S}_0) \subseteq \mathfrak{A}, \ (t;t_0) \in \mathfrak{T}_0 \times \mathfrak{T}_i$.

A time-varying set $\mathfrak{S}(t;t_0;\mathfrak{S}_0)$ *is* ***bounded on*** $\mathfrak{T}_0 \times \mathfrak{T}_i$ *if and only if there is a constant bounded nonempty set* $\mathfrak{A}$ *such that*

$$\mathfrak{S}(t;t_0;\mathfrak{S}_0) \subseteq \mathfrak{A}, \ \forall (t;t_0) \in \mathfrak{T}_0 \times \mathfrak{T}_i. \tag{3.28}$$

Definition 67 *A variable set* $\mathfrak{B}(t;t_0;\mathfrak{B}_0)$ *is a* ***subset*** *of a set* $\mathfrak{S}(t;t_0;\mathfrak{S}_0)$ *at a moment* $t \in \mathfrak{T}_0$ *for any* $t_0 \in \mathfrak{T}_i$ *if and only if the set* $\mathfrak{S}(t;t_0;\mathfrak{S}_0)$ *contains all elements of the set* $\mathfrak{B}(t;t_0;\mathfrak{B}_0)$*, which is denoted by* $\mathfrak{B}(t;t_0;\mathfrak{B}_0) \subseteq \mathfrak{S}(t;t_0;\mathfrak{S}_0)$*, or equivalently* $\mathfrak{S}(t;t_0;\mathfrak{S}_0) \supseteq \mathfrak{B}(t;t_0;\mathfrak{B}_0)$.

The empty set ϕ *is a subset of every nonempty variable set.*

A set $\mathfrak{B}(t;t_0;\mathfrak{B}_0)$ *is a* ***proper subset*** *of a set* $\mathfrak{S}(t;t_0;\mathfrak{S}_0)$ *at a moment* $t \in \mathfrak{T}_0$ *for any* $t_0 \in \mathfrak{T}_i$ *if and only if it is a subset of the set* $\mathfrak{S}(t;t_0;\mathfrak{S}_0)$ *at a moment* t *for any* $t_0 \in \mathfrak{T}_i$ *and there are elements of* $\mathfrak{S}(t;t_0;\mathfrak{S}_0)$ *which do not belong to the set* $\mathfrak{B}(t;t_0;\mathfrak{B}_0)$ *at any* $t_0 \in \mathfrak{T}_i$*, which is denoted by* $\mathfrak{B}(t;t_0;\mathfrak{B}_0) \subset \mathfrak{S}(t;t_0;\mathfrak{S}_0)$*, or equivalently* $\mathfrak{S}(t;t_0;\mathfrak{S}_0) \supset \mathfrak{B}(t;t_0;\mathfrak{B}_0)$.

Definition 68 ***The distance of a vector*** $\mathbf{S} \in \mathfrak{R}^n$ ***from a time-varying set*** $\mathfrak{S}(t) \subseteq \mathfrak{R}^n$ ***at a moment*** $t \in \mathfrak{T}$ *is denoted by* $d[\mathbf{S}, \mathfrak{S}(t)]$ *and determined by*

$$d[\mathbf{S}, \mathfrak{S}(t)] = inf\left[\|\mathbf{S} - \mathbf{W}\| : \mathbf{W} \in \mathfrak{S}(t)\right], t \in \mathfrak{T}.$$

The distance of a vector $\mathbf{S} \in \mathfrak{R}^n$ ***from a time-varying set*** $\mathfrak{S}(t;t_0;\mathfrak{S}_0) \subseteq \mathfrak{R}^n$ ***at any*** $(t;t_0) \in \mathfrak{T}_0 \times \mathfrak{T}_i$ *is denoted by* $d[\mathbf{S}, \mathfrak{S}(t;t_0;\mathfrak{S}_0)]$ *and determined by*

$$d[\mathbf{S}, \mathfrak{S}(t;t_0;\mathfrak{S}_0)] = inf\left[\|\mathbf{S} - \mathbf{W}\| : \mathbf{W} \in \mathfrak{S}(t;t_0;\mathfrak{S}_0)\right], \ \forall (t;t_0) \in \mathfrak{T}_0 \times \mathfrak{T}_i. \tag{3.29}$$

Definition 69 ***The distance between two time-varying sets*** $\mathfrak{S}_k(t) \subseteq \mathfrak{R}^n$, $k = 1, 2$, ***at any moment*** $t \in \mathfrak{T}$ *is denoted by* $d[\mathfrak{S}_1(t), \mathfrak{S}_2(t)]$ *and determined by*

$$d[\mathfrak{S}_1(t), \mathfrak{S}_2(t)] = inf\left[\|\mathbf{S} - \mathbf{W}\| : \mathbf{S} \in \mathfrak{S}_1(t), \ \mathbf{W} \in \mathfrak{S}_2(t)\right], \forall t \in \mathfrak{T}. \tag{3.30}$$

The distance between two time-varying sets $\mathfrak{S}_k(t;t_0;\mathfrak{S}_{0k}) \subseteq \mathfrak{R}^n$ ***at*** $(t;t_0) \in \mathfrak{T}_0 \times \mathfrak{T}_i$ ***for any*** $t_0 \in \mathfrak{T}_i$, $k = 1, 2$, *is denoted by* $d[\mathfrak{S}_1(t;t_0;\mathfrak{S}_{10}), \mathfrak{S}_2(t;t_0;\mathfrak{S}_{20})]$,

$$d[\mathfrak{S}_1(t;t_0;\mathfrak{S}_{10}), \mathfrak{S}_2(t;t_0;\mathfrak{S}_{20})] =$$
$$inf\left[\|\mathbf{S} - \mathbf{W}\| : \mathbf{S} \in \mathfrak{S}_1(t;t_0;\mathfrak{S}_{10}), \ \mathbf{W} \in \mathfrak{S}_2(t;t_0;\mathfrak{S}_{20})\right], \ (t,t_0) \in \mathfrak{T}_0 \times \mathfrak{T}_i. \tag{3.31}$$

Definition 70 *The point* $\mathbf{S} = \mathbf{S}(t;t_0)$ *is an* ***interior point*** *of the set* $\mathfrak{S}(t;t_0;\mathfrak{S}_0)$ *at any* $(t;t_0) \in \mathfrak{T}_0 \times \mathfrak{T}_i$*, if and only if there is a real number* $\mu \in \mathfrak{R}^+$, $\mu = \mu[\mathbf{S}(t;t_0;\mathbf{S}_0)]$*, such that* $\mathfrak{B}_\mu(t;t_0;\mathbf{S}_0) \subseteq \mathfrak{S}(t;t_0;\mathfrak{S}_0)$.

The interior $In\mathfrak{S}(t;t_0;\mathfrak{S}_0)$ *of the set* $\mathfrak{S}(t;t_0;\mathfrak{S}_0)$ *at any* $(t;t_0) \in \mathfrak{T}_0 \times \mathfrak{T}_i$ *is the set of all points* $\mathbf{S} = \mathbf{S}(t;t_0;\mathbf{S}_0) \in \mathfrak{S}(t;t_0;\mathfrak{S}_0)$*, for which there is a positive real number* $\mu = \mu[\mathbf{S}(t;t_0;\mathbf{S}_0)] = \mu(t;t_0) \in \mathfrak{R}^+$*, determined by* $\mathbf{S} = \mathbf{S}(t;t_0;\mathbf{S}_0)$*, such that the open hyperball* $\mathfrak{B}_\mu(t;t_0;\mathbf{S})$ *centered at* $\mathbf{S} = \mathbf{S}(t;t_0;\mathbf{S}_0)$ *with the radius* $\mu(t;t_0)$ *at* $(t;t_0) \in \mathfrak{T}_0 \times \mathfrak{T}_i$ *is a subset of the set* $\mathfrak{S}(t;t_0;\mathfrak{S}_0)$, $\mathfrak{B}_\mu(t;t_0;\mathbf{S}) \subseteq \mathfrak{S}(t;t_0;\mathfrak{S}_0)$,

$$In\mathfrak{S}(t;t_0;\mathfrak{S}_0) = \left\{ \begin{array}{c} \mathbf{S} : \mathbf{S} \in \mathfrak{S}(t;t_0;\mathfrak{S}_0) \Longrightarrow \\ \exists \mu(t;t_0;\mathbf{S}) = \mu[\mathfrak{S}(t;t_0;\mathfrak{S}_0), \mathbf{S}] \in \mathfrak{R}^+ \Longrightarrow \\ \mathfrak{B}_\mu(t;\mathbf{S}) \subseteq \mathfrak{S}(t;t_0;\mathfrak{S}_0) \end{array} \right\}, \ \forall (t;t_0) \in \mathfrak{T}_0 \times \mathfrak{T}_i. \tag{3.32}$$

Equivalently, the interior $In\mathfrak{S}(t;t_0;\mathfrak{S}_0)$ *of the set* $\mathfrak{S}(t;t_0;\mathfrak{S}_0)$ *at any* $(t;t_0) \in \mathfrak{T}_0 \times \mathfrak{T}_i$ *is the set of all the interior points of the set* $\mathfrak{S}(t;t_0;\mathfrak{S}_0)$ *at* $(t;t_0)$.

Definition 71 *A variable set $\mathfrak{S}(t;t_0;\mathfrak{S}_0)$ is an* ***open set*** *at $(t;t_0) \in \mathfrak{T}_0 \times \mathfrak{T}_i$ if and only if it is its own interior $In\mathfrak{S}(t;t_0;\mathfrak{S}_0)$, $\mathfrak{S}(t;t_0;\mathfrak{S}_0) = In\mathfrak{S}(t;t_0;\mathfrak{S}_0)$, at $(t;t_0) \in \mathfrak{T}_0 \times \mathfrak{T}_i$.*

Definition 72 *$\partial\mathfrak{S}(t;t_0;\mathfrak{S}_0)$ is* ***the boundary of the set*** *$\mathfrak{S}(t;t_0;\mathfrak{S}_0)$ at $(t;t_0) \in \mathfrak{T}_0 \times \mathfrak{T}_i$, which is the set of all points $\mathbf{S}$ such that in every open hyperball $\mathfrak{B}_\eta(t;t_0;\mathbf{S})$ centered at $\mathbf{S}$ at $(t;t_0)$ there is a point $\mathbf{Y} \in \mathfrak{B}_\eta(t;t_0;\mathbf{S})$ belonging to the set $\mathfrak{S}(t;t_0;\mathfrak{S}_0)$, $\mathbf{Y} \in \mathfrak{B}_\eta(t;t_0;\mathbf{S}) \cap \mathfrak{S}(t;t_0;\mathfrak{S}_0)$, and a point $\mathbf{W} \in \mathfrak{B}_\eta(t;t_0;\mathbf{S})$ that is not in the set $\mathfrak{S}(t;t_0;\mathfrak{S}_0)$, $\mathbf{W} \in \mathfrak{B}_\eta(t;t_0;\mathbf{S}) \backslash \mathfrak{S}(t;t_0;\mathfrak{S}_0)$. The point $\mathbf{S}$ is* ***the boundary point*** *of the set $\mathfrak{S}(t;t_0;\mathfrak{S}_0)$ at $(t;t_0) \in \mathfrak{T}_0 \times \mathfrak{T}_i$.*

Note 73 ***Boundary and unbounded set***

A set $\mathfrak{S}(t;t_0;\mathfrak{S}_0)$ can have the boundary $\partial\mathfrak{S}(t;t_0;\mathfrak{S}_0)$ at every $(t;t_0) \in \mathfrak{T}_0 \times \mathfrak{T}_i$ but need not be bounded on $\mathfrak{T}_0 \times \mathfrak{T}_i$, i.e., it can be unbounded on $\mathfrak{T}_0 \times \mathfrak{T}_i$. For example, the set $\mathfrak{S}_{ex}(t;t_0;\mathfrak{S}_{ex0})$ defined by

$$\mathfrak{S}_{ex}(t;t_0;\mathfrak{S}_{ex0}) = \left\{\mathbf{S} : \|\mathbf{S}\| < e^{|t-t_0|}\right\},\ \forall(t;t_0) \in \mathfrak{T}_0 \times \mathfrak{T}_i,\ \mathfrak{S}_{ex0} = \{\mathbf{S} : \|\mathbf{S}\| < 1\},$$

is bounded at every $(t;t_0) \in \mathfrak{T}_0 \times \mathfrak{T}_i$ with the boundary $\partial\mathfrak{S}_{ex}(t;t_0)$,

$$\partial\mathfrak{S}_{ex}(t;t_0) = \left\{\mathbf{S} : \|\mathbf{S}\| = e^{|t-t_0|}\right\},\ \forall(t;t_0) \in \mathfrak{T}_0 \times \mathfrak{T}_i,$$

but it is not bounded on $\mathfrak{T}_0 \times \mathfrak{T}_i$ because there does not exist a constant bounded set $\mathfrak{A}_b$ such that the set $\mathfrak{S}_{ex}(t;t_0;\mathfrak{S}_{ex0})$ is subset of the set $\mathfrak{A}_b$ on $\mathfrak{T}_0 \times \mathfrak{T}_i$,

$$\begin{array}{c} \forall\mathfrak{A}_b \subset \mathfrak{R}^n \Longrightarrow \exists(t;t_0) \in \mathfrak{T}_0 \times \mathfrak{T}_i,\ \Longrightarrow |t-t_0| > 0\wedge \\ [\mathfrak{S}_{ex}(t;t_0)\backslash\mathfrak{A}_b] \neq \phi. \end{array}$$

Definition 74 *$Cl\mathfrak{S}(t;t_0;\mathfrak{S}_0)$ is* ***the closure of the set*** *$\mathfrak{S}(t;t_0;\mathfrak{S}_0)$ at $(t;t_0) \in \mathfrak{T}_0 \times \mathfrak{T}_i$, which is the set of all interior and boundary points of the set $\mathfrak{S}(t;t_0;\mathfrak{S}_0)$ at $(t;t_0) \in \mathfrak{T}_0 \times \mathfrak{T}_i$,*

$$Cl\mathfrak{S}(t;t_0;\mathfrak{S}_0) = \partial\mathfrak{S}(t;t_0;\mathfrak{S}_0) \cup In\mathfrak{S}(t;t_0;\mathfrak{S}_0),\ (t;t_0) \in \mathfrak{T}_0 \times \mathfrak{T}_i. \tag{3.33}$$

Definition 75 *A set $\mathfrak{S}(t;t_0;\mathfrak{S}_0)$ is a* ***closed set*** *at a moment $t \in \mathfrak{T}$ if and only if it is its own closure $Cl\mathfrak{S}(t;t_0;\mathfrak{S}_0)$, $\mathfrak{S}(t;t_0;\mathfrak{S}_0) = Cl\mathfrak{S}(t;t_0;\mathfrak{S}_0)$, at the moment t,*

$$\mathfrak{S}(t;t_0;\mathfrak{S}_0) = Cl\mathfrak{S}(t;t_0;\mathfrak{S}_0),\ t \in \mathfrak{T}. \tag{3.34}$$

Definition 76 *A set $\mathfrak{S}(t;t_0;\mathfrak{S}_0)$ is a* ***compact set*** *at $(t;t_0) \in \mathfrak{T}_0 \times \mathfrak{T}_i$ if and only if it is both closed set and bounded set at $(t;t_0) \in \mathfrak{T}_0 \times \mathfrak{T}_i$.*

It is compact on $\mathfrak{T}_0 \times \mathfrak{T}_i$ if and only if it is compact at every $(t;t_0) \in \mathfrak{T}_0 \times \mathfrak{T}_i$.

For stability properties, for tracking properties [232], and for their studies, it is inherent to clarify whether they are valid on *time*-invariant sets or on *time*-varying sets. A short description of several important phenomena of *time*-varying sets [115], [195, pp. 33-42] will be explained.

Definition 77 ***The set difference*** *between the set $\mathfrak{A}(t)$ and the set $\mathfrak{B}(t)$ at a moment $t \in \mathfrak{T}_0$ is the set $\mathfrak{A}(t)\backslash\mathfrak{B}(t)$. It is the set of all vectors $\mathbf{S}$ in $\mathfrak{A}(t)$, which do not belong to $\mathfrak{B}(t)$ at the moment t,*

$$\mathfrak{A}(t)\backslash\mathfrak{B}(t) = \{\mathbf{S} :\ \mathbf{S} \in \mathfrak{A}(t),\ \mathbf{S} \notin \mathfrak{B}(t)\},\ t \in \mathfrak{T}_0. \tag{3.35}$$

Definition 78 *$\mathfrak{A}(t)\Delta\mathfrak{B}(t)$ is* ***the symmetric set difference*** *between the set $\mathfrak{A}(t)$ and the set $\mathfrak{B}(t)$ at a moment $t\in\mathfrak{T}_0$. It is the set of all vectors $\mathbf{S}$ in $\mathfrak{A}(t)$, which do not belong to $\mathfrak{B}(t)$, and of all vectors $\mathbf{S}$ that are in $\mathfrak{B}(t)$ but not in $\mathfrak{A}(t)$ at the moment t,*

$$\mathfrak{A}(t)\Delta\mathfrak{B}(t) = [\mathfrak{A}(t)\backslash\mathfrak{B}(t)] \cup [\mathfrak{B}(t)\backslash\mathfrak{A}(t)]. \tag{3.36}$$

Set $\mathfrak{D}(t_0) \subseteq \mathfrak{R}^n$ is *the set of the appropriate initial states* $\mathbf{S}_0 \in \mathfrak{R}^n$ at *an initial moment* $t_0 \in \mathfrak{T}_i$,

$$\mathfrak{D}(t_0) = \{\mathbf{S} : \mathbf{S} = \mathbf{S}(t_0) = \mathbf{S}_0 \in \mathfrak{R}^n\} \subseteq \mathfrak{R}^n. \tag{3.37}$$

Definition 79 *Set $\mathfrak{D}(t;t_0) \subseteq \mathfrak{R}^n$ is the* ***set of all system instantaneous state values $\mathbf{S}(t)$ at a moment*** *$t \in \mathfrak{T}_0$* ***of the system motion*** *$\mathcal{S}(.;t_0;\mathbf{S}_0;\mathbf{I})$* ***starting from*** *$\mathbf{S}_0 \in \mathfrak{D}(t_0;t_0) = \mathfrak{D}(t_0)$* ***at an initial moment*** *$t_0 \in \mathfrak{T}_i$,*

$$\mathfrak{D}(t;t_0) = \{\mathbf{S} : \mathbf{S} = \mathbf{S}(t) = \mathbf{S}(t;t_0;\mathbf{S}_0;\mathbf{I}),\ \mathbf{S}_0 \in \mathfrak{D}(t_0)\},\ \mathfrak{D}(t_0;t_0) = \mathfrak{D}(t_0). \tag{3.38}$$

It is ***the time development of the set*** *$\mathfrak{D}(t_0)$* ***along system motions*** *$\mathcal{S}(.;t_0;\mathbf{S};\mathbf{I})$* ***at a moment*** *$t \in \mathfrak{T}_0$.*

Definition 80 *Set $\mathfrak{D}_m(\mathfrak{T}_i)$ is* ***the intersection,*** *i.e.,* ***the lower limit set, of the set*** *$\mathfrak{D}(t_0)$* ***on*** *$\mathfrak{T}_i$,*

$$\mathfrak{D}_m(\mathfrak{T}_i) = \cap[\mathfrak{D}(t_0) : t_0 \in \mathfrak{T}_i], \tag{3.39}$$

and set $\mathfrak{D}_m(t;\mathfrak{T}_i) \subseteq \mathfrak{R}^n$ is ***the intersection of*** *$\mathfrak{D}(t;t_0)$* ***in*** *t_0 over $\mathfrak{T}_i$ at a moment $t \in \mathfrak{T}_0$,*
:

$$\mathfrak{D}_m(t;\mathfrak{T}_i) = \cap[\mathfrak{D}(t;t_0) : t_0 \in \mathfrak{T}_i],\ t \in \mathfrak{T}_0. \tag{3.40}$$

Definition 81 *The intersection set $\mathfrak{S}_m(\mathfrak{T}_0)$, of $\mathfrak{S}(t)$ in t over $\mathfrak{T}_0$ is* ***the lower set limit of the set*** *$\mathfrak{S}(t)$* ***on*** *$\mathfrak{T}_i$,*

$$\mathfrak{S}_m(\mathfrak{T}_0) = \cap[\mathfrak{S}(t) : t \in \mathfrak{T}_0]. \tag{3.41}$$

Analogously is defined $\mathfrak{S}_m(\mathfrak{T})$,

$$\mathfrak{S}_m(\mathfrak{T}) = \cap[\mathfrak{S}(t) : t \in \mathfrak{T}]. \tag{3.42}$$

Besides,

$$\mathfrak{S}_{0m}(\mathfrak{T}_i) \equiv \cap[\mathfrak{S}(t_0) : t_0 \in \mathfrak{T}_i],\ \mathfrak{S}_{0m} \equiv \cap[\mathfrak{S}(t_0) : t_0 \in \mathfrak{T}]. \tag{3.43}$$

The intersection set $\mathfrak{S}_m(t;\mathfrak{T}_i)$, of $\mathfrak{S}(t;t_0;\mathfrak{S}_0)$ in t_0 over $\mathfrak{T}_i$ at the moment $t \in \mathfrak{T}_0$ is ***the lower set limit of the set*** *$\mathfrak{S}(t;t_0;\mathfrak{S}_0)$* ***on*** *$\mathfrak{T}_i$* ***at the moment*** *$t \in \mathfrak{T}_0$,*

$$\begin{aligned}
\mathfrak{S}_m(t;\mathfrak{T}_i;\mathfrak{S}_0) &\equiv \cap[\mathfrak{S}(t;t_0;\mathfrak{S}_0) : t_0 \in \mathfrak{T}_i],\ t \in \mathfrak{T}_0,\\
\mathfrak{S}_m(\mathfrak{T}_i;t_0;\mathfrak{S}_0) &\equiv \cap[\mathfrak{S}(t;t_0;\mathfrak{S}_0) : t \in \mathfrak{T}_i]\\
\mathfrak{S}_m(\mathfrak{T}_i;\mathfrak{S}_0) = \mathfrak{S}_m(\mathfrak{T}_i;\mathfrak{T}_i;\mathfrak{S}_0) &\equiv \cap[\mathfrak{S}(t_0;t_0;\mathfrak{S}_0) : t_0 \in \mathfrak{T}_i] =\\
&= \cap[\mathfrak{S}(t_0) : t_0 \in \mathfrak{T}_i] = \cap[\mathfrak{S}_0 : t_0 \in \mathfrak{T}_i].
\end{aligned} \tag{3.44}$$

Definition 82 *Set $\mathfrak{S}_M(\mathfrak{T}_i)$ is the union set of $\mathfrak{S}_t \equiv \mathfrak{S}(t;t;\mathfrak{S}_t)$ on $\mathfrak{T}_i$:*
$\mathfrak{S}_M(\mathfrak{T}_i) = \cup[\mathfrak{S}(t;t;\mathfrak{S}_t) : t \in \mathfrak{T}_i]$. It is ***the upper set limit of the set*** *$\mathfrak{S}(t)$* ***on*** *$\mathfrak{T}_I$.*

Set $\mathfrak{S}_M(t;\mathfrak{T}_i;\mathfrak{S}_0)$ is the union set of $\mathfrak{S}(t;t_0;\mathfrak{S}_0)$ on $\mathfrak{T}_i$ at the moment $t \in \mathfrak{T}_0$. It is ***the upper set limit of the set*** *$\mathfrak{S}(t;t_0;\mathfrak{S}_0)$* ***on*** *$\mathfrak{T}_i$* ***at the moment*** *$t \in \mathfrak{T}_0$.*

$$\begin{aligned}
\mathfrak{S}_M(t;\mathfrak{T}_i;\mathfrak{S}_0) &\equiv \cup[\mathfrak{S}(t;t_0;\mathfrak{S}_0) : t_0 \in \mathfrak{T}_i],\ t \in \mathfrak{T}_0,\\
\mathfrak{S}_M(\mathfrak{T}_i;\mathfrak{S}_0) = \mathfrak{S}_M(\mathfrak{T}_i;\mathfrak{T}_i;\mathfrak{S}_0) &\equiv \cup[\mathfrak{S}(t_0;t_0;\mathfrak{S}_0) : t_0 \in \mathfrak{T}_i] =\\
&= \cup[\mathfrak{S}(t_0) : t_0 \in \mathfrak{T}_i] = \cup[\mathfrak{S}_0 : t_0 \in \mathfrak{T}_i].
\end{aligned} \tag{3.45}$$

Definition 83 *$Ev_{\mathfrak{T}_i}\mathfrak{S}(t)$ is* ***the time evolution of the set*** *$\mathfrak{S}(t) \equiv \mathfrak{S}(t;t)$* ***in*** *$t$* ***over*** *$\mathfrak{T}_i$, i.e., the set $\mathfrak{S}(t)$ evolution in t over $\mathfrak{T}_i$, which is composed of all pairs $(t;\mathbf{S}) \in \mathfrak{T}_i \times \mathfrak{S}(t)$.*

$$Ev_{\mathfrak{T}_i}\mathfrak{S}(t) = \{(t,\mathbf{S}) \in \mathfrak{T} \times \mathfrak{R}^n : \ (t,\mathbf{S}) \in \mathfrak{T}_i \times \mathfrak{S}(t)\} \subseteq \mathfrak{T} \times R^n, \tag{3.46}$$

$Ev_{\mathfrak{T}_i}\mathfrak{S}(t)$ is a subset of **the integral state space** $\mathfrak{I}$ defined by (3.47) (see in the sequel Definition 115, Equations (4.41), Section 4.2.1):

$$\mathfrak{I} = \mathfrak{T} \times \mathfrak{R}^n. \tag{3.47}$$

If and only if the set $\mathfrak{S}(t_0)$ is constant on $\mathfrak{T}_i$, $\mathfrak{S}(t_0) = \mathfrak{S}$ for all $t_0 \in \mathfrak{T}_i$, then $Ev_{\mathfrak{T}_i}\mathfrak{S}(t_0) = \mathfrak{T}_i \times \mathfrak{S}$.

Analogously,

$$Ev_{\mathfrak{T}_0}\mathfrak{D}(t;\mathfrak{T}_i) = \{(t,\mathbf{S}) \in \mathfrak{T}_0 \times \mathfrak{R}^n : \ (t,\mathbf{S}) \in \mathfrak{T}_0 \times \mathfrak{D}(t;\mathfrak{T}_i)\} \subseteq \mathfrak{T}_0 \times \mathfrak{R}^n, \tag{3.48}$$

Let $\mathfrak{B}_\xi(\mathbf{a};t;t_0)$ be an open hyperball with the radius $\xi > 0$ centered at the point $\mathbf{a}(t;t_0)$ in the space $\mathfrak{R}^n$ at $t \in \mathfrak{T}_0$,

$$\mathfrak{B}_\xi(\mathbf{a};t;t_0) = \{\mathbf{S} : \mathbf{S} \in \mathfrak{R}^n, \ \|\mathbf{S} - \mathbf{a}(t;t_0)\| < \xi\} \subseteq .\mathfrak{R}^n, \ t \in \mathfrak{T}_0, \ t_0 \in \mathfrak{T}_i.$$

If and only if $\mathbf{a}(t;t_0) \equiv \mathbf{0}_n$, then we omit $(\mathbf{0}_n)$ from the notation $\mathfrak{B}_\xi(\mathbf{0}_n)$,

$$\mathfrak{B}_\xi = \mathfrak{B}_\xi(\mathbf{0}_n) = \{\mathbf{S} : \mathbf{S} \in \mathfrak{R}^n, \ \|\mathbf{S}\| < \xi\}. \tag{3.49}$$

3.2.1 Neighborhood of a time-varying set

Fraktur letter $\mathfrak{N}$ denotes a neighborhood.

Definition 84 ***Neighborhood of a time-varying nonempty set*** *$\mathfrak{S}(t)$*

A nonempty variable set $\mathfrak{N}[\mathfrak{S}(t)]$, $\mathfrak{N}[\mathfrak{S}(t)] \subseteq \mathfrak{R}^n$, is a ***neighborhood of a time-varying nonempty set*** *$\mathfrak{S}(t)$* ***at a moment*** *$t \in \mathfrak{T}$ if and only if there exists $\mu(t)$, $\mu(t) \in \mathfrak{R}^+$ or $\mu(t) = \infty$, such that the set $\{\mathbf{S} : \mathbf{S} \in \mathfrak{R}^n, \ d[\mathbf{S}, \mathfrak{S}(t)] < \mu(t)\}$ is a subset of the set $\mathfrak{N}[\mathfrak{S}(t)]$* ***at the moment*** *$t \in \mathfrak{T}$:*

$$\begin{gathered} t \in \mathfrak{T}, \ \exists \mathfrak{N}[\mathfrak{S}(t)] \Longleftrightarrow \exists \mu(t) \in \mathfrak{R}^+ \cup \{\infty\} \Longrightarrow \\ \mathfrak{S}(t) \subset \ \{\mathbf{S} : \mathbf{S} \in \mathfrak{R}^n, \ d[\mathbf{S},\mathfrak{S}(t)] < \mu(t)\} \subseteq \mathfrak{N}[\mathfrak{S}(t)]. \end{gathered} \tag{3.50}$$

If, and only if, this holds for every $t \in \mathfrak{T}$ ($t \in \mathfrak{T}_0$), then the set $\mathfrak{N}[\mathfrak{S}(t)]$ is a ***variable neighborhood of a time-varying nonempty set*** *$\mathfrak{S}(t)$* ***at every moment*** *$t \in \mathfrak{T}$ ($t \in \mathfrak{T}_0$), respectively.*

Every neighborhood $\mathfrak{N}[\mathfrak{S}(t)]$ of a time-varying nonempty set $\mathfrak{S}(t)$ at $t \in \mathfrak{T}$ is also a nonempty set at $t \in \mathfrak{T}$. If the set $\mathfrak{S}(t)$ is additionally connected set at $t \in \mathfrak{T}$, then its neighborhood $\mathfrak{N}[\mathfrak{S}(t)]$ is also connected at $t \in \mathfrak{T}$.

A nonempty constant set $\mathfrak{N}[\mathfrak{T}_i, \mathfrak{S}_M(\mathfrak{T}_i)]$, $\mathfrak{N}[\mathfrak{T}_i, \mathfrak{S}_M(\mathfrak{T}_i)] \subseteq \mathfrak{R}^n$, is a ***neighborhood of a time-varying set*** *$\mathfrak{S}(t)$* ***on*** *$\mathfrak{T}_i$ if and only if there exists $\mu(\mathfrak{T}_i)$, $\mu(\mathfrak{T}_i) \in \mathfrak{R}^+$ or $\mu(\mathfrak{T}_i) = \infty$, such that the set $\{\mathbf{S} : \mathbf{S} \in \mathfrak{R}^n, \ d[\mathbf{S}, \mathfrak{S}_M(\mathfrak{T}_i)] < \mu(\mathfrak{T}_i)\}$ is a subset of the set $\mathfrak{N}[\mathfrak{T}_i, \mathfrak{S}_M(\mathfrak{T}_i)]$ at every moment $t \in \mathfrak{T}_i$,*

$$\begin{gathered} \exists \mathfrak{N}[\mathfrak{T}_i, \mathfrak{S}_M(\mathfrak{T}_i)] \Longleftrightarrow \exists \mu(\mathfrak{T}_i) \in \mathfrak{R}^+ \cup \{\infty\} \Longrightarrow \\ \mathfrak{Cl}\mathfrak{S}_M(\mathfrak{T}_i) \subset \ \{\mathbf{S} : \mathbf{S} \in \mathfrak{R}^n, \ d[\mathbf{S}, \mathfrak{S}_M(\mathfrak{T}_i)] < \mu(\mathfrak{T}_i)\} \subseteq \mathfrak{N}[\mathfrak{T}_i, \mathfrak{S}_M(\mathfrak{T}_i)], \ \forall t \in \mathfrak{T}_i. \end{gathered} \tag{3.51}$$

Comment 85 *The notation $\mathfrak{N}[\mathfrak{S}(t)]$ clarifies that the set $\mathfrak{N}[\mathfrak{S}(t)]$ is a neighborhood of a set $\mathfrak{S}(t)$ and that $\mathfrak{N}[\mathfrak{S}(t)]$ varies with time if and only if $\mathfrak{S}(t)$ is a time-varying set.*

Comment 86 *Every neighborhood of a nonempty set $\mathfrak{S}(t)$ at a moment $t \in \mathfrak{T}$ is also a nonempty set at the same moment $t \in \mathfrak{T}$.*

Lemma 87 ***Neighborhood and time evolutions of sets***

Let $\mathfrak{N}_k[\mathfrak{S}(t)]$, $k = 1, 2$, be two different neighborhoods of a set $\mathfrak{S}(t)$ at $t \in \mathfrak{T}_\omega$. Then:

1) The intersection $\mathfrak{I}\{t; \mathfrak{N}_k[\mathfrak{S}(t)]\}$ of two neighborhoods $\mathfrak{N}_k[\mathfrak{S}(t)]$, $k = 1, 2$, of the set $\mathfrak{S}(t)$ at any $t \in \mathfrak{T}_\omega$ is also a neighborhood of $\mathfrak{S}(t)$ at any $t \in \mathfrak{T}_\omega$.

2) The intersection of the time evolutions $Ev_{\mathfrak{T}_\omega}\mathfrak{N}_k[\mathfrak{S}(t)]$, $k = 1, 2$, of the neighborhoods $\mathfrak{N}_k[\mathfrak{S}(t)]$ of the set $\mathfrak{S}(t)$ on $\mathfrak{T}_\omega$ is also time evolution of some neighborhood $\mathfrak{N}[\mathfrak{S}(t)]$ of the set $\mathfrak{S}(t)$ on $\mathfrak{T}_\omega$:

$$\begin{gathered}\exists \mathfrak{N}[\mathfrak{S}(t)] \subseteq \mathfrak{N}_k[\mathfrak{S}(t)], \ k = 1, 2, \ \forall t \in \mathfrak{T}_\omega \Longrightarrow \\ Ev_{\mathfrak{T}_\omega}\mathfrak{N}_1[\mathfrak{S}(t)] \cap Ev_{\mathfrak{T}_\omega}\mathfrak{N}_2[\mathfrak{S}(t)] = Ev_{\mathfrak{T}_\omega}\mathfrak{N}[\mathfrak{S}(t)].\end{gathered} \tag{3.52}$$

Proof. Let $\mathfrak{N}_k[\mathfrak{S}(t)], k = 1, 2$, be neighborhoods of the set $\mathfrak{S}(t)$ at $t \in \mathfrak{T}_\omega$, and let $\mathfrak{N}_\cap(t; \mathfrak{S})$ be their intersection,

$$\mathfrak{N}_\cap(t; \mathfrak{S}) = \mathfrak{N}_1[\mathfrak{S}(t)] \cap \mathfrak{N}_2[\mathfrak{S}(t)].$$

The set $\mathfrak{S}(t)$ is subset of its neighborhood $\mathfrak{N}_k[\mathfrak{S}(t)]$, $k = 1, 2$, Definition 84.

1) Let $t \in \mathfrak{T}_\omega$ be arbitrarily chosen and let $\mu(t) > 0$ be such that

$$d\{\partial\mathfrak{S}(t), \partial\mathfrak{N}_k[\mathfrak{S}(t)]\} > \mu(t) > 0, \ \forall k = 1, 2. \tag{3.53}$$

Such $\mu(t) > 0$ exists due to (3.50), Definition 84. The neighborhood $\mathfrak{N}_{\mu(t)/2}[\mathfrak{S}(t)]$ of the set $\mathfrak{S}(t)$ is simultaneously subset of both $\mathfrak{N}_1[\mathfrak{S}(t)]$ and $\mathfrak{N}_2[\mathfrak{S}(t)]$

$$\mathfrak{N}_\cap[\mathfrak{S}(t)] = \mathfrak{N}_1[t; \mathfrak{S}(t)] \cap \mathfrak{N}_2[t; \mathfrak{S}(t)] \supseteq \mathfrak{N}_{\mu(t)/2}[\mathfrak{S}(t)] \neq \phi, \ \forall t \in \mathfrak{T}_\omega.$$

The intersection $\mathfrak{N}_\cap(t; \mathfrak{S})$ of the two neighborhoods $\mathfrak{N}_1[\mathfrak{S}(t)]$ and $\mathfrak{N}_2[\mathfrak{S}(t)]$ of a set $\mathfrak{S}(t)$ at $t \in \mathfrak{T}_\omega$ is also some neighborhood $\mathfrak{N}[\mathfrak{S}(t)]$ of $\mathfrak{S}(t)$ at any $t \in \mathfrak{T}_\omega$.

2) Definition 84 and the result under 1) guarantee the existence of the neighborhood $\mathfrak{N}[\mathfrak{S}(t)]$ that obeys $\mathfrak{N}[\mathfrak{S}(t)] \subseteq \mathfrak{N}_1[\mathfrak{S}(t)]$ and $\mathfrak{N}[\mathfrak{S}(t)] \subseteq \mathfrak{N}_2[\mathfrak{S}(t)]$ at every $t \in \mathfrak{T}_\omega$. Let $\mathfrak{N}[\mathfrak{S}(t)] = \cap\{\mathfrak{N}_k[\mathfrak{S}(t)], k = 1, 2\}, \forall t \in \mathfrak{T}_\omega$. This implies

$$Ev_{\mathfrak{T}_T}\mathfrak{N}_1[\mathfrak{S}(t)] \cap Ev_{\mathfrak{T}_T}\mathfrak{N}_2[\mathfrak{S}(t)] = Ev_{\mathfrak{T}_T}\mathfrak{N}[\mathfrak{S}(t)]$$

and completes the proof. ■

Definition 88 ***The equivalent definition of a neighborhood of a time-varying nonempty set $\mathfrak{S}(t)$ on $\mathfrak{T}_\omega$***

A time-varying set $\mathfrak{N}[\mathfrak{S}(t)]$ is ***a neighborhood of the time-varying nonempty set $\mathfrak{S}(t)$ on $\mathfrak{T}_\omega$*** *if and only if the evolution $Ev_{\mathfrak{T}_\omega}\mathfrak{N}[\mathfrak{S}(t)]$ of $\mathfrak{N}[\mathfrak{S}(t)]$ in $t \in \mathfrak{T}_\omega$ is a neighborhood of the evolution $Ev_{\mathfrak{T}_\omega}\mathfrak{S}(t)$ of $\mathfrak{S}(t)$ in $t \in \mathfrak{T}_\omega$,*

$$Ev_{\mathfrak{T}_\omega}\mathfrak{S}(t) \subset Ev_{\mathfrak{T}_\omega}\mathfrak{N}[\mathfrak{S}(t)].$$

Let

$$\mathfrak{N}_{\mu(t)}[\mathfrak{S}(t)] = \{\mathbf{S} : d[\mathbf{S}, \mathfrak{S}(t)] < \mu(t)\}. \tag{3.54}$$

Definition 89 ***$\mu(t)$-neighborhood of a time-varying set $\mathfrak{S}(t)$***

If and only if $\mu(t) \in \mathfrak{R}^+$or $\mu(t) = \infty$ and

$$\mathfrak{N}[\mathfrak{S}(t)] = \{\mathbf{S} : d[\mathbf{S},\mathfrak{S}(t)] < \mu(t)\}, \tag{3.55}$$

then the set $\mathfrak{N}[\mathfrak{S}(t)]$ is ***the $\mu(t)$-neighborhood $\mathfrak{N}_{\mu(t)}[\mathfrak{S}(t)]$ of the time-varying set $\mathfrak{S}(t)$ at a moment*** *$t \in \mathfrak{T}$,*

$$t \in \mathfrak{T},\ \mu(t) \in \mathfrak{R}^+ or\ \mu(t) = \infty,$$
$$\mathfrak{N}[\mathfrak{S}(t)] = \mathfrak{N}_{\mu(t)}[\mathfrak{S}(t)] \Longleftrightarrow \{\mathbf{S} : d[\mathbf{S},\mathfrak{S}(t)] < \mu(t)\} = \mathfrak{N}[\mathfrak{S}(t)]. \tag{3.56}$$

If and only if (3.56) holds for every $t \in \mathfrak{T}_\omega$, then the $\mu(t)$-neighborhood $\mathfrak{N}_{\mu(t)}[\mathfrak{S}(t)]$ of the set $\mathfrak{S}(t)$ is ***the $\mu(t)$-neighborhood $\mathfrak{N}_{\mu(t)}[\mathfrak{S}(t);\mathfrak{T}_\omega]$ of the set $\mathfrak{S}(t)$ on $\mathfrak{T}_\omega$.***

Note 90 *The set $\mathfrak{S}(t)$ is a proper subset of its $\mu(t)$-neighborhood $\mathfrak{N}_{\mu(t)}[\mathfrak{S}(t);\mathfrak{T}_\omega]$ for every $t \in \mathfrak{T}_\omega$,*

$$\mu(t) \in \mathfrak{R}^+ \cup \{\infty\} \Longrightarrow \mathfrak{S}(t) \subset \mathfrak{N}_{\mu(t)}[\mathfrak{S}(t);\mathfrak{T}_\omega]\ \forall t \in \mathfrak{T}_\omega. \tag{3.57}$$

Definition 91 ***Permanently following neighborhood of a time-varying nonempty set $\mathfrak{S}(t)$ with a nonempty boundary $\partial\mathfrak{S}(t)$ on $\mathfrak{T}_\omega$***

A neighborhood $\mathfrak{N}[\mathfrak{S}(t)]$ of the set $\mathfrak{S}(t)$ with a nonempty boundary $\partial\mathfrak{S}(t)$ is ***a permanently following (PF) neighborhood $\mathfrak{N}_{PF}[\mathfrak{S}(t);\mathfrak{T}_\omega]$ on $\mathfrak{T}_\omega$ of the set $\mathfrak{S}(t)$*** *if and only if the closure $Cl\mathfrak{S}(t)$ of $\mathfrak{S}(t)$ is a proper subset of the interior $In\mathfrak{N}[\mathfrak{S}(t)]$ of $\mathfrak{N}[\mathfrak{S}(t)]$ for every $t \in \mathfrak{T}_\omega$ and there is $\mu(t) > 0$ for every $t \in \mathfrak{T}_\omega$ such that the smallest distance $d(.)$ between the boundaries $\partial\mathfrak{S}(t)$ of $\mathfrak{S}(t)$ and $\partial\mathfrak{N}[\mathfrak{S}(t)]$ of $\mathfrak{N}[\mathfrak{S}(t)]$ on $\mathfrak{T}_\omega$ is not less than $\mu(t)$ at any moment $t \in \mathfrak{T}_\omega$,*

$$\mathfrak{N}[\mathfrak{S}(t)] = \mathfrak{N}_{PF}[\mathfrak{S}(t);\mathfrak{T}_\omega] \Longleftrightarrow Cl\mathfrak{S}(t) \subset In\mathfrak{N}(t)\ \forall t \in \mathfrak{T}_\omega \wedge$$
$$\exists\mu(t) > 0\ \forall t \in \mathfrak{T}_\omega \Longrightarrow d\{\partial\mathfrak{S}(t), \partial\mathfrak{N}(t)\} \geq \mu(t),\ \forall t \in \mathfrak{T}_\omega. \tag{3.58}$$

If and only if (3.58) holds for a constant $\mu > 0$, $\mu(t) = \mu > 0\ \forall t \in \mathfrak{T}_\omega$, then the time-varying set $\mathfrak{N}[\mathfrak{S}(t)]$ is ***a uniform permanently following (UPF) neighborhood $\mathfrak{N}_{UPF}[\mathfrak{S}(t);\mathfrak{T}_\omega]$ of the set $\mathfrak{S}(t)$ on $\mathfrak{T}_\omega$.***

Definition 92 ***Fixed following neighborhood of a time-varying nonempty set $\mathfrak{S}(t)$ on $\mathfrak{T}_\omega$***

A time-varying set $\mathfrak{N}[\mathfrak{S}(t)]$ is ***a fixed following (FX) neighborhood $\mathfrak{N}_{FX}[\mathfrak{S}(t);\mathfrak{T}_\omega]$ of the set $\mathfrak{S}(t)$ on $\mathfrak{T}_\omega$*** *if and only if the set $\mathfrak{S}(t)$ is a proper subset of the set $\mathfrak{N}[\mathfrak{S}(t)]$ on $\mathfrak{T}_\omega$ and for a known given constant $\mu \in \mathfrak{R}^+$ the smallest distance between the boundaries $\partial\mathfrak{S}(t)$ of $\mathfrak{S}(t)$ and $\partial\mathfrak{N}[\mathfrak{S}(t)]$ of $\mathfrak{N}[\mathfrak{S}(t)]$ on $\mathfrak{T}_\omega$ is not less than μ,*

$$\mathfrak{N}[\mathfrak{S}(t)] = \mathfrak{N}_{Fx}[\mathfrak{S}(t);\mathfrak{T}_\omega] \Longleftrightarrow \forall t \in \mathfrak{T}_\omega \Longrightarrow \mathfrak{S}(t) \subset \mathfrak{N}[\mathfrak{S}(t)]$$
$$\wedge for\ a\ known\ given\ \mu \in \mathfrak{R}^+ \Longrightarrow inf\{d[\partial\mathfrak{S}(t), \partial\mathfrak{N}(t)] : t \in \mathfrak{T}_\omega\} \geq \mu. \tag{3.59}$$

Note 93 ***On "following neighborhoods"***

The term "following neighborhood" specifies that the "following neighborhood" of a movable set $\mathfrak{S}(t)$ follows the set $\mathfrak{S}(t)$ movement by preserving to be a neighborhood of the set $\mathfrak{S}(t)$ at every moment.

Note 94 ***On "following neighborhoods of a constant set"***

If a set is constant, then its "following neighborhood" is also a constant set.

Comment 95 ***Permanent, uniform and fixed following neighborhoods of a set***
Time variations of the variable set $\mathfrak{S}(t)$ induce time variations of both the fixed following neighborhood and the (uniform) permanently following neighborhood of the set $\mathfrak{S}(t)$.

A time-varying fixed following neighborhood $\mathfrak{N}_{FX}\left[\mathfrak{S}\left(t\right);\mathfrak{T}_{\omega}\right]$ of a time-varying nonempty set $\mathfrak{S}(t)$ on $\mathfrak{T}_{\omega}$ is also a permanently following neighborhood $\mathfrak{N}_{PF}\left[\mathfrak{S}\left(t\right);\mathfrak{T}_{\omega}\right]$ and a uniform permanently following neighborhood $\mathfrak{N}_{UPF}\left[\mathfrak{S}\left(t\right);\mathfrak{T}_{\omega}\right]$ of the set $\mathfrak{S}(t)$ on $\mathfrak{T}_{\omega}$. This does not mean that they are equal.

Comment 96 ***Variable sets, their neighborhoods and systems***
Let a constant $\mu \in \mathfrak{R}^{+}$ or $\mu = \infty$.

1. If a set is time-varying, i.e., variable, e.g., the set $\mathfrak{S}(t;t_0;\mathfrak{S}_0)$, then for every constant $\mu > 0$ its μ-neighborhood $\mathfrak{N}_{\mu}\left[\mathfrak{S}(t;t_0;\mathfrak{S}_0)\right]$ is time-varying, i.e., it is implicitly dependent on both the current moment t and the initial moment t_0 through the set $\mathfrak{S}(t;t_0;\mathfrak{S}_0)$.

2. If a set is time-varying and a specific set of a time-invariant system, e.g., the set $\mathfrak{S}(t;t_0;\mathfrak{S}_0)$, then for any constant $\mu \in \mathfrak{R}^{+}$ its μ-neighborhood is also time-varying, i.e., implicitly dependent on both the current moment t and the initial moment t_0 through the set $\mathfrak{S}(t;t_0;\mathfrak{S}_0) : \mathfrak{N}_{\mu}\left[\mathfrak{S}(t;t_0;\mathfrak{S}_0)\right]$.

3) If a set is time-varying and a specific set of a time-varying system, e.g., the set $\mathfrak{S}(t;t_0;\mathfrak{S}_0)$, then for any constant $\mu \in \mathfrak{R}^{+}$ its μ-neighborhood is also time-varying, i.e., explicitly and implicitly dependent on both the current moment t and the initial moment $t_0 : \mathfrak{N}_{\mu}\left[t;t_0;\mathfrak{S}(t;t_0;\mathfrak{S}_0)\right]$, due to temporal variations of both the system properties and the set $\mathfrak{S}(t;t_0;\mathfrak{S}_0)$. Then,

$$\mathfrak{N}_{\mu}\left[t_0;t_0;\mathfrak{S}\left(t_0;t_0\right)\right] \equiv \mathfrak{N}_{\mu}\left(t_0;t_0;\mathfrak{S}_0\right). \tag{3.60}$$

4. If a set should be such a neighborhood $\mathfrak{N}\left[\mu;.;\mathfrak{S}(t;t_0;\mathfrak{S}_0)\right]$ of a time-varying set $\mathfrak{S}(t;t_0;\mathfrak{S}_0)$ that the system motions starting from $\mathfrak{N}\left[\mu;.;\mathfrak{S}(t;t_0;\mathfrak{S}_0)\right]$ at an initial moment $t_0 \in \mathfrak{T}_i$ rest in the μ-neighborhood $\mathfrak{N}_{\mu}\left[\mathfrak{S}(t;t_0;\mathfrak{S}_0)\right]$ of the time-varying set $\mathfrak{S}(t;t_0;\mathfrak{S}_0)$ for all $t \in \mathfrak{T}_0$, then, for any constant $\mu \in \mathfrak{R}^{+}$:

a) the neighborhood $\mathfrak{N}\left[\mu;....;\mathfrak{S}(t;t_0;\mathfrak{S}_0)\right]$ of the set $\mathfrak{S}(t;t_0;\mathfrak{S}_0)$ is time-varying and depends implicitly on both the current moment t and the initial moment t_0 despite the system is time-invariant: $\mathfrak{N}\left[\mu;.;\mathfrak{S}(t;t_0;\mathfrak{S}_0)\right] = \mathfrak{N}\left[\mu;\mathfrak{S}(t;t_0;\mathfrak{S}_0)\right]$,

b) the neighborhood $\mathfrak{N}\left[\mu;...;\mathfrak{S}(t;t_0;\mathfrak{S}_0)\right]$ of the set $\mathfrak{S}(t;t_0;\mathfrak{S}_0)$ depends explicitly and implicitly on the initial moment t_0 if the system is time-varying: $\mathfrak{N}\left[\mu;...;\mathfrak{S}(t;t_0;\mathfrak{S}_0)\right] = \mathfrak{N}\left[\mu;t_0;\mathfrak{S}(t;t_0;\mathfrak{S}_0)\right]$.

Comment 97 *Let $\mu \in \mathfrak{R}^{+}$. The μ-neighborhood $\mathfrak{N}_{\mu}\left[\mathfrak{S}(t;t_0;\mathfrak{S}_0);\mathfrak{T}_{\omega}\right]$ of the set $\mathfrak{S}(t;t_0;\mathfrak{S}_0)$ on $\mathfrak{T}_{\omega}$ is also the fixed, the uniform and the permanent time-varying neighborhood of the set $\mathfrak{S}(t;t_0;\mathfrak{S}_0)$ on $\mathfrak{T}_{\omega}$. While the boundary of the permanent, of the uniform and of the fixed neighborhood of the set $\mathfrak{S}(t;t_0;\mathfrak{S}_0)$ is not defined, the boundary $\partial\mathfrak{N}_{\mu}\left[\mathfrak{S}(t;t_0;\mathfrak{S}_0);\mathfrak{T}_{\omega}\right]$ of the $\mu-$neighborhood $\mathfrak{N}_{\mu}\left[\mathfrak{S}(t;t_0;\mathfrak{S}_0);\mathfrak{T}_{\omega}\right]$ of the set $\mathfrak{S}(t;t_0;\mathfrak{S}_0)$ on $\mathfrak{T}_{\omega}$ is well defined,*

$$\partial\mathfrak{N}_{\mu}\left[\mathfrak{S}(t;t_0;\mathfrak{S}_0);\mathfrak{T}_{\omega}\right] = \left\{\mathbf{S} : d\left[\mathbf{S},\mathfrak{S}\left(t\right)\right] = \mu\right\},\ \forall t \in \mathfrak{T}_{\omega}. \tag{3.61}$$

If the set is time-invariant set $\mathfrak{S}$, then its μ-neighborhood $\mathfrak{N}_{\mu}\left(\mathfrak{S};\mathfrak{T}_{\omega}\right)$ becomes its time-invariant, i.e., constant, μ-neighborhood $\mathfrak{N}_{\mu}\left(\mathfrak{S}\right)$, $\mathfrak{N}_{\mu}\left(\mathfrak{S};\mathfrak{T}_{\omega}\right) \equiv \mathfrak{N}_{\mu}\left(\mathfrak{S}\right)$. However, this need not hold for the permanent and uniform neighborhoods of a constant set $\mathfrak{S}$ if the set $\mathfrak{S}$ is a specific set (e.g., the uniform stability domain) of a time-varying system.

Definition 98 ***The permanent following neighborhood of the vector value $\mathbf{Z}(t)$ of a vector function $\mathbf{Z}(.)$***

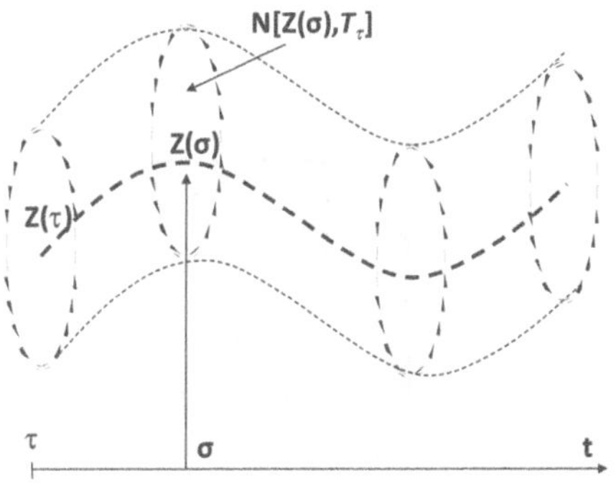

Figure 3.2: A fixed neighborhood $\mathfrak{N}[\mathbf{S}(t);\mathfrak{T}_\tau]$ of a vector function $\mathbf{S}(.)$ on $\mathfrak{T}_\tau$.

*a) A time-varying set $\mathfrak{N}[\mathbf{Z}(t)]$, for short $\mathfrak{N}(t;\mathbf{Z})$, is **a connected permanent following neighborhood of the vector value $\mathbf{Z}(t)\in\mathfrak{R}^n$ of a continuous vector function $\mathbf{Z}(.)$ at** $t\in\mathfrak{T}_\omega$ if and only if the singleton $\mathfrak{Z}(t)=\{\mathbf{Z}(t)\}$ is a proper subset of the $\mathfrak{N}[\mathbf{Z}(t)]$ of $\mathfrak{N}[\mathbf{Z}(t)]$ at $t\in\mathfrak{T}_\omega$,*

$$\mathfrak{Z}(t)=\{\mathbf{Z}(t)\}\subset In\mathfrak{N}[\mathbf{Z}(t)],\quad t\in\mathfrak{T}_\omega, \tag{3.62}$$

and there is $\eta(t)>0$ such that the distance of $\mathbf{Z}(t)$ from the boundary $\partial\mathfrak{N}[\mathbf{Z}(t)]$ of $\mathfrak{N}[\mathbf{Z}(t)]$ is not less than $\eta(t)$ at $t\in\mathfrak{T}_\omega$,

$$\begin{gathered}\mathfrak{Z}(t)=\{\mathbf{Z}(t)\}\subset\mathfrak{N}[\mathbf{Z}(t)],\\ \exists\eta(t)>0\Longrightarrow d\{\mathbf{Z}(t),\partial\mathfrak{N}[t,\mathbf{Z}(t)]\}\geq\eta(t),\quad t\in\mathfrak{T}_\omega,\end{gathered} \tag{3.63}$$

see Figure 3.2.

*If and only if this holds for $\eta(t)\equiv\eta=const.>0$ for every $t\in\mathfrak{T}_\omega$ then $\mathfrak{N}[\mathbf{Z}(t)]$ is **a uniform permanently following neighborhood of the vector value $\mathbf{Z}(t)$ of a vector function $\mathbf{Z}(.)$ on** $\mathfrak{T}_0$, which is denoted by $\mathfrak{N}_u[\mathfrak{T}_\omega;\mathbf{Z}(t)]$.*

Definition 99 ***The μ-neighborhood of the value $\mathbf{Z}(t)$ of a vector function $\mathbf{Z}(.)$: $\mathfrak{T}_\omega\longrightarrow\mathfrak{R}^n$***

*A set $\mathfrak{N}_\mu[\mathbf{Z}(t)]$ is **the μ-neighborhood of the value $\mathbf{Z}(t)$ of a continuous vector function $\mathbf{Z}(.):\mathfrak{T}\longrightarrow\mathfrak{R}^n$ at** $t\in\mathfrak{T}_\omega$ if and only if (3.63) holds for a fixed positive real number μ, $\eta(t)\equiv\mu$, or for $\eta(t)\equiv\mu=\infty$.*

*If and only if this holds for every $t\in\mathfrak{T}_\omega$, then $\mathfrak{N}_\mu[\mathbf{Z}(t);\mathfrak{T}_\omega]$ is **the μ-neighborhood of the vector value $\mathbf{Z}(t)$ of a vector function $\mathbf{Z}(.)$ on** $\mathfrak{T}_\omega$.*

Let $\mathbf{V}\in\mathfrak{R}^n$, $\mathbf{V}=[V_1\ \ V_2...V_n]^T$ and $\mathbf{W}\in\mathfrak{R}^n$, $\mathbf{W}=[W_1\ \ W_2...W_n]^T$. The vector absolute value $|\mathbf{V}|$ of the vector $\mathbf{V}$ is elementwise,

$$|\mathbf{V}|=[|V_1|\ \ |V_2|\ \ ...\ \ |V_n|]^T\in\mathfrak{R}^n_+.$$

Definition 100 ***The elementwise v-neighborhood of the vector value $\mathbf{Z}(t)$ of a vector function $\mathbf{Z}(.):\mathfrak{T}\longrightarrow\mathfrak{R}^n$ at $t\in\mathfrak{T}_0$***

*Let $\mathbf{V}\in\mathfrak{R}^{+n}$, $\mathbf{V}=[V_1\ \ V_2\ ...V_n]^T$. **The elementwise v-neighborhood of the vector value $\mathbf{Z}(t)$ of a vector function $\mathbf{Z}(.):\mathfrak{T}\longrightarrow\mathfrak{R}^n$ at** $t\in\mathfrak{T}_0$ is the vector neighborhood $\mathfrak{N}_{\mathbf{v}}[\mathbf{Z}(t)]$ of $\mathbf{Z}(t)$ at $t\in\mathfrak{T}_0$ defined by*

$$\begin{aligned}&\mathfrak{N}_{\mathbf{v}}[\mathbf{Z}(t)]=\{\mathbf{W}:\mathbf{W}\in\mathfrak{R}^n,\ |\mathbf{W}-\mathbf{Z}(t)|<\mathbf{V}\}\\ &=\left\{\mathbf{W}:\mathbf{W}\in\mathfrak{R}^n,\ |W_i-Z_i(t)|<V_i,\ V_i\in\mathfrak{R}^+,\ \forall i=1,2,\ ...,n\right\},\ t\in\mathfrak{T}_0,\end{aligned} \tag{3.64}$$

see Figure 3.3.

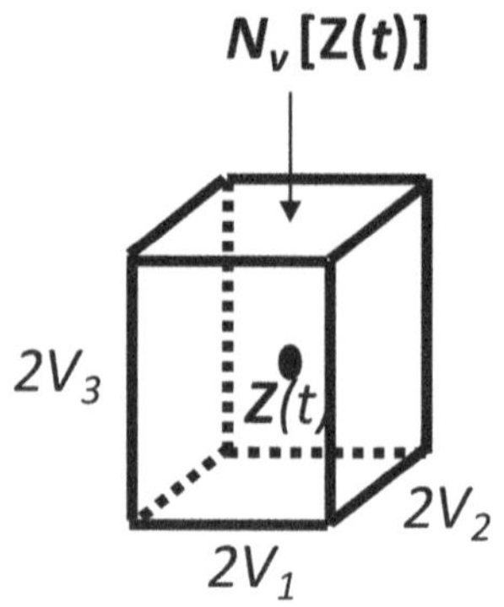

Figure 3.3: The elementwise **v**-neighborhood $\mathfrak{N}_{\mathbf{v}}\left[\mathbf{Z}\left(t\right)\right]$ of the vector value $\mathbf{Z}\left(t\right)$ of a vector function $\mathbf{Z}\left(.\right)$ at $t\in\mathfrak{T}_0$.

3.2.2 Time-varying uniquely bounded sets

Definition 101 *A time-varying set $\mathfrak{S}\left(t\right)$ is* ***uniquely bounded at a moment*** $t\in\mathfrak{T}$ *if and only if:*

1. it is a bounded set at the moment $t\in\mathfrak{T}$,

and

2. there is a point $Z(t)$ at a moment $t\in\mathfrak{T}$ such that every ray from it in any direction passes through the set boundary exactly once at the moment t.

Definition 102 *A time-varying set $\mathfrak{S}\left(t\right)$ is* ***uniquely bounded on*** $\mathfrak{T}_0$ *if and only if:*

1. it is a uniquely bounded set at every moment $t\in\mathfrak{T}_0$,

and

2. there is a point $Z(t)$ at every moment $t\in\mathfrak{T}o$ such that every ray from it in any direction passes through the boundary exactly once at every moment $t\in\mathfrak{T}$.

Definition 103 *A time-varying set $\mathfrak{S}\left(t\right)$ is* ***O-uniquely bounded on*** $\mathfrak{T}_0$ *if and only if:*

1. it is O-uniquely bounded set at every moment $t\in\mathfrak{T}_0$,

and

2. there is in its interior a point $Z(t)$ at every moment $t\in\mathfrak{T}_0$ such that every ray from it in any direction passes through the set boundary exactly once at every moment $t\in\mathfrak{T}_0$.

3.2.3 Continuity of sets

The set operator Δ denotes *the symmetric set difference*, Equation (3.36), between two sets $\mathfrak{A}\left(t\right)$ and $\mathfrak{B}\left(t\right)$, which is the set of all vectors $\mathbf{S}$ in $\mathfrak{A}\left(t\right)$, which do not belong to $\mathfrak{B}\left(t\right)$, and of all vectors that are in $\mathfrak{B}\left(t\right)$ but not in $\mathfrak{A}\left(t\right)$ at a moment $t\in\mathfrak{T}$,

$$\mathfrak{A}\left(t\right)\Delta\mathfrak{B}\left(t\right)=\left[\mathfrak{A}\left(t\right)\backslash\mathfrak{B}\left(t\right)\right]\cup\left[\mathfrak{B}\left(t\right)\backslash\mathfrak{A}\left(t\right)\right],\ t\in\mathfrak{T}. \tag{3.65}$$

It should be distinguished from the ordinary set difference $\mathfrak{A}\left(t\right)\backslash\mathfrak{B}\left(t\right)$, which is the set of all vectors $\mathbf{S}$ in $\mathfrak{A}\left(t\right)$ that do not belong to $\mathfrak{B}\left(t\right)$,

$$\mathfrak{A}\left(t\right)\backslash\mathfrak{B}\left(t\right)=\left\{\mathbf{S}:\ \mathbf{S}\in\mathfrak{A}\left(t\right),\ \mathbf{S}\notin\mathfrak{B}\left(t\right)\right\},\ t\in\mathfrak{T}. \tag{3.66}$$

The following definition refines [232, Definition 43 in Section 2.3, p. 18] by correcting the typographical error.

Definition 104 ***Set continuity in time***

Let a time-varying set $\mathfrak{S}(t)$ *be connected and bounded on* $\mathfrak{T}_0$ *and let* $\tau \in \mathfrak{T}_0$*. Then:*

i) There exists lim $[\mathfrak{S}(t) : t \longrightarrow \tau]$ *if and only if there is a connected bounded constant set* $\mathfrak{L}$ *and for every time sequence* $\mathfrak{T}_{\longrightarrow\tau}$ *converging to* τ*,*

$$\mathfrak{T}_{\longrightarrow\tau} = \{t_k : t_k \in \mathfrak{T},\ k \longrightarrow \infty \Longrightarrow t_k \longrightarrow \tau\}, \tag{3.67}$$

and for every $\varepsilon \in \mathfrak{R}^+$ *there exists a natural number* $\mu = \mu(\tau, \varepsilon)$ *such that*

$$[\mathfrak{S}(t_k)\Delta\mathfrak{L}] \subset \mathfrak{N}(\partial\mathfrak{L};\varepsilon),\quad t_k \in \mathfrak{T}_{\longrightarrow\tau},\ \forall k \in]\mu, \infty[,$$

which is expressed by

$$\lim[\mathfrak{S}(t) : t \longrightarrow \tau] = \mathfrak{L}, \tag{3.68}$$

and the set $\mathfrak{L}$ *is* ***the limit set*** *of* $\mathfrak{S}(t)$ *at* τ*.*

ii) The set $\mathfrak{S}(t)$ *is* ***continuous in*** $t \in \mathfrak{T}_0$, $\mathfrak{T}_0 \subseteq \mathfrak{T}$, $t_0 \in \mathfrak{T}_i$, *if and only if*

$$\lim[\mathfrak{S}(t) : t \longrightarrow \tau] = \mathfrak{S}(\tau),\quad \forall\tau \in \mathfrak{T}_0, \tag{3.69}$$

which is denoted by $\mathfrak{S}(t) \in \mathfrak{C}(\mathfrak{T}_0)$*. This is* ***set continuity on*** $\mathfrak{T}_0$*.*

*The expressions "****on*** $\mathfrak{T}_0$*" and "*$\mathfrak{T}_0$*" should be omitted if and only if the preceding conditions hold also when* $\mathfrak{T}_0$ *and* $\mathfrak{T}_i$ *are replaced by* $\mathfrak{T}$*.*

The distance of $\mathbf{S} \in \mathfrak{R}^n$ from a *time*-varying set $\mathfrak{S}(t) \subset \mathfrak{R}^n$, (68), permits us to introduce *two set distances* $d_{(.)}[\mathfrak{S}_1(t), \mathfrak{S}_2(t)]$, $(.) \in \{m, M\}$, between *time*-varying sets $\mathfrak{S}_1(t) \subset \mathfrak{R}^n$ and $\mathfrak{S}_2(t) \subset \mathfrak{R}^n$ at a moment $t \in \mathfrak{T}$, [324].

The inner set distance $d_m[\mathfrak{S}_1(t), \mathfrak{S}_2(t)]$ *between the sets* $\mathfrak{S}_1(t)$ *and* $\mathfrak{S}_2(t)$ *at* $t \in \mathfrak{T}$:

$$\begin{gathered} d_m[\mathfrak{S}_1(t), \mathfrak{S}_2(t)] = inf\{d(\mathbf{x}, \mathbf{y}) : \mathbf{x} \in \mathfrak{S}_1(t),\ \mathbf{y} \in \mathfrak{S}_2(t)\} = \\ = min\left\{ \begin{array}{c} inf\langle d[\mathbf{S}, \mathfrak{S}_1(t)] : \mathbf{S} \in \mathfrak{S}_2(t)\rangle, \\ inf\langle d[\mathbf{S}, \mathfrak{S}_2(t)] : \mathbf{S} \in \mathfrak{S}_1(t)\rangle \end{array} \right\}, \end{gathered} \tag{3.70}$$

and *the outer set distance* $d_M[\mathfrak{S}_1(t), \mathfrak{S}_2(t)]$ *between the sets* $\mathfrak{S}_1(t)$ *and* $\mathfrak{S}_2(t)$ *at* $t \in \mathfrak{T}$:

$$\begin{gathered} d_M[\mathfrak{S}_1(t), \mathfrak{S}_2(t)] = sup\{d(\mathbf{x}, \mathbf{y}) : \mathbf{x} \in \mathfrak{S}_1(t),\ \mathbf{y} \in \mathfrak{S}_2(t)\} = \\ = max\left\{ \begin{array}{c} sup\langle d[\mathbf{S}, \mathfrak{S}_1(t)] : \mathbf{S} \in \mathfrak{S}_2(t)\rangle, \\ sup\langle d[\mathbf{S}, \mathfrak{S}_2(t)] : \mathbf{S} \in \mathfrak{S}_1(t)\rangle \end{array} \right\}. \end{gathered} \tag{3.71}$$

The notions of the set distances (3.70) and (3.71) enabled Grujic [Gruyitch] to define in [115, Definition 3, pp. 530, 531], [195, Definition 17, pp. 33, 34] continuity of a *time*-varying set equivalently to Definition 104.

Definition 105 ***Set continuity in time (the equivalent definition)***

i) A set $\mathfrak{L}$ *is* ***the limit set*** *of* $\mathfrak{S}(t)$ *at* $\tau \in \mathfrak{T}$*, denoted by (3.68), if and only if*

$$lim\{d_M[\mathfrak{S}(t), \mathfrak{L}] : t \longrightarrow \tau\} = 0. \tag{3.72}$$

ii) A time-varying set is ***continuous in*** $t \in \mathfrak{T}_0$, $\mathfrak{T}_0 \subseteq \mathfrak{T}$, $t_0 \in \mathfrak{T}_i$, *if and only if (3.69) holds. This is* ***set continuity on*** $\mathfrak{T}_0$*.*

The following definition links sets and system motions:

Definition 106 ***Positive invariance of time-varying set relative to system motions***

A time-continuous set $\mathfrak{S}(t; t_0; \mathfrak{S}_0)$ *is positive invariant on* $\mathfrak{T}_0 \times \mathcal{J}^i$ *with respect to the system motions* $\mathcal{S}(.; t_0; \mathbf{S}_0; \mathbf{I})$ *if and only if* $\mathbf{S}_0 \in \mathfrak{S}(t_0; t_0)$ *implies* $\mathbf{S}(t; t_0; \mathbf{S}_0; \mathbf{I}) \in \mathfrak{S}(t; t_0; \mathfrak{S}_0)$, $\forall [t, \mathbf{I}(.)] \in \mathfrak{T}_0 \times \mathcal{J}^i$*.*

This is valid also for time-invariant sets $\mathfrak{S} \in 2^{\mathfrak{R}^n}$*. Then* $\mathfrak{S}$ *replaces* $\mathfrak{S}(t_0; t_0)$*.*

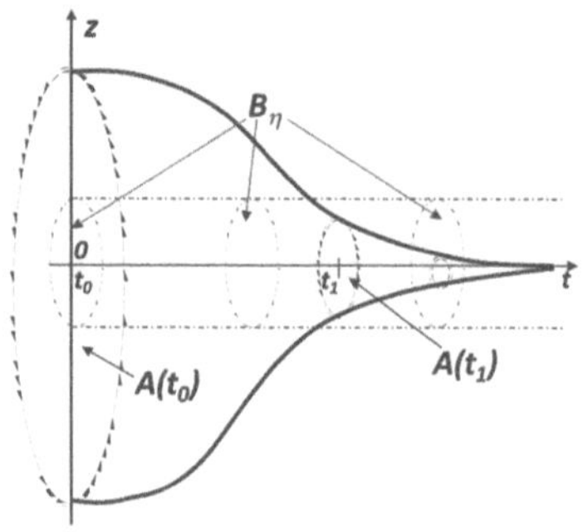

Figure 3.4: Asymptotically contractive set $\mathfrak{A}(t)$. The Fraktur letter $\mathfrak{A}$ is represented in this figure by the Roman letter A.

3.2.4 Differentiability of variable sets

The basis for what follows are Dini derivatives B.3 of a continuous set valued function $\mathfrak{S}(.)$ of $(t, \mathbf{z}) \in \mathfrak{T} \times \mathfrak{R}^K$, $\mathfrak{S}(t, \mathbf{z}) \in \mathfrak{C}\left(\mathfrak{T} \times \mathfrak{R}^K\right)$. They are defined in Section B.3. In what follows we refer to the forward-time Eulerian derivative simply as the Eulerian derivative, Equation B.5.

Definition 107 *1) A continuous set valued function $\mathfrak{S}(.)$ of $(t, \mathbf{z}) \in \mathfrak{T} \times \mathfrak{R}^K$, $\mathfrak{S}(t, \mathbf{z}) \in \mathfrak{C}\left(\mathfrak{T} \times \mathfrak{R}^K\right)$, is differentiable at $t \in \mathfrak{T}$ if and only if it has the (forward time) Eulerian derivative at the moment $t \in \mathfrak{T}$:*

$$\exists \mathfrak{S}^{(1)}(t, \mathbf{z}) \Leftrightarrow \exists \mathfrak{S}_f^{(1)}(t, \mathbf{z}) \Longrightarrow \mathfrak{S}^{(1)}(t, \mathbf{z}) = \frac{d\mathfrak{S}(t, \mathfrak{B})}{dt} = \mathfrak{S}_f^{(1)}(t, \mathbf{z}). \tag{3.73}$$

2) A continuous set valued function $\mathfrak{S}(.)$ of $(t, \mathbf{z}) \in \mathfrak{T} \times \mathfrak{R}^K$, $\mathfrak{S}(t, \mathbf{z}) \in \mathfrak{C}\left(\mathfrak{T} \times \mathfrak{R}^K\right)$, is differentiable on $\mathfrak{T}_\tau$ if and only if it is differentiable at every moment $t \in \mathfrak{T}_\tau$, i.e. if and only if it has the Eulerian derivative at every moment $t \in \mathfrak{T}_\tau$:

$$\exists \mathfrak{S}^{(1)}(t, \mathbf{z}) \text{ on } \mathfrak{T}_\tau \Leftrightarrow \exists \mathfrak{S}^{(1)}(t, \mathbf{z}) \ \forall t \in \mathfrak{T}_\tau. \tag{3.74}$$

3.2.5 Set contraction

The supremum of the *time*-set $\mathfrak{T}_0$ (2.2) is ∞, but the supremum of the *time*-set $\mathfrak{T}_i$ (Section 2.2) can be ∞ but need not (2.3). However, in the framework of the Lyapunov stability it is ∞ (2.4):

$$\mathfrak{T}_0 = \{t : t \in \mathfrak{T},\ num\ t \in [num\ t_0, \infty[\},\ sup\mathfrak{T}_0 = \infty \tag{3.75}$$
$$\mathfrak{T}_i = \{t : t[T]\langle s\rangle,\ t \in \mathfrak{T}\} \subseteq \mathfrak{T},\ inf\ \mathfrak{T}_i \geq -\infty,\ sup\mathfrak{T}_i \leq \infty. \tag{3.76}$$

The hyperballs $\mathfrak{B}_{\delta(t_0,\varepsilon)}$ and $\mathfrak{B}_{\Delta(t_0)}$ vary in general with t_0 for *time*-varying systems. For strongly nonstationary systems, their closures can contract to the singleton $\mathfrak{O}$ as t_0 goes to infinity [115]. This justified the introduction and the use of the new family of the sets called *asymptotically contractive sets;* see Figure *3.4* (Grujic [Gruyitch] in *[115, Definition 3, p. 531], [170, Theorem 3, pp. 46-48, Theorem 4, pp. 51-53], [195, Definition 18, p. 34]).* Let the set $\mathfrak{S}(t)$ be a nonempty connected subset of $\mathfrak{R}^n$ with a nonempty boundary $\partial\mathfrak{S}(t)$, $\mathfrak{S}(t) \subset \mathfrak{R}^n$, $\partial\mathfrak{S}(t) \neq \phi$. Let $d[\mathbf{S},\mathfrak{S}(t)]$ be the distance of $\mathbf{S}$ from $\mathfrak{S}(t)$.

Definition 108 ***Asymptotically contractive sets***

a) A time-varying set $\mathfrak{A}(t)$ ***asymptotically contracts as*** *$t \longrightarrow sup\mathfrak{T}_i$* ***to the constant set*** *$\mathfrak{S}$, i.e., it is* ***asymptotically contractive relative to*** *$\mathfrak{S}$* ***as*** *$t \longrightarrow sup\mathfrak{T}_i$ if and only if both*

$$In\mathfrak{A}(t) \supset \mathfrak{S}, \quad \forall t \in \mathfrak{T}_i, \tag{3.77}$$

and

$$lim\left[Cl\mathfrak{A}(t) : t \longrightarrow sup\mathfrak{T}_i\right] = Cl\ \mathfrak{S}, \tag{3.78}$$

hold. This is ***the asymptotic contraction relative to*** *$\mathfrak{S}$* ***as*** *$t \to sup\mathfrak{T}_i$; see Figure 3.4.*

Otherwise, the set $\mathfrak{A}(t)$ ***does not asymptotically contract to the constant set*** *$\mathfrak{S}$* ***as*** *$t \to sup\mathfrak{T}_i$.*

A time-varying set $\mathfrak{A}(t)$ ***asymptotically contracts on*** *$\mathfrak{T}_0$* ***to the constant set*** *$\mathfrak{S}$, i.e., it is* ***asymptotically contractive relative to*** *$\mathfrak{S}$ if and only if both*

$$In\mathfrak{A}(t) \supset \mathfrak{S}, \quad \forall t \in \mathfrak{T}_0, \tag{3.79}$$

and

$$lim\left[Cl\mathfrak{A}(t) : t \longrightarrow \infty\right] = Cl\ \mathfrak{S}, \tag{3.80}$$

hold. This is the asymptotic contraction relative to $\mathfrak{S}$ as $t \to \infty$.

Otherwise, the set $\mathfrak{A}(t)$ ***does not asymptotically contract on*** *$\mathfrak{T}_0$* ***to the constant set*** *$\mathfrak{S}$.*

*The expression "****relative to*** *$\mathfrak{S}$" is to be omitted if and only if $Cl\mathfrak{S} = \mathfrak{O} = \{\mathbf{0}_n\}$.*

b) A time-varying set $\mathfrak{A}(t)$ ***asymptotically contracts as*** *$t \longrightarrow sup\mathfrak{T}_i$* ***to a time-varying set*** *$\mathfrak{S}(t)$, i.e., it is* ***asymptotically contractive relative to*** *$\mathfrak{S}(t)$* ***as*** *$t \longrightarrow sup\mathfrak{T}_i$ if and only if both*

$$In\mathfrak{A}(t) \supset \mathfrak{S}(t), \quad \forall t \in \mathfrak{T}_i, \tag{3.81}$$

and

$$lim\left[Cl\mathfrak{A}(t) : t \longrightarrow sup\mathfrak{T}_i\right] = lim\left[Cl\ \mathfrak{S}(t) : t \longrightarrow sup\mathfrak{T}_i\right], \tag{3.82}$$

hold. This is ***the asymptotic contraction relative to*** *$\mathfrak{S}(t)$* ***as*** *$t \to sup\mathfrak{T}_i$.*

Otherwise, the set $\mathfrak{A}(t)$ ***does not asymptotically contract to the time-varying set*** *$\mathfrak{S}(t)$* ***as*** *$t \to sup\mathfrak{T}_i$.*

A time-varying set $\mathfrak{A}(t)$ ***asymptotically contracts on*** *$\mathfrak{T}_0$* ***to the time-varying set*** *$\mathfrak{S}(t)$, i.e., it is* ***asymptotically contractive relative to*** *$\mathfrak{S}(t)$* ***on*** *$\mathfrak{T}_0$ if and only if both*

$$In\mathfrak{A}(t) \supset \mathfrak{S}(t), \quad \forall t \in \mathfrak{T}_0, \tag{3.83}$$

and

$$lim\left[Cl\mathfrak{A}(t) : t \longrightarrow \infty\right] = lim\left[Cl\mathfrak{S}(t) : t \longrightarrow \infty\right] = Cl\mathfrak{S}(\infty), \tag{3.84}$$

hold. This is the asymptotic contraction relative to $\mathfrak{S}(t)$ as $t \to \infty$.

Otherwise, the set $\mathfrak{A}(t)$ ***does not asymptotically contract on*** *$\mathfrak{T}_0$* ***to the time-varying set*** *$\mathfrak{S}(t)$.*

*The expression "****relative to*** *$\mathfrak{S}(t)$" is to be omitted if and only if $Cl\mathfrak{S}(t) \equiv \mathfrak{O} = \{\mathbf{0}_n\}$.*

Example 109 *The following time varying set is asymptotically contractive:*

$$\mathfrak{A}(t) = \left\{\mathbf{S} : \|\mathbf{S}\| \left(1 + \|\mathbf{S}\|\right)^{-1} < e^{-t}\right\} = \left\{\mathbf{S} : \|\mathbf{S}\| < e^{-t}\left(1 - e^{-t}\right)^{-1}\right\} \Longrightarrow$$

$$In\mathfrak{A}(t) = \mathfrak{A}(t) = \left\{\mathbf{S} : \|\mathbf{S}\| \left(1 + \|\mathbf{S}\|\right)^{-1} < e^{-t}\right\} \supset \mathfrak{O}, \quad \forall t \in \mathfrak{T}_0, \quad (t_0 \geq 0) \in \mathfrak{T},$$

$$lim\left[Cl\mathfrak{A}(t) : t \longrightarrow \infty\right] = lim\left[\left\{\mathbf{S} : \|\mathbf{S}\| \left(1 + \|\mathbf{S}\|\right)^{-1} \leq e^{-t}\right\} : t \longrightarrow \infty\right] = \mathfrak{O} = \{\mathbf{0}_n\}.$$

Lemma 110 ***Asymptotically contractive sets and fixed neighborhoods***

a) Let a time-varying set $\mathfrak{A}(t)$ *asymptotically contract on* $\mathfrak{T}_0$ *to the constant set* $\mathfrak{S}$. *Then, for any constant neighborhood* $\mathfrak{N}(\mathfrak{S})$ *of* $\mathfrak{S}$ *there is a moment* $\tau \in \mathfrak{T}_0$ *such that* $Cl\mathfrak{A}(t) \subset Cl\mathfrak{N}(\mathfrak{S})$ *for all* $(t > \tau) \in \mathfrak{T}_0 \cup \{\infty\}$.

b) Let a time-varying set $\mathfrak{A}(t)$ *be not asymptotically contractive on* $\mathfrak{T}_0$ *relative to the constant set* $\mathfrak{S}$ *and* $\mathfrak{A}(t) \supset \mathfrak{S}$, $\forall t \in \mathfrak{T}_0$. *Then, there are a constant neighborhood* $\mathfrak{N}(\mathfrak{S})$ *of* $\mathfrak{S}$ *and a moment* $\tau \in \mathfrak{T}_0$ *such that* $Cl\mathfrak{A}(t) \supset Cl\mathfrak{N}(\mathfrak{S})$ *for all* $(t > \tau) \in \mathfrak{T}_0 \cup \{\infty\}$.

c) Let a time-varying set $\mathfrak{A}(t)$ *asymptotically contract on* $\mathfrak{T}_0$ *to the time-varying set* $\mathfrak{S}(t)$. *Then, for any fixed known* $\eta \in \mathfrak{R}^+$ *determining the fixed* $\eta-$*neighborhood* $\mathfrak{N}_\eta[\mathfrak{S}(t)]$ *of* $\mathfrak{S}(t)$ *there is a moment* $\tau \in \mathfrak{T}_0$ *such that* $Cl\mathfrak{A}(t) \subset Cl\mathfrak{N}_\eta[\mathfrak{S}(t)]$ *for all* $(t > \tau) \in \mathfrak{T}_0 \cup \{\infty\}$.

d) Let a time-varying set $\mathfrak{A}(t)$ *be not asymptotically contractive on* $\mathfrak{T}_0$ *relative to the time-varying set* $\mathfrak{S}(t)$ *and* $\mathfrak{A}(t) \supset \mathfrak{S}$, $\forall t \in \mathfrak{T}_0$. *Then, there are a fixed known* $\eta \in \mathfrak{R}^+$ *determining the fixed neighborhood* $\mathfrak{N}_\eta[\mathfrak{S}(t)]$ *of* $\mathfrak{S}(t)$ *and a moment* $\tau \in \mathfrak{T}_0$ *such that* $Cl\mathfrak{A}(t) \supset Cl\mathfrak{N}_\eta[\mathfrak{S}(t)]$ *for all* $(t > \tau) \in \mathfrak{T}_0 \cup \{\infty\}$.

Proof. a) Let a *time*-varying set $\mathfrak{A}(t)$ asymptotically contract on $\mathfrak{T}_0$ to the constant set $\mathfrak{S}$. Let the statement under a) be wrong, i.e., let there exist $\tau \in \mathfrak{T}_0$ and a constant neighborhood $\mathfrak{N}(\mathfrak{S})$ of $\mathfrak{S}$ such that $\mathfrak{A}(t) \supseteq \mathfrak{N}(\mathfrak{S})$ for all $(t > \tau) \in \mathfrak{T}_0 \cup \{\infty\}$. This contradicts (3.80). The contradiction is the consequence of the assumption on the existence of $\tau \in \mathfrak{T}_0$ and a constant neighborhood $\mathfrak{N}(\mathfrak{S})$ of $\mathfrak{S}$ such that $\mathfrak{A}(t) \supseteq \mathfrak{N}(\mathfrak{S})$ for all $(t > \tau) \in \mathfrak{T}_0 \cup \{\infty\}$. The failure of the assumption proves the statement under a).

b) The statement under b) results from the fact that the *time*-varying set $\mathfrak{A}(t)$ is not asymptotically contractive on $\mathfrak{T}_0$ relative to the constant set $\mathfrak{S}$ and from a).

c) Let a *time*-varying set $\mathfrak{A}(t)$ asymptotically contract on $\mathfrak{T}_0$ to the *time*-varying set $\mathfrak{S}(t)$. Let the statement under c) be wrong, i.e., let there exist a moment $\tau \in \mathfrak{T}_0$ and a fixed known $\eta \in \mathfrak{R}^+$ determining the fixed $\eta-$neighborhood $\mathfrak{N}_\eta[\mathfrak{S}(t)]$ of $\mathfrak{S}(t)$ such that $\mathfrak{A}(t) \supseteq \mathfrak{N}_\eta[\mathfrak{S}(t)]$ for all $(t > \tau) \in \mathfrak{T}_0 \cup \{\infty\}$. This contradicts (3.84). The contradiction is the consequence of the assumption on the existence of a moment $\tau \in \mathfrak{T}_0$ and a fixed $\eta-$neighborhood $\mathfrak{N}_\eta[\mathfrak{S}(t)]$ of $\mathfrak{S}(t)$ such that $\mathfrak{A}(t) \supseteq \mathfrak{N}_\eta[\mathfrak{S}(t)]$ for all $(t > \tau) \in \mathfrak{T}_0 \cup \{\infty\}$. The failure of the assumption proves the statement under c).

d) The statement under d) results from the fact that the *time*-varying set $\mathfrak{A}(t)$ is not asymptotically contractive on $\mathfrak{T}_0$ relative to *time*-varying set $\mathfrak{S}(t)$ and from c). ■

Note 111 *The outer diameter* $\mathfrak{d}_M(t;\mathfrak{S})$ *of an asymptotically contractive set* $\mathfrak{S}(t)$ *obeys:*

$$t \longrightarrow \infty \Longrightarrow \mathfrak{d}_M(t;\mathfrak{S}) \longrightarrow 0,\ \ \mathfrak{d}_M(\mathfrak{T}_0;\mathfrak{S}) = 0,\ \ \forall \mathfrak{T}_0 \subseteq \mathfrak{T}.$$

Chapter 4

Systems

4.1 Physical dynamical systems and state

Definition 112 ***Physical dynamical system***

A physical system is ***dynamical*** *if and only if its behavior depends on its physical situation at the initial instant, i.e., depends on its initial physical situation. Otherwise, a physical system is* ***static****.*

Natural sciences study dynamical physical systems directly by investigating them as they are in reality. System science, control science, and control engineering use mathematical descriptions, i.e., mathematical models, of the physical dynamical systems in order to study, to design, to construct them. The mathematical models represent mathematical systems containing the crucial physical characteristics of the original physical dynamical systems.

There are three substantial characteristic groups of variables that are associated with the dynamical system in general. Their definitions follow.

The physical nature of a system is anything.

Definition 113 ***Input variables, input vector and input space***

A variable that acts on the system and whose influence is essential, crucial, and inherent for the system behavior is the ***system input variable*** *denoted by* $I \in \mathfrak{R}$. *The system can be under the action of several mutually independent input variables* $I_1, I_2, ..., I_M$. *They compose* ***the system input vector*** *(for short,* ***input****)*

$$\mathbf{I} = [I_1 \ \ I_2 \ \ ... \ \ I_M]^T \in \mathfrak{R}^M, \tag{4.1}$$

which is an element of ***the input space*** $\mathfrak{R}^M$.

The instantaneous values *of the input variables* I_i *and of the input vector* $\mathbf{I}$ ***at an instant*** $t \in \mathfrak{T}$ *are* $I_i(t)$ *and* $\mathbf{I}(t)$*, respectively.*

The capital letters I *and* $\mathbf{I}$ *denote the total (scalar, vector) values of the variable* I *and of the vector* $\mathbf{I}$ *relative to their total zero (scalar, vector) value, if it exists, or relative to their accepted zero (scalar, vector) value, respectively.*

The time evolution $\mathbf{I}(t)$ *of the input vector* $\mathbf{I}$ *takes place, i.e., the input vector* $\mathbf{I}$ *propagates, in* ***the integral input space*** $\mathcal{I}$,

$$\mathcal{I} = \mathfrak{T} \times \mathfrak{R}^M. \tag{4.2}$$

The system output behavior is determined by the temporal evolution of its *output variables and their derivatives,* in the sense of the following definition:

DOI: 10.1201/9781003744528-7

Definition 114 ***Output variables, output vector, output space and response***

A variable $Y \in \mathfrak{R}$ is an ***output variable*** *of the system if and only if its values result from the system behavior, they are (directly or indirectly) measurable, and we are interested in them.*

The number N is the maximal number of mutually (linearly, algebraically, differentially and integrally) independent output variables Y_1, Y_2, . . ., Y_N on $\mathfrak{T}$ of the system. They form the ***output vector*** $\mathbf{Y}$ *of the system, which is element of* ***the output space*** $\mathfrak{R}^N$*:*

$$\mathbf{Y} = [Y_1 \ Y_2 \ ...Y_N]^T \in \mathfrak{R}^N. \tag{4.3}$$

The time evolution $\mathbf{Y}(t)$ of the output vector $\mathbf{Y}$ takes place, i.e., the output vector $\mathbf{Y}$ propagates, in ***the integral output space*** $\mathcal{I}$*,*

$$\mathcal{I} = \mathfrak{T} \times \mathfrak{R}^N. \tag{4.4}$$

The instantaneous values of the variables Y_i and of the vector $\mathbf{Y}$ at an instant $t \in \mathfrak{T}$ are $Y_i(t)$ and $\mathbf{Y}(t)$, respectively.

The time variation $\mathbf{Y}(t)$ of the system output vector $\mathbf{Y}$ is the ***system (output) response****.*

The system desired (reference) output behavior *is denoted by $\mathbf{Y}_d(t)$.*

A (physical and a mathematical) dynamical system can be subjected to the action of the input vector derivatives $\mathbf{I}^{(\mu)}(t)$, $\mu \in \{1, 2, ...\}$. The system internal and output dynamical behavior depend then not only on the input vector $\mathbf{I}(t)$ but also on all its derivatives acting on the system. This is reality that inspires us, justifies and demands us to generalize the concept of the dynamical system *state* as follows in view of [65, Definition 3-6, p. 83], [313, p. 105], [389, 2. Definition, p. 380], [390, 2. Definition, p. 380], [407, p. 4], [408, p. 664].

Definition 115 ***Physical state of a physical dynamical system***

The (internal, output) physical state of a physical dynamical system at a moment $\tau \in \mathfrak{T}$ *is, respectively, the system (internal, output) dynamical physical situation at the moment τ, which, together with the input vector and all its derivatives acting on the system at any moment $(t \geq \tau) \in \mathfrak{T}$, determines uniquely the system behavior—i.e., the system (internal, output) state and the system output response—for all $(t > \tau) \in \mathfrak{T}$, respectively.*

The minimal number n of linearly independent physical variables S_i on $\mathfrak{T}$, $i = 1, 2, ..., n$, the values $S_i(\tau)$ of which are at every moment $\tau \in \mathfrak{T}$ in the biunivoque correspondence with the system physical state at the same moment τ, is ***the physical state dimension*** *and the variables S_i, $i = 1, 2,. . ., n$, are* ***physical state variables of the system.*** *They compose* ***the physical state vector*** S ***of the system, respectively,***

$$\mathbf{S} = [S_1 \ \ S_2 \ ...S_n]^T \in \mathfrak{R}^n, \ (\mathbf{S}_I = [S_{I1} \ \ S_{I2} \ ...S_{In_I}]^T \in \mathfrak{R}^{n_I},$$
$$\mathbf{S}_O = [S_{O1} \ \ S_{O2} \ ...S_{On_O}]^T \in \mathfrak{R}^{n_O}). \tag{4.5}$$

The space $\mathfrak{R}^n$ is ***the physical state space of the system.*** ***The integral state space*** $\mathcal{I}$ *is the space product $\mathfrak{T} \times \mathfrak{R}^n$,*

$$\mathcal{I} = \mathfrak{T} \times \mathfrak{R}^n. \tag{4.6}$$

Note 116 Existence of the output state

The system output state exists if and only if the system creates the output dynamics despite both the internal state and the input vector being constant. It is described by a differential equation of the system output vector, that contains the internal state vector and can contain also the input vector (together with its derivatives). Then the system state vector $\mathbf{S}$ *(4.5) is composed of both the system internal state* $\mathbf{S}_I = \widehat{\mathbf{S}} \in \mathfrak{R}^{n_I} = \mathfrak{R}^{\widehat{n}}$ *and the system output state vector* $\mathbf{S}_O = \widetilde{\mathbf{S}} \in \mathfrak{R}^{n_O} = \mathfrak{R}^{\widetilde{n}}$,

$$\mathbf{S} = \begin{bmatrix} \mathbf{S}_I \\ \mathbf{S}_O \end{bmatrix} = \begin{bmatrix} \widehat{\mathbf{S}} \\ \widetilde{\mathbf{S}} \end{bmatrix} \in \mathfrak{R}^n, \ n = n_I + n_O = \widehat{n} + \widetilde{n}. \tag{4.7}$$

The system output state does not exist if and only if the system output dynamical behavior is induced exclusively by a variation of the internal state and/or by a variation of the input vector. It is described by an algebraic equation of the system output vector, that contains the internal state vector and can contain also the input vector. Then the system internal state vector $\mathbf{S}_I = \mathbf{S} \in \mathfrak{R}^{n_I} = \mathfrak{R}^n$ *is its state vector* $\mathbf{S}$ *(4.5),*

$$\mathbf{S_I} = \mathbf{S} \in \mathfrak{R}^{n_I} = \mathfrak{R}^n, \ n = n_I. \tag{4.8}$$

4.2 Mathematical dynamical systems

4.2.1 Smooth dynamical system

This section refers to [232, Section 3.1: General dynamical systems, pp. 21-23].

Michel, Hou, and Liu [379, pp. 1, 2] presented the general quadruple determination of a dynamical system, the internal dynamical behavior of which, hence the dynamical properties of which, are independent of the initial instant t_0. Such determination treats only the uniform internal dynamical behavior of the system. It is appropriate for uniform Lyapunov stability studies, but not for stability study in general.

By following [379, pp. 1, 2] let us extend the determination of dynamical systems to the systems for which the initial moment t_0 can have a crucial influence on their internal dynamics and output behavior and properties. It is a characteristic of *time*-varying systems.

Definition 117 ***Smooth mathematical dynamical system in general means smooth time-varying mathematical dynamical system***

A smooth mathematical dynamical system $\mathcal{S}$, *for short:* ***a variable dynamical system*** $\mathcal{S}$, *is a set of mappings (vector functions)* $\mathcal{S}(.)$ *and* $\mathcal{Y}(.)$, *which are dependent on the initial moment* t_0, *from the Cartesian product space* $\mathfrak{T}_0 \times \mathfrak{T}_i \times \mathfrak{L}^* \times \mathfrak{S}^*(t_0) \times \mathcal{J}$ *into the Cartesian product space* $\mathfrak{R}^n \times \mathfrak{R}^N$,

$$(\mathcal{S}, \mathcal{Y}) : \ \mathfrak{T}_0 \ \times \ \mathfrak{T}_i \ \times \ \mathfrak{L}^* \ \times \ \mathfrak{S}^*(t_0) \ \times \ \mathcal{J} \longrightarrow \mathfrak{R}^n \ \times \ \mathfrak{R}^N.$$

They are called ***system motion*** *and* ***system output response,*** *the vector values of which at any moment* $t \in \mathfrak{T}_0$ *are* ***the system state vector*** $\mathbf{S}(t) \in \mathfrak{R}^n$ *and* ***the system output vector*** $\mathbf{Y}(t) \in \mathfrak{R}^N$, *respectively,*

$$\mathbf{S}(t) = \mathbf{S}(t; t_0; \mathbf{S}_0; \mathbf{I}) \in \mathfrak{R}^n, \ \left\{ \begin{array}{c} \forall (t, t_0, \mathbf{S}_0, \mathbf{I}) \\ \in \mathfrak{T}_0 \ \times \ \mathfrak{T}_i \ \times \ \mathfrak{L}^* \ \times \ \mathfrak{S}^*(t_0) \ \times \ \mathcal{J} \end{array} \right\}, \tag{4.9}$$

$$\mathbf{Y}(t) = \mathbf{Y}(t; t_0; \mathbf{S}_0; \mathbf{I}) \in \mathfrak{R}^N, \ \left\{ \begin{array}{c} \forall (t, t_0, \mathbf{S}_0, \mathbf{I}) \\ \in \mathfrak{T}_0 \ \times \ \mathfrak{T}_i \ \times \ \mathfrak{L}^* \ \times \ \mathfrak{S}^*(t_0) \ \times \ \mathcal{J} \end{array} \right\}, \tag{4.10}$$

and which have the following properties:

P1 ***Existence:*** *Every system motion $\mathcal{S}(.)$ is defined in $\mathfrak{R}^n$ on the product set $\mathfrak{T}_0 \times \mathfrak{T}_i \times \mathfrak{L}^* \times \mathfrak{S}^*(t_0) \times \mathcal{J}$ and every system output response $\mathcal{Y}(.)$ is defined in $\mathfrak{R}^N$ on the product set $\mathfrak{T}_0 \times \mathfrak{T}_i \times \mathfrak{L}^* \times \mathfrak{S}^*(t_0) \times \mathcal{J}$,:*

$$\mathbf{S}(t;t_0;\mathbf{S}_0;\mathbf{I}) \in \mathfrak{R}^n, \left\{ \begin{array}{c} \forall(t,t_0,\mathbf{S}_0,\mathbf{I}) \\ \in \mathfrak{T}_0 \times \mathfrak{T}_i \times \mathfrak{L}^* \times \mathfrak{S}^*(t_0) \times \mathcal{J} \end{array} \right\}, \tag{4.11}$$

$$\mathbf{Y}(t;t_0;\mathbf{S}_0;\mathbf{I}) \in \mathfrak{R}^N, \left\{ \begin{array}{c} \forall(t,t_0,\mathbf{S}_0,\mathbf{I}) \\ \in \mathfrak{T}_0 \times \mathfrak{T}_i \times \mathfrak{L}^* \times \mathfrak{S}^*(t_0) \times \mathcal{J} \end{array} \right\}. \tag{4.12}$$

P2 ***Initial condition:*** *Every system motion $\mathcal{S}(.)$ and every system output response $\mathcal{Y}(.)$ satisfy the initial conditions:*

$$\mathbf{S}(t_0;\mathbf{S}_0;\mathbf{I}) = \mathbf{S}_0, \ \forall(t_0,\mathbf{S}_0,\mathbf{I}) \in \mathfrak{S}^* \times \mathcal{J}, \tag{4.13}$$

$$\mathbf{Y}(t_0;t_0;\mathbf{S}_0;\mathbf{I}) = \mathbf{Y}_0, \ \forall(t_0,\mathbf{S}_0,\mathbf{I}) \in \mathfrak{L}^* \times \mathfrak{S}^*(t_0) \times \mathcal{J}, \tag{4.14}$$

P3 ***Continuous dependence:*** *Every system motion $\mathcal{S}(.)$, its first derivative and every system output response $\mathcal{Y}(.)$ are continuous in $(t,t_0,\mathbf{S}_0,\mathbf{I}) \in \mathfrak{T}_0 \times \mathfrak{T}_i \times \mathfrak{L}^* \times \mathfrak{S}^*(t_0) \times \mathcal{J}$:*

$$\mathbf{S}(t;t_0;\mathbf{S}_0;\mathbf{I}) \in \mathfrak{C}(\mathfrak{T}_0 \times \mathfrak{T}_i \times \mathfrak{L}^* \times \mathfrak{S}^*(t_0) \times \mathcal{J}), \tag{4.15}$$

$$\mathbf{S}^{(1)}(t;t_0;\mathbf{S}_0;\mathbf{I}) \in \mathfrak{C}(\mathfrak{T}_0 \times \mathfrak{T}_i \times \mathfrak{L}^* \times \mathfrak{S}^*(t_0) \times \mathcal{J}) \tag{4.16}$$

$$\mathbf{Y}(t;t_0;\mathbf{S}_0;\mathbf{I}) \in \mathfrak{C}(\mathfrak{T}_0 \times \mathfrak{T}_i \times \mathfrak{L}^* \times \mathfrak{S}^*(t_0) \times \mathcal{J}), \tag{4.17}$$

P4 ***Uniqueness:*** *For every $(t,t_0,\tau,\mathbf{S}_0,\mathbf{I}) \in \mathfrak{T}_0 \times \mathfrak{L}^* \times \mathfrak{L}^* \times \mathfrak{S}^*(t_0) \times \mathcal{J}$ every system motion $\mathcal{S}(.)$ and every system output response $\mathcal{Y}(.)$ satisfy*

$$\mathbf{S}\left[t;t_0+\tau;\mathcal{S}(t_0+\tau;t_0;\mathbf{S}_0;\mathbf{I});\mathbf{I}\right] = \mathcal{S}(t;t_0;\mathbf{S}_0;\mathbf{I}), \tag{4.18}$$

$$\mathbf{Y}\left[t;t_0+\tau;\mathcal{S}(t_0+\tau;t_0;\mathbf{S}_0;\mathbf{I});\mathbf{I}\right] = \mathcal{Y}(t;t_0;\mathbf{S}_0;\mathbf{I}). \tag{4.19}$$

Comment 118 *Property P4 is the systems counterpart to the set Property 63, Section 3.2.*

Definition 119 ***Smooth time-invariant mathematical dynamical system means a mathematical dynamical system independent of the initial instant t_0***

A smooth time-invariant mathematical dynamical system *$\mathcal{S}$, for short:* ***a (constant) dynamical system*** *$\mathcal{S}$, is a set of mappings (vector functions) $\mathcal{S}(.)$ and $\mathcal{Y}(.)$, which are independent of the initial moment t_0, from the Cartesian product space $\mathfrak{T} \times \mathfrak{S}^* \times \mathcal{J}$ into the Cartesian product space $\mathfrak{R}^n \times \mathfrak{R}^N$,*

$$(\mathcal{S},\mathcal{Y}) : \mathfrak{T} \times \mathfrak{S}^* \times \mathcal{J} \longrightarrow \mathfrak{R}^n \times \mathfrak{R}^N,$$

which are called the ***system motion*** *and* ***system output response,*** *the vector values of which at any moment $t \in \mathfrak{T}$ are* ***the system state vector*** *$\mathbf{S}(t) \in \mathfrak{R}^n$ and* ***the system output vector*** *$\mathbf{Y}(t) \in \mathfrak{R}^N$, respectively,*

$$\mathbf{S}(t) = \mathbf{S}(t;\mathbf{S}_0;\mathbf{I}) \in \mathfrak{R}^n, \ \{\forall(t,\mathbf{S}_0,\mathbf{I}) \in \mathfrak{T} \times \mathfrak{S}^* \times \mathcal{J}\}, \tag{4.20}$$

$$\mathbf{Y}(t) = \mathbf{Y}(t;\mathbf{S}_0;\mathbf{I}) \in \mathfrak{R}^N, \ \{\forall(t,\mathbf{S}_0,\mathbf{I}) \in \mathfrak{T} \times \mathfrak{S}^* \times \mathcal{J}\}, \tag{4.21}$$

and which have the following properties:

P1 ***Existence:*** *Every system motion $\mathcal{S}(.)$ is defined in $\mathfrak{R}^n$ on the product set $\mathfrak{T} \times \mathfrak{S}^* \times \mathcal{J}$ and every system output response $\mathcal{Y}(.)$ is defined in $\mathfrak{R}^N$ on the product set $\mathfrak{T}_0 \times \mathfrak{T}_i \times \mathfrak{L}^* \times \mathfrak{S}^*(t_0) \times \mathcal{J}$:*

$$\mathbf{S}(t;\mathbf{S}_0;\mathbf{I}) \in \mathfrak{R}^n, \ \{\forall(t,\mathbf{S}_0,\mathbf{I}) \in \mathfrak{T} \times \mathfrak{S}^* \times \mathcal{J}\}, \tag{4.22}$$

$$\mathbf{Y}(t;\mathbf{S}_0;\mathbf{I}) \in \mathfrak{R}^N, \ \{\forall(t,\mathbf{S}_0,\mathbf{I}) \in \mathfrak{T} \times \mathfrak{S}^* \times \mathcal{J}\}. \tag{4.23}$$

P2 ***Initial condition:*** *Every system motion* $\mathcal{S}(.)$ *and every system output response* $\mathcal{Y}(.)$ *satisfy the initial conditions:*

$$\mathbf{S}(t_0;\mathbf{S}_0;\mathbf{I}) = \mathbf{S}_0,\ \forall(t_0,\mathbf{S}_0,\mathbf{I}) \in \mathfrak{T} \times \mathfrak{S}^* \times \mathcal{J}, \tag{4.24}$$

$$\mathbf{Y}(t_0;\mathbf{S}_0;\mathbf{I}) = \mathbf{Y}_0,\ \forall(t_0,\mathbf{S}_0,\mathbf{I}) \in \mathfrak{T} \times \mathfrak{S}^* \times \mathcal{J}, \tag{4.25}$$

P3 ***Continuous dependence:*** *Every system motion* $\mathcal{S}(.)$*, its first derivative and every system output response* $\mathcal{Y}(.)$ *are continuous in* $(t,\mathbf{S}_0,\mathbf{I}) \in \mathfrak{T}\times\mathfrak{S}^*\times\mathcal{J}$*:*

$$\mathbf{S}(t;\mathbf{S}_0;\mathbf{I}) \in \mathfrak{C}\left(\mathfrak{T} \times \mathfrak{S}^* \times \mathcal{J}\right), \tag{4.26}$$

$$\mathbf{S}^{(1)}(t;\mathbf{S}_0;\mathbf{I}) \in \mathfrak{C}\left(\mathfrak{T} \times \mathfrak{S}^* \times \mathcal{J}\right) \tag{4.27}$$

$$\mathbf{Y}(t;\mathbf{S}_0;\mathbf{I}) \in \mathfrak{C}\left(\mathfrak{T}_0 \times \mathfrak{S}^* \times \mathcal{J}\right), \tag{4.28}$$

P4 ***Uniqueness:*** *For every* $(t,\mathbf{S}_0,\mathbf{I}) \in \mathfrak{T}_0 \times \mathfrak{S}^* \times \mathcal{J}$ *every system motion* $\mathcal{S}(.)$ *and every system output response* $\mathcal{Y}(.)$ *satisfy*

$$\mathbf{S}\left[t;\tau;\mathcal{S}(\tau;\mathbf{S}_0;\mathbf{I});\mathbf{I}\right] = \mathcal{S}(t;\mathbf{S}_0;\mathbf{I}), \tag{4.29}$$

$$\mathbf{Y}\left[t;\tau;\mathcal{S}(\tau;\mathbf{S}_0;\mathbf{I});\mathbf{I}\right] = \mathcal{Y}(t;t_0;\mathbf{S}_0;\mathbf{I}). \tag{4.30}$$

Comment 120 ***On Property P4***

Property 4 of time-invariant dynamical systems, which is related only to the internal dynamics of the systems, reads [247, Definition 3.20, p. 81], [322, p. 28]:

$$\mathcal{S}\left[t;\mathcal{S}(\tau;\mathbf{S}_0;\mathbf{I});\mathbf{I}\right] = \mathcal{S}(t+\tau;\mathbf{S}_0;\mathbf{I}). \tag{4.31}$$

which results in $t_0 = 0$ *from the group property [247, Definition 3.20, p. 81] of the motions of the time-invariant dynamical systems expressed by:*

$$\mathcal{S}\left[t;t_0+\tau;\mathcal{S}(t_0+\tau;t_0;\mathbf{S}_0;\mathbf{I});\mathbf{I}\right] = \mathcal{S}(t+\tau;t_0;\mathbf{S}_0;\mathbf{I}) \tag{4.32}$$

because the initial moment does not influence motions of the time-invariant systems so that it can be accepted equal to zero: $t_0 = 0$*. Equation (4.32) for* $t_0 = 0$ *reduces to (4.31).*

However, the initial moment can have a substantial influence on the character of motions of a time-varying system. A motion starting from $\mathbf{S}_0$ *at an initial moment* t_0 *can converge to the origin* $\mathbf{0}_n$*, but a motion starting from the same initial state* $\mathbf{S}_0$ *taken as the initial state at another initial moment* t_0^* *can diverge from the origin* $\mathbf{0}_n$*. Therefore, equations (4.31) and (4.32) do not hold for time-varying systems.*

We use the notion of the output variables in the classical sense. To be precise:

Definition 121 ***Output variables and output vector***

A variable $Y \in \mathfrak{R}$ *is an* ***output variable*** *of the system if and only if its values result from the system behavior, they are measurable, and somebody is interested in them. The capital letter denotes the total value of the variable relative to its total zero value, if it exists, or relative to its accepted zero value.*

The number N *is the maximal number of the mutually independent output variables* $Y_1,\ Y_2,\ ...\ Y_N$ *of the system. They form the* ***output vector*** $\mathbf{Y}$ *of the system:* $\mathbf{Y} = \left[Y_1\ Y_2\ ...\ Y_N\right]^T \in\mathfrak{R}^N$

An adequate mathematical model of the physical dynamical system also contains an adequate description of the state of the physical dynamical system, which should be, if possible, also the state of the (mathematical) dynamical system (see Definition 115 in Section and Definition 124 in Section 4.3).

4.2.2 Generalized dynamical system

What follows broadens the concepts of both the motions to the generalized motions and the dynamical systems to the generalized dynamical systems established in the framework of the *time*-invariant nonlinear systems in [247, Section 3.2, pp. 54-73].

Let $\mathfrak{T}_0^0$ be the largest subset of $\mathfrak{T}_0$ with the zero measure. "**Almost always on** $\mathfrak{T}_0$" means "**for every** $t \in (\mathfrak{T}_0 \backslash \mathfrak{T}_0^0)$".

For Dini derivatives, see Chapter B in Appendix VII.

Definition 122 ***Generalized motions (solutions) of the system*** *(5.22), (5.23)*

A vector function $\mathcal{S}(.) : \mathfrak{T}_0 \times \mathfrak{T}_i \times \mathfrak{R}^n \times \mathcal{J} \longrightarrow \mathfrak{R}^n$ *is a* ***generalized motion (generalized solution)*** *of the system (5.22), (5.23) through* $\mathbf{S}_0$ *at* $t = t_0$ *if and only if* $\mathcal{S}(.)$, $\mathbf{S}_0 \in \mathfrak{R}^n$, *and* $t_0 \in \mathfrak{T}$ *obey the following for fixed* $\mathbf{I}(.) \in \mathcal{J}$.

- $(G1)$ ***Existence:*** *The vector function* $\mathcal{S}(.)$ *is defined in* $\mathfrak{R}^n$, *on* $\mathfrak{T}_0$,

$$\mathcal{S}(t; t_0, \mathbf{S}_0; \mathbf{I}) \equiv \mathbf{S}(t) \in \mathfrak{R}^n, \ \forall t \in \mathfrak{T}_0,$$

if and only if $t_0 \in \mathfrak{L}$, $\mathfrak{L} \subseteq \mathfrak{T}_i \subseteq \mathfrak{T}$, *and* $\mathbf{S}_0 \in \mathfrak{S}$.

- $(G2)$ ***Initial condition:*** *The vector function* $\mathcal{S}(.)$ *satisfies the initial condition:*

$$\mathcal{S}(t_0; t_0; \mathbf{S}_0; \mathbf{I}) \equiv \mathbf{S}_0.$$

- $(G3)$ ***Continuous dependence:*** *The vector function* $\mathcal{S}(.)$ *is continuous on* $\mathfrak{T}_0$ *for* $t_0 \in \mathfrak{L}$,

$$\mathcal{S}(t; t_0, \mathbf{S}_0; \mathbf{I}) \in \mathfrak{C}(\mathfrak{T}_0), \ \ \forall t_0 \in \mathfrak{L},$$

and its first derivative is continuous almost everywhere on $\mathfrak{T}_0$,

$$\mathcal{S}^{(1)}(t; t_0, \mathbf{S}_0; \mathbf{I}) \in \mathfrak{C}(\mathfrak{T}_0 \backslash \mathfrak{T}_0^0), \ \ t_0 \in \mathfrak{L}.$$

At every t^* *at which* $\mathcal{S}(t; t_0, \mathbf{S}_0; \mathbf{I})$ *is not continuously differentiable,* $t^* \in \mathfrak{T}_0^0$, *it has derivatives on both sides*

$$\exists D_l \mathcal{S}(t^*; t_0, \mathbf{S}_0; \mathbf{I}) \text{ and } \exists D_r \mathcal{S}(t^*; t_0, \mathbf{S}_0; \mathbf{I})$$

if t^* *is an interior point of* $\mathfrak{T}_0$.

If t^* *is a lower boundary point of the time interval* $\mathfrak{T}_0$,

$$t^* = min\ \mathfrak{T}_0 = t_0,$$

then $\mathcal{S}(t; t_0, \mathbf{S}_0; \mathbf{I})$ *has the right-hand derivative,*

$$\exists D_r \mathcal{S}(t^*; t_0, \mathbf{S}_0; \mathbf{I}).$$

If t^* *is an upper boundary point of the time interval* $\mathfrak{T}_0$,

$$t^* = sup\ \mathfrak{T}_0 = \infty,$$

then $\mathcal{S}(t; t_0, \mathbf{S}_0; \mathbf{I})$ *has the left-hand derivative,*

$$\exists D_l \mathcal{S}(t^*; t_0, \mathbf{S}_0; \mathbf{I}).$$

- (G4) ***Uniqueness:*** *For every* $(t, t_0, \tau, \mathbf{S}_0, \mathbf{I}) \in \mathfrak{T}_0 \times \mathfrak{L}^* \times \mathfrak{L}^* \times \mathfrak{S}^*(t_0) \times \mathcal{J}$, *every system motion* $\mathcal{S}(.)$ *and every system output response* $\mathcal{Y}(.)$ *satisfy*

$$\mathbf{S}[t; t_0 + \tau; \mathcal{S}(t_0 + \tau; t_0; \mathbf{S}_0; \mathbf{I}); \mathbf{I}] = \mathcal{S}(t; t_0; \mathbf{S}_0; \mathbf{I}), \tag{4.33}$$
$$\mathbf{Y}[t; t_0 + \tau; \mathcal{S}(t_0 + \tau; t_0; \mathbf{S}_0; \mathbf{I}); \mathbf{I}] = \mathcal{Y}(t; t_0; \mathbf{S}_0; \mathbf{I}). \tag{4.34}$$

This definition extends Definition 3.8 of [247, pp. 57, 58] to *time*-varying nonlinear systems and represents a special case of [232, Definition 72, pp. 32, 33].

Definition 123 ***A (generalized) dynamical system***
A (generalized) dynamical system $\mathcal{S}$ *is a set of mappings (vector functions)* $\mathcal{S}(.)$ *and* $\mathcal{Y}(.)$ *from the Cartesian product space* $\mathfrak{T}_0 \times \mathfrak{L} \times \mathfrak{S}$ *for fixed* $\mathbf{I}(.) \in \mathcal{J}$ *into the Cartesian product space* $\mathfrak{R}^n \times \mathfrak{R}^N$,

$$\mathcal{S} : \mathfrak{T}_0 \times \mathfrak{L} \times \mathfrak{S} \to \mathfrak{R}^n \times \mathfrak{R}^N, \ \mathbf{I}(.) \in \mathcal{J},$$

which have the following properties.

P1 ***Existence:*** *Every vector function* $\mathcal{S}(.)$ *called* **the system (generalized) motion** *is defined in* $\mathfrak{R}^n$ *on* $\mathfrak{T}_0 \times \mathfrak{L} \times \mathfrak{S}$ *for fixed* $\mathbf{I}(.) \in \mathcal{J}$, *and every vector function* $\mathcal{Y}(.)$ *called* **the system output response** *is defined in* $\mathfrak{R}^N$ *on* $\mathfrak{T}_0 \times \mathfrak{L} \times \mathfrak{S}$ *for fixed* $\mathbf{I}(.) \in \mathcal{J}$; *i.e.,*

$$\mathcal{S}(t; t_0; \mathbf{S}_0; \mathbf{I}) \in \mathfrak{R}^n, \ \mathcal{Y}(t; t_0; \mathbf{S}_0; \mathbf{I}) \in \mathfrak{R}^N,$$
$$\forall (t, t_0, \mathbf{S}_0) \in \mathfrak{T}_0 \times \mathfrak{L} \times \mathfrak{S}^*(t_0), \ \mathbf{I}(.) \in \mathcal{J}. \tag{4.35}$$

P2 ***Initial condition:*** *Every system (generalized) motion* $\mathcal{S}(.)$ *and every system output response* $\mathcal{Y}(.)$ *satisfy the initial conditions,*

$$\mathcal{S}(t_0; t_0; \mathbf{S}_0; \mathbf{I}) = \mathbf{S}_0, \ \forall (t_0, \mathbf{S}_0) \in \mathfrak{L} \times \mathfrak{S}, \ \mathbf{I}(.) \in \mathcal{J},$$
$$\mathcal{Y}(t_0; t_0; \mathbf{S}_0; \mathbf{I}) \in \mathbf{Y}_0, \ \forall (t_0, \mathbf{S}_0) \in \mathfrak{L} \times \mathfrak{S}, \ \mathbf{I}(.) \in \mathcal{J}.. \tag{4.36}$$

P3 ***Continuous dependence:*** *Every system (generalized) motion* $\mathcal{S}(.)$ *and every output response* $\mathcal{Y}(.)$ *of the system are continuous in* $(t, t_0, \mathbf{S}_0)$:

$$\mathcal{S}(t; t_0; \mathbf{S}_0; \mathbf{I}) \in \mathfrak{C}(\mathfrak{T}_0 \times \mathfrak{L} \times \mathfrak{S}), \mathbf{I}(.) \in \mathcal{J},$$
$$\mathcal{Y}(t; t_0; \mathbf{S}_0; \mathbf{I}) \in \mathfrak{C}(\mathfrak{T}_0 \times \mathfrak{L} \times \mathfrak{S}), \mathbf{I}(.) \in \mathcal{J}. \tag{4.37}$$

P4 ***Uniqueness:*** *For every* $(t, t_0, \tau, \mathbf{S}_0) \in \mathfrak{T}_0 \times \mathfrak{L} \times \mathfrak{T}_0 \times \mathfrak{S}$ *and for fixed* $\mathbf{I}(.) \in \mathcal{J}$, *every system extended (generalized) motion* $\mathcal{S}(.)$ *and every system output response* $\mathcal{Y}(.)$ *satisfy*

$$\mathcal{S}[t; t_0 + \tau; \mathcal{S}(t_0 + \tau; t_0; \mathbf{S}_0; \mathbf{I}); \mathbf{I}] = \mathcal{S}(t; t_0; \mathbf{S}_0; \mathbf{I}), \tag{4.38}$$
$$\mathcal{Y}[t; t_0 + \tau; \mathcal{S}(t_0 + \tau; t_0; \mathbf{S}_0; \mathbf{I}); \mathbf{I}] = \mathcal{Y}(t; t_0; \mathbf{S}_0; \mathbf{I}). \tag{4.39}$$

4.3 Mathematical and physical state

Definition 115 (Section 4.2.1) determines the physical state, physical state variables and their state space.

Definition 124 ***Mathematical state of a physical and/or of a mathematical dynamical system***

The (internal, output) mathematical state of a physical and/or of a mathematical dynamical system at a moment $\tau \in \mathfrak{T}$ *is the minimal amount of information about the system at the moment* τ*, which, together with information about the action on the system (about the system input vector and all its derivatives acting on the system) at any moment* $(t \geq \tau) \in \mathfrak{T}$*, determines uniquely the system behavior—i.e., the system (internal, output) state and its output response—for all* $(t > \tau) \in \mathfrak{T}$*, respectively.*

The minimal number n *of linearly independent physical and/or mathematical variables* S_i (S_{Ii}, S_{Oi}) *on* $\mathfrak{T}$, $i = 1, 2, \ldots, n$, $(i = 1, 2, \ldots, n_I, i = 1, 2, \ldots, n_O)$ *the values* $S_i(\tau)$ $\{S_{Ii}(\tau), S_{Oi}(\tau)\}$ *of which are at every moment* $\tau \in \mathfrak{T}$ *in the biunivoque correspondence with the system (internal, output) physical and/or mathematical state at the same moment* τ*, is* ***the mathematical state dimension*** *and the variables* S_i, $i = 1, 2, \ldots, n$, *are* ***physical and/or mathematical state variables of the system,*** *respectively. They compose* ***the physical and/or mathematical (internal, output) state vector*** S (S_I, S_O) ***of the system,***

$$\mathbf{S} = [S_1 \; S_2 \; ... S_n]^T \in \mathfrak{R}^n, \; \mathbf{S}_I = [S_{I1} \; S_{I2} \; ... S_{In_I}]^T \in \mathfrak{R}^{n_I}, \mathbf{S}_O = [S_{O1} \; S_{O2} \; ... S_{On_O}]^T \in \mathfrak{R}^{n_O}, \tag{4.40}$$

respectively. The space $\mathfrak{R}^n$ $(\mathfrak{R}^{n_I}, \mathfrak{R}^{n_O})$ *is* ***the physical and/or mathematical (internal, output) state space of the system,*** *respectively.* ***The integral (internal, output) space*** $\mathcal{I}$ $(\mathcal{I}_I, \mathcal{I}_O)$ *is the (internal, output) space product:*

$$\mathcal{I} = \mathfrak{T} \times \mathfrak{R}^n \; (\mathcal{I}_I = \mathfrak{T} \times \mathfrak{R}^{n_I}, \; \mathcal{I}_O = \mathfrak{T} \times \mathfrak{R}^{n_O}), \tag{4.41}$$

respectively.

If and only if the (internal, output) state variables and the (internal, output) state vector are both physical and mathematical (internal, output) state variables and physical and mathematical (internal, output) state vector, then they are called ***the (internal, output) state variables*** *and* ***the (internal, output) state vector****, respectively.*

The ***(internal, output)*** *state vector function* $\mathcal{S}(.) : \mathfrak{T} \longrightarrow \mathfrak{R}^n$ $(\mathcal{S}_I(.) : \mathfrak{T} \longrightarrow \mathfrak{R}^{n_I}$, $\mathcal{S}_O(.) : \mathfrak{T} \longrightarrow \mathfrak{R}^{n_O})$ *is* ***the (internal, output) motion of the system,*** *respectively.*

The instantaneous (internal, output) value of the (internal, output) state vector function $\mathcal{S}(.)$ *at an instant* $t \in \mathfrak{T}$ *is* ***the instantaneous (internal, output) state vector*** $\mathbf{S}(t)$ $(\mathbf{S}_I(t), \mathbf{S}_O(t))$ ***at the instant*** t,

$$\mathbf{S}(t) \equiv \mathcal{S}(t; t_0; \mathbf{S}_0; \mathbf{I}) = \mathbf{S}(t; t_0; \mathbf{S}_0; \mathbf{I}) \Longrightarrow \mathbf{S}(t_0) \equiv \mathbf{S}(t_0; t_0; \mathbf{S}_0; \mathbf{I}) \equiv \mathbf{S}_0, \tag{4.42}$$

$$\mathbf{S}_I(t) \equiv \mathcal{S}_I(t; t_0; \mathbf{S}_{I0}; \mathbf{I}) = \mathbf{S}_I(t; t_0; \mathbf{S}_{I0}; \mathbf{I}) \Longrightarrow \mathbf{S}_I(t_0) \equiv \mathbf{S}_I(t_0; t_0; \mathbf{S}_{I0}; \mathbf{I}) \equiv \mathbf{S}_{I0}, \tag{4.43}$$

$$\mathbf{S}_O(t) \equiv \mathcal{S}_O(t; t_0; \mathbf{S}_{O0}; \mathbf{I}) = \mathbf{S}_I(t; t_0; \mathbf{S}_{O0}; \mathbf{I}) \Longrightarrow \mathbf{S}_I(t_0) \equiv \mathbf{S}_I(t_0; t_0; \mathbf{S}_{O0}; \mathbf{I}) \equiv \mathbf{S}_{O0}. \tag{4.44}$$

The system desired state behavior *is denoted by* $\mathbf{S}_d(t)$ $[\mathbf{S}_{Id}(t), \mathbf{S}_{Od}(t)]$*.*

This definition broadens and generalizes the well-known and commonly accepted definition of the state of the dynamical systems in general, of the control systems in particular.

Note 125 ***Omission of the input vector notation***

The input vector notation $\mathbf{I}$ *will be omitted from the notations of the motion* $\mathcal{S}(.; t_0; \mathbf{S}_0; \mathbf{I})$ *and of the state* $\mathbf{S}(t; t_0; \mathbf{S}_0; \mathbf{I})$ *if and only if* $\mathbf{I}(.)$ *is well defined so that* $\mathbf{I}(t)$ *is known on* $\mathfrak{T}_0$ *for every* $t_0 \in \mathfrak{T}_i$ *and fixed, e.g.:*

$$\mathcal{S}(.; t_0; \mathbf{S}_0; \mathbf{I}) \equiv \mathcal{S}(.; t_0; \mathbf{S}_0) \text{ and } \mathbf{S}(t; t_0; \mathbf{S}_0; \mathbf{I}) \equiv \mathbf{S}(.; t_0; \mathbf{S}_0). \tag{4.45}$$

For more details, see in the sequel Condition 180 and Note 182 in Section 5.6, Chapter 9.

4.4 Classes of the systems and states

In what follows, the term *mathematical system* denotes the accepted mathematical model (description) of the corresponding physical system.

The system explicit internal dynamics variable is its *internal state variable* S_I. This is typical for ISO, $EISO$ and $HISO$ systems.

This book treats the five general classes of nonlinear systems.

Note 126 ***System classes***

The new notation enables us to unify the qualitative (e.g., controllability, observability, stability, trackability, tracking) theory of all five classes of the dynamical systems studied herein (see Task 1 in Preface II).

The properties of the system determine the form and the character of the system state vector $\mathbf{S}$:

- **The Input-Output** (IO) **system** is described by the *m-th* order nonlinear vector differential *input-output, i.e., (internal and output state),* equation of the output vector $\mathbf{Y} \in \mathfrak{R}^N$,

$$\mathbf{Y} = \begin{bmatrix} Y_1 \vdots Y_2 \vdots ... \vdots Y_N \end{bmatrix}^T \in \mathfrak{R}^N, \ Y_i \in \mathfrak{R}, \ i = 1, 2, ..., N. \tag{4.46}$$

  The extended output vector $\mathbf{Y}^{m-1}$,

$$\mathbf{Y}^{m-1} = \begin{bmatrix} \mathbf{Y}^T \vdots \mathbf{Y}^{(1)^T} \vdots ... \vdots \mathbf{Y}^{(m-1)^T} \end{bmatrix}^T \in \mathfrak{R}^{mN}, \ n = mN, \tag{4.47}$$

  is **the state vector** $\mathbf{S}_{IO}$, which is also the **internal state vector** $\mathbf{S}_{IOI}$, **the output state vector** $\mathbf{S}_{IOO}$ and **the full state vector** $\mathbf{S}_F$ of the IO system,

$$\mathbf{S}_{IOI} = \mathbf{S}_{IOO} = \mathbf{S}_{IOF} = \mathbf{S}_{IO} = \mathbf{Y}^{m-1} \in \mathfrak{R}^n, \ n = mN. \tag{4.48}$$

- **The Input-State-Output** (ISO) **system** is determined by the first order nonlinear vector differential equation in the vector $\mathbf{X}$ - *the (internal) state equation,* and by *the algebraic output vector equation* of the output vector $\mathbf{Y}$. The only derivative in them is the first derivative of the state vector $\mathbf{X}$ in the state equation. **The state vector** $\mathbf{S}_{ISO}$ is the vector $\mathbf{X}$, which is also **the internal state vector** $\mathbf{S}_{ISOI}$ and **the full state vector** $\mathbf{S}_{ISOF}$ of the ISO system:

$$\mathbf{S}_{ISOI} = \mathbf{S}_{ISOF} = \mathbf{S}_{ISO} = \mathbf{X} \in \mathfrak{R}^n. \tag{4.49}$$

  The ISO system does not possess the output state vector $\mathbf{S}_O$ because it does not possess an independent output dynamics. Its output equation does not contain any derivative of the output vector.

- **The Higher order-Input-State-Output** ($HISO$) **system** is characterized by the *m-th* order nonlinear vector differential equation, i.e., *the m-th order (internal) state equation,* in *the substate vector* $\mathbf{X}$,

$$\mathbf{X} = \begin{bmatrix} X_1 \vdots X_2 \vdots ... \vdots X_\rho \end{bmatrix}^T \in \mathfrak{R}^\rho, \ X_i \in \mathfrak{R}, \ i = 1, 2, ..., \rho, \tag{4.50}$$

and is additionally determined by *the algebraic output vector equation* of the output vector $\mathbf{Y}$. **The internal state vector $\mathbf{S}_{HISOI}$** is the extended vector $\mathbf{X}^{m-1}$,

$$\mathbf{X}^{m-1} = \left[\mathbf{X}^T \vdots \mathbf{X}^{(1)^T} \vdots ... \vdots \mathbf{X}^{(m-1)^T}\right]^T \in \mathfrak{R}^{m\rho},\ n = m\rho, \tag{4.51}$$

which is also its **full state vector $\mathbf{S}_f$**,

$$\mathbf{S}_{HISOI} = \mathbf{S}_{HISOf} = \mathbf{S}_{HISO} = \mathbf{X}^{m-1} \in \mathfrak{R}^n,\ n = m\rho. \tag{4.52}$$

They do not possess **the output state vector $\mathbf{S}_O$**. The derivatives of the input vector can exist only in the state equation.

- **The Input-Internal and Output state** (IIO) **system** is characterized by the *m-th* order nonlinear vector differential equation, i.e., by *the m-th order internal state equation*, in *the substate vector* $\mathbf{X}$, and by the nonlinear output vector $\nu - th$ order differential equation, i.e., by *the output state equation* of the output vector $\mathbf{Y}$. The extended vector $\mathbf{X}^{m-1}$ (4.52) is **the IIO system internal state vector $\mathbf{S}_{IIOI}$** (4.53),

$$\mathbf{S}_{IIOI} = \mathbf{X}^{m-1} \in \mathfrak{R}^{n_I},\ n_I = m\rho, \tag{4.53}$$

and the system **output state vector $\mathbf{S}_{IIOO}$** is the extended output vector $\mathbf{Y}^{\nu-1}$,

$$\mathbf{S}_{IIOO} = \mathbf{Y}^{\nu-1} = \left[\mathbf{Y}^T \vdots \mathbf{Y}^{(1)^T} \vdots ... \vdots \mathbf{Y}^{(\nu-1)^T}\right]^T \in \mathfrak{R}^{n_O},\ n_O = \nu N. \tag{4.54}$$

The full state vector $\mathbf{S}_{IIOf}$, which is the state vector $\mathbf{S}_{IIO}$, is composed of the internal state vector $\mathbf{S}_{IIOI}= \mathbf{X}^{m-1}$ and of the output state vector $\mathbf{S}_{IIO} = \mathbf{Y}^{\nu-1}$,

$$\mathbf{S}_{IIOf}= \begin{bmatrix} \mathbf{S}_{IIOI} \\ \mathbf{S}_{IIOO} \end{bmatrix} = \begin{bmatrix} \mathbf{X}^{m-1} \\ \mathbf{Y}^{\nu-1} \end{bmatrix} = \mathbf{S}_{IIO} \in \mathfrak{R}^n,\ n = m\rho + \nu N. \tag{4.55}$$

The IIO system can have derivatives of the input vector in both equations.

- **The Extended Input-State-Output** ($EISO$) **system** is determined by the first order nonlinear vector differential equation in the state vector $\mathbf{S}$ (4.56),

$$\mathbf{S} = \left[S_1 \vdots S_2 \vdots ... \vdots S_n\right]^T \in \mathfrak{R}^n,\ S_i \in \mathfrak{R},\ \forall i = 1, 2, ..., n, \tag{4.56}$$

which is *the (internal) state equation,* by *the algebraic output vector equation* of the output vector $\mathbf{Y}$, and, in addition to the first derivative of the state vector, there are derivatives of the input vector only in the state equation. **The state vector $\mathbf{S}_{EISO}$** is the vector $\mathbf{S}$ (4.56), which is also **the internal state vector $\mathbf{S}_{EISOI}$**, and **the full state vector $\mathbf{S}_{EISOF}$** of the $EISO$ system

$$\mathbf{S}_{EISOI} = \mathbf{S}_{EISOF} = \mathbf{S}_{EISO} = \mathbf{X} \in \mathfrak{R}^n, \tag{4.57}$$

They do not possess the output state vector $\mathbf{S}_O$ for the same reason that the ISO systems do not have the output state vector.

Note 127 ***On the highest derivative of the input vector***
In order to avoid the problem of the appearance of impulse discontinuities in the system behavior, systems theory and control theory restrict the order of the highest derivative of the

input vector to be at most equal to the system order. However, the problem of the appearance of impulse discontinuities in the system behavior does not exist if the input vector function is defined and continuously differentiable $\mu-$times, where μ is the order of the highest input vector derivative acting on the system. For its physical origin in the system mathematical model, see Sections 4.7, 4.9, 4.10.

Comment 128 *The state variables and the state vectors defined by (4.46)-(4.55) have the full physical sense.*

4.5 Invariant sets of dynamical systems

Note 129 *Invariant sets represent a characteristic phenomenon of dynamical systems. They are tied to system motions. This book deals only with connected (negatively, positively) invariant sets of dynamical systems.*

Let us recall Equations (2.6)-(2.6) (Section 2.2):

$\mathfrak{T}_0^-$ is *a nonempty connected subset of* $\mathfrak{T}$,

$$\mathfrak{T}_0^- = \{t : \ t \in \mathfrak{T}, \ -\infty < t \leq t_0 \ < \infty\} \subset \mathfrak{T}, \tag{4.58}$$

$\mathfrak{T}_0$ is *a nonempty connected subset of* $\mathfrak{T}$,

$$\mathfrak{T}_0 = \{t : \ t \in \mathfrak{T}, \ t_0 \leq t \ < \infty\} \subset \mathfrak{T}. \tag{4.59}$$

Equations (4.58) and (4.59) show that

$$t_0 \in \ T_0^- \cap T_0 = \{t_0\} \ \ and \ \ T_0^- \cup T_0 = \ T. \tag{4.60}$$

Definition 130 *A time-varying, i.e., variable, set $\mathfrak{S}(t)$ is:*

a) ***a positively invariant set at*** $t_0 \in \mathfrak{T}_i$ ***of a dynamical system****, which is denoted by $\mathfrak{I}^+(t;t_0)$, if and only if the system state $\mathbf{S}_0$ from the set $\mathfrak{S}(t_0)$ at the moment t_0 implies that the system motion $\mathcal{S}(.;t_0;\mathbf{S}_0;\mathbf{I})$ in the set $\mathfrak{S}(t)$ since t_0 forever,*

$$t_0 \in \mathfrak{T}_i, \ \mathfrak{S}(t) = \mathfrak{I}^+(t;t_0) \Longleftrightarrow$$
$$\forall \mathbf{S}_0 \in \mathfrak{S}(t_0) = \mathfrak{S}_0 \Longrightarrow \mathbf{S}(t;t_0;\mathbf{S}_0;\mathbf{I}) \in \mathfrak{S}(t), \ \forall t \in \mathfrak{T}_0. \tag{4.61}$$

If and only if additionally the set $\mathfrak{S}(t)$ is constant, $\mathfrak{S}(t) = \mathfrak{I}^+(t) \equiv \mathfrak{S}$ then: $\mathfrak{I}^+(t) \equiv \mathcal{I}^{\shortmid}$.

b) ***a positively invariant set on*** $\mathfrak{T}_i$ ***of a dynamical system****, which is denoted by $\mathfrak{I}^+(t;\mathfrak{T}_i)$, if and only if the system state $\mathbf{S}_0$ from the set $\mathfrak{S}(t_0)$ at any initial moment $t_0 \in \mathfrak{T}_i$ implies that the system motion $\mathcal{S}(.;t_0;\mathbf{S}_0;\mathbf{I})$ rests in the set $\mathfrak{S}(t) = \mathfrak{I}^+(t;\mathfrak{T}_i)$ since t_0 forever,*

$$\mathfrak{S}(t) = \mathfrak{I}^+(t;\mathfrak{T}_i) \Longleftrightarrow$$
$$\forall t_0 \in \mathfrak{T}_i \wedge \forall \mathbf{S}_0 \in \mathfrak{S}(t_0) = \mathfrak{S}_0 \Longrightarrow \mathbf{S}(t;t_0;\mathbf{S}_o;\mathbf{I}) \in \mathfrak{S}(t), \ \forall t \in \mathfrak{T}_0. \tag{4.62}$$

If and only if additionally the set $\mathfrak{S}(t)$ is constant, $\mathfrak{S}(t) \equiv \mathfrak{S} = \mathfrak{I}^+(t;\mathfrak{T}_i)$, then for short: $\mathfrak{I}^+(t;\mathfrak{T}_i) \equiv \mathcal{I}^+$.

c) ***a negatively invariant set at*** $t_0 \in \mathfrak{T}_i$ ***of a dynamical system****, which is denoted by $\mathfrak{I}^-(t;t_0)$, if and only if the system state $\mathbf{S}_0$ from the set $\mathfrak{S}(t_0)$ at the moment t_0 implies that the system motion $\mathcal{S}(.;t_0;\mathbf{S}_0;\mathbf{I})$ was in the set $\mathfrak{S}(t)$ up to the moment t_0,*

$$t_0 \in \mathfrak{T}, \ \mathfrak{S}(t) = \mathfrak{I}^-(t;t_0) \Longleftrightarrow$$
$$\forall \mathbf{S}_0 \in \mathfrak{S}(t_0) = \mathfrak{S}_0 \Longrightarrow \mathbf{S}(t;t_0;\mathbf{S}_0;\mathbf{I}) \in \mathfrak{S}(t), \ \forall t \in \mathfrak{T}_0^-. \tag{4.63}$$

If and only if additionally the set $\mathfrak{S}(t)$ *is constant,* $\mathfrak{S}(t) = \mathfrak{I}^-(t) \equiv \mathfrak{S}$, *then for short:* $\mathfrak{I}^-(t) \equiv \mathcal{I}^-$.

d) ***a negatively invariant set on*** $\mathfrak{T}_i$ ***of a dynamical system****, which is denoted by* $\mathfrak{I}^-(t;\mathfrak{T}_i)$*, if and only if the system state* $\mathbf{S}_0$ *from the set* $\mathfrak{S}(t_0)$ *at any initial moment* $t_0 \in \mathfrak{T}_i$ *implies that the system motion* $\mathcal{S}(.;t_0;\mathbf{S}_0;\mathbf{I})$ *was in the set* $\mathfrak{S}(t) = \mathfrak{I}^-(t;\mathfrak{T}_i)$ *up to the moment* t_0*,*

$$\mathfrak{S}(t) = \mathfrak{I}^-(t;\mathfrak{T}_i) \Longleftrightarrow$$
$$\forall t_0 \in \mathfrak{T}_i \wedge \forall \mathbf{S}_0 \in \mathfrak{S}(t_0) = \mathfrak{S}_0 \Longrightarrow \mathbf{S}(t;t_0;\mathbf{S}_0;\mathbf{I}) \in \mathfrak{S}(t),\ \forall t \in \mathfrak{T}_0^-. \tag{4.64}$$

If and only if additionally the set $\mathfrak{S}(t)$ *is constant,* $\mathfrak{S}(t) \equiv \mathfrak{S} = \mathfrak{I}^-(t;\mathfrak{T}_i)$*, then for short:* $\mathfrak{I}^-(t;\mathfrak{T}_i) \equiv \mathcal{I}^-$.

e) ***an invariant set of a dynamical system at*** $t_0 \in \mathfrak{T}_i$*, which is denoted by* $\mathfrak{I}\left(t;t_0^-\right)$*, if and only if it is both a negatively invariant set at* t_0 *and a positively invariant set at* t_0 *of the dynamical system,*

$$t_0 \in \mathfrak{T}_i,\ \mathfrak{S}(t) = \mathfrak{I}\left(t;t_0^-\right) \Longleftrightarrow$$
$$\forall \mathbf{S}_0 \in \mathfrak{S}(t_0) = \mathfrak{S}_0 \Longrightarrow \mathbf{S}(t;t_0;\mathbf{S}_0;\mathbf{I}) \in \mathfrak{S}(t),\ \forall t \in \mathfrak{T}. \tag{4.65}$$

If and only if additionally the set $\mathfrak{S}(t)$ *is constant,* $\mathfrak{S}(t) = \mathfrak{I}(t) \equiv \mathfrak{S}$*, then for short:* $\mathfrak{I}\left(t;t_0^-\right) \equiv \mathcal{I}$.

f) ***an invariant set of a dynamical system on*** $\mathfrak{T}_i$*, which is denoted by* $\mathfrak{I}(t;\mathfrak{T}_i)$*, if and only if it is both a negatively invariant set on* $\mathfrak{T}_i^-$ *and a positively invariant set on* $\mathfrak{T}_i$ *of the dynamical system,*

$$\mathfrak{S}(t) = \mathfrak{I}(t;\mathfrak{T}_i) \Longleftrightarrow$$
$$\forall t_0 \in \mathfrak{T}_i \wedge \forall \mathbf{S}_0 \in \mathfrak{S}(t_0) = \mathfrak{S}_0 \Longrightarrow \mathbf{S}(t;t_0;\mathbf{S}_0;\mathbf{I}) \in \mathfrak{S}(t),\ \forall t \in \mathfrak{T}. \tag{4.66}$$

If and only if additionally the set $\mathfrak{S}(t)$ *is constant,* $\mathfrak{S}(t) = \mathfrak{I}(t;\mathfrak{T}_i) \equiv \mathfrak{S}$*, then for short:* $\mathfrak{I}(t;\mathfrak{T}_i) \equiv \mathcal{I}$.

Definition 131 *A set* $\mathfrak{S}(t;t_0;\mathfrak{S}_0)$ *is:*

a) ***a positively invariant set at*** $t_0 \in \mathfrak{T}_i$ ***of a dynamical system****, which is denoted by* $\mathfrak{I}^+\left(t;t_0;\mathfrak{I}_0^+\right)_i$*, if and only if the system state* $\mathbf{S}_0$ *from the set* $\mathfrak{S}(t_0;t_0;\mathfrak{S}_0)$ *at the moment* t_0 *implies that the system motion* $\mathcal{S}(.;t_0;\mathbf{S}_0;\mathbf{I})$ *rests in the set* $\mathfrak{S}(t;t_0;\mathfrak{S}_0)$ *since* t_0 *forever,*

$$t_0 \in \mathfrak{T}_i,\ \mathfrak{S}(t;t_0;\mathfrak{S}_0) = \mathfrak{I}^+\left(t;t_0;\mathfrak{I}_0^+\right) \Longleftrightarrow$$
$$\forall \mathbf{S}_0 \in \mathfrak{S}(t_0;t_0;\mathfrak{S}_0) = \mathfrak{S}_0 \Longrightarrow \mathbf{S}(t;t_0;\mathbf{S}_0;\mathbf{I}) \in \mathfrak{S}(t;t_0;\mathfrak{S}_0),\ \forall t \in \mathfrak{T}_0. \tag{4.67}$$

If and only if additionally the set $\mathfrak{S}(t;t_0;\mathfrak{S}_0)$ *is constant,* $\mathfrak{S}(t;t_0;\mathfrak{S}_0) = \mathfrak{I}^+\left(t;t_0;\mathfrak{I}_0^+\right) \equiv \mathfrak{S}$*, then for short:* $\mathfrak{I}^+\left(t;t_0;\mathfrak{I}_0^+\right) \equiv \mathcal{I}^+$.

b) ***a positively invariant set on*** $\mathfrak{T}_i$ ***of a dynamical system****, which is denoted by* $\mathfrak{I}^+(t;\mathfrak{T}_i)$*, if and only if the system state* $\mathbf{S}_0$ *from the set* $\mathfrak{S}(t_0;t_0;\mathfrak{S}_0)$ *at any moment* $t_0 \in \mathfrak{T}_i$ *implies that the system motion* $\mathcal{S}(.;t_0;\mathbf{S}_0;\mathbf{I})$ *rests in the set* $\mathfrak{S}(t;t_0;\mathfrak{S}_0)$ *since* t_0 *forever,*

$$\mathfrak{S}(t;t_0;\mathfrak{S}_0) = \mathfrak{I}^+(t;\mathfrak{T}_i) \Longleftrightarrow$$
$$\forall t_0 \in \mathfrak{T}_i \wedge\ \forall \mathbf{S}_0 \in \mathfrak{S}(t_0) = \mathfrak{S}_0 \Longrightarrow \mathbf{S}(t;t_0;\mathbf{S}_0;\mathbf{I}) \in \mathfrak{S}(t;t_0;\mathfrak{S}_0),\ \forall t \in \mathfrak{T}_0. \tag{4.68}$$

If and only if additionally the set $\mathfrak{S}(t;t_0;\mathfrak{S}_0)$ *is constant,* $\mathfrak{S}(t;t_0;\mathfrak{S}_0) = \mathfrak{I}^+(t;\mathfrak{T}_i) \equiv \mathfrak{S}$*, then for short:* $\mathfrak{I}^+(t;\mathfrak{T}_i) \equiv \mathcal{I}^+$.

*c) **a negatively invariant set at** $t_0 \in \mathfrak{T}_i$ **of a dynamical system**, which is denoted by $\mathfrak{I}^-\left(t;;t_0;\mathfrak{I}_0^-\right)$, if and only if the system state $\mathbf{S}_0$ from the set $\mathfrak{S}(t_0;t_0;\mathfrak{S}_0)$ at the moment t_0 implies that the system motion $\mathcal{S}(.;t_0;\mathbf{S}_0;\mathbf{I})$ was in the set $\mathfrak{S}(t;t_0;\mathfrak{S}_0)$ up to the moment t_0,*

$$t_0 \in \mathfrak{T}_i,\ \mathfrak{S}(t;t_0;\mathfrak{S}_0) = \mathfrak{I}^-\left(t;;t_0;\mathfrak{I}_0^-\right) \Longleftrightarrow$$
$$\forall \mathbf{S}_0 \in \mathfrak{S}(t_0;t_0;\mathfrak{S}_0) = \mathfrak{S}_0 \Longrightarrow \mathbf{S}(t;t_0;\mathbf{S}_0;\mathbf{I}) \in \mathfrak{S}(t;t_0;\mathfrak{S}_0),\ \forall t \in \mathfrak{T}_0^-. \tag{4.69}$$

If and only if additionally the set $\mathfrak{S}(t;t_0;\mathfrak{S}_0)$ is constant, $\mathfrak{S}(t;t_0;\mathfrak{S}_0) = \mathfrak{I}^-\left(t;t_0;\mathfrak{I}_0^-\right) \equiv \mathfrak{S}$, then for short: $\mathfrak{I}^-\left(t;t_0;\mathfrak{I}_0^-\right) \equiv \mathcal{I}^-$.

*d) **a negatively invariant set on** $\mathfrak{T}_i$ **of a dynamical system**, which is denoted by $\mathfrak{I}^-(t;\mathfrak{T}_i)$, if and only if the system state $\mathbf{S}_0$ from the set $\mathfrak{S}(t_0;t_0;\mathfrak{S}_0)$ at any moment $t_0 \in \mathfrak{T}_i$ implies that the system motion $\mathcal{S}(.;t_0;\mathbf{S}_0;\mathbf{I})$ was also in the set $\mathfrak{S}(t;t_0;\mathfrak{S}_0)$ up to the moment t_0,*

$$\mathfrak{S}(t;t_0;\mathfrak{S}_0) = \mathfrak{I}^-(t;\mathfrak{T}_i) \Longleftrightarrow$$
$$\forall t_0 \in \mathfrak{T}_i \wedge \forall \mathbf{S}_0 \in \mathfrak{S}(t_0;t_0;\mathfrak{S}_0) = \mathfrak{S}_0 \Longrightarrow \mathbf{S}(t;t_0;\mathbf{S}_0;\mathbf{I}) \in \mathfrak{S}(t;t_0;\mathfrak{S}_0),\ \forall t \in \mathfrak{T}_0^-. \tag{4.70}$$

If and only if additionally the set $\mathfrak{S}(t;t_0;\mathfrak{S}_0)$ is constant, $\mathfrak{S}(t;t_0;\mathfrak{S}_0) \equiv \mathfrak{S} \equiv \mathfrak{I}^-(t;\mathfrak{T}_i)$, then for short: $\mathfrak{I}^-(t;\mathfrak{T}_i) \equiv \mathcal{I}^-$.

*e) **an invariant set of a dynamical system at** $t_0 \in \mathfrak{T}_i$, which is denoted by $\mathfrak{I}(t;t_0;\mathfrak{I}_0)$, if and only if it is both a negatively invariant set at t_0 and a positively invariant set at t_0 of the dynamical system,*

$$t_0 \in \mathfrak{T}_i,\ \mathfrak{S}(t;t_0;\mathfrak{S}_0) = \mathfrak{I}(t;t_0;\mathfrak{I}_0) \Longleftrightarrow$$
$$\forall \mathbf{S}_0 \in \mathfrak{S}(t_0;t_0;\mathfrak{S}_0) = \mathfrak{S}_0 \Longrightarrow \mathbf{S}(t;t_0;\mathbf{S}_0;\mathbf{I}) \in \mathfrak{S}(t;t_0;\mathfrak{S}_0),\ \forall t \in \mathfrak{T}. \tag{4.71}$$

If and only if additionally the set $\mathfrak{S}(t;t_0;\mathfrak{S}_0)$ is constant, $\mathfrak{S}(t;t_0;\mathfrak{S}_0) = \mathfrak{I}(t;t_0;\mathfrak{I}_0) \equiv \mathfrak{S}$, then for short: $\mathfrak{I}(t;t_0;\mathfrak{I}_0) \equiv \mathcal{I}$.

*f) **an invariant set of a dynamical system on** $\mathfrak{T}_i$, which is denoted by $\mathfrak{I}(t;\mathfrak{T}_i)$, if and only it is both a negatively invariant set on $\mathfrak{T}_i^-$ and a positively invariant set on $\mathfrak{T}_i$ of the dynamical system,*

$$\mathfrak{S}(t;t_0;\mathfrak{S}_0) = \mathfrak{I}(t;\mathfrak{T}_i) \Longleftrightarrow$$
$$\forall t_0 \subset \mathfrak{T}_i \wedge \forall \mathbf{S}_0 \in \mathfrak{S}(t_0;t_0;\mathfrak{S}_0) = \mathfrak{S}_0 \Longrightarrow \mathbf{S}(t;t_0;\mathbf{S}_0;\mathbf{I}) \in \mathfrak{S}(t;t_0;\mathfrak{S}_0),\ \forall t \in \mathfrak{T}. \tag{4.72}$$

If and only if additionally the set $\mathfrak{S}(t;t_0;\mathfrak{S}_0)$ is constant, $\mathfrak{S}(t;t_0;\mathfrak{S}_0) = \mathfrak{I}(t;\mathfrak{T}_i) \equiv \mathfrak{S}$, then for short: $\mathfrak{I}(t;\mathfrak{T}_i) \equiv \mathcal{I}$.

The uniqueness of the system motions guarantees the following:

Comment 132 *Let $\mathfrak{S}_d(t)$ be a singleton, the single element of which is the system instantaneous desired state $\mathbf{S}_d(t)$ at every $t \in \mathfrak{T}$:*

$$\mathfrak{S}_d(t) = \{\mathbf{S}_d(t)\},\ \forall t \in \mathfrak{T}. \tag{4.73}$$

The time-varying set $\mathfrak{S}_d(t)$, Equation 4.73, is an invariant set with respect to system motions $\mathcal{S}_d(.;t_0;\mathbf{S}_{0d};\mathbf{I}_N)$,

$$\mathfrak{S}_d(t) = \mathfrak{I}_d(t). \tag{4.74}$$

Comment 133 *Let the $\mathfrak{S}(t;t_0;\mathfrak{S}_0)$ be time-invariant, $\mathfrak{S}(t;t_0;\mathfrak{S}_0) = \mathfrak{S}$ for all $t \in \mathfrak{T}$. It can be a limit cycle $\mathfrak{S}_{LC}$, $\mathfrak{S} = \mathfrak{S}_{LC}$ on $\mathfrak{T}$. It is an invariant set. It can be an equilibrium set $\mathfrak{S}_E = \{\mathbf{S}_E\}$ on $\mathfrak{T}_0$, which is positively invariant on $\mathfrak{T}_0$, where $\mathbf{S}_E$ is a system equilibrium state on $\mathfrak{T}_0$,*

$$\mathfrak{S}_{LC} = \mathfrak{I}_{LC}(\mathfrak{T}) = \mathcal{I}_{LC} \text{ and } \mathfrak{S}_E = \mathfrak{I}_E^+(\mathfrak{T}_0) = \mathcal{I}_E^+. \tag{4.75}$$

4.6 System regimes

Analysis of system regimes is omitted due to the space limitation and the fact that it is presented in the books [217, Chapter 3, pp. 33-85], [230, Chapter 3, pp. 37-88], [244, Chapter 3, pp. 33-51] on linear systems, which is valid also for nonlinear systems as shown in [232, Chapter 4: Desired regime, pp. 41-51].

4.7 *IO* systems

4.7.1 Original *IO* system description

We will use the compact notation introduced in (1.9), (1.10) and (1.11), Subsection 1.1.3 of Section 1.1, e.g.,

$$\mathbf{Y}^{m-1} = \left[\mathbf{Y}^{T} \vdots \mathbf{Y}^{T(1)} \vdots \mathbf{Y}^{T(2)} \vdots \vdots \mathbf{Y}^{T(m-2)} \vdots \mathbf{Y}^{T(m-1)} \right]^{T} \in \mathfrak{R}^{mN}, \ \mathbf{Y} \in \mathfrak{R}^{N}, \ mN > 0,$$

$$A^{(m-1)} = \left[A_0 \vdots A_1 \vdots ... \vdots A_{m-1} \right] \in \mathfrak{R}^{N \times mN}, \ A_i \in \mathfrak{R}^{N \times N}, \ i = 0, 1, .., m-1.$$

The original mathematical models of some physical plants, control systems, or systems in general, result directly in the form of the *time*-varying nonlinear vector *Input-Output* (*IO*) differential equation 4.76 that we present in the general compact form:

$$Q\left[t, \mathbf{Y}^{m-1}(t)\right] \mathbf{Y}^{(m)}(t) + A^{(m-1)} \mathbf{Y}^{m-1}(t) + \mathbf{q}\left[t, \mathbf{Y}^{m-1}(t)\right] = \mathbf{h}\left[t, \mathbf{I}^{\mu}(t)\right], \ \forall t \in \mathfrak{T},$$
$$m \geq 1, \ 0 \leq \mu \leq m-1, \ \mathbf{I} \in \mathcal{J}, \ \mathcal{J} \subset \mathfrak{C}^{(\mu+1)M}, \tag{4.76}$$

where the matrix function $Q(.)$ satisfies:

$$Q(.) : \mathfrak{T} \times \mathfrak{R}^{mN} \longrightarrow \mathfrak{R}^{N \times N}, \tag{4.77}$$

and the nonlinear vector functions $\mathbf{q}(.)$ and $\mathbf{h}(.)$,

$$\mathbf{q}(.) : \mathfrak{T} \times \mathfrak{R}^{mN} \longrightarrow \mathfrak{R}^{N}, \ \mathbf{h}(.) : \mathfrak{T} \times \mathfrak{R}^{(\mu+1)M} \longrightarrow \mathfrak{R}^{N}, \tag{4.78}$$

are all defined and continuously differentiable in all their arguments, if not stated otherwise.

The desired system output behavior $\mathbf{Y}_d(t)$ is either determined in advance or during the system operation. It is known for the control synthesis and implementation.

The matrix $A^{(m-1)}$, the matrix function $Q(.)$ and the vector function $\mathbf{q}(.)$ determine the internal dynamics of the *IO* system (4.76).

The vector function $\mathbf{h}(.)$ defines how the system transmits the influence of the extended input vector $\mathbf{I}^{\mu}$ on the system behavior.

Property 134 ***Nonsingularity property of the*** *IO* ***system***
The matrix $Q(t, \mathbf{Y}^{m-1})$ of the IO system (4.76) is non-singular,

$$detQ(t, \mathbf{Y}^{m-1}) \neq 0, \ \forall \left(t, \mathbf{Y}^{m-1}\right) \in \mathfrak{T} x \mathfrak{R}^{mN}. \tag{4.79}$$

The *IO* system (4.76) that possesses Property 134 is called **regular** (i.e., **nonsingular**) *IO* **system** (4.76).

The condition (4.79) ensures solvability of Equation (4.76) in $\mathbf{Y}^{(m)}(t)$, respectively,

$$\mathbf{Y}^{(m)}(t) = Q^{-1}\left[t, \mathbf{Y}^{m-1}(t)\right] \left\{ \mathbf{h}\left[t, \mathbf{I}^{\mu}(t)\right] - A^{(m-1)} \mathbf{Y}^{m-1}(t) - \mathbf{q}\left[t, \mathbf{Y}^{m-1}(t)\right] \right\}. \tag{4.80}$$

Note 135 *Property 134 is sufficient, but not necessary, for all output variables of the system (4.76) to have the same order m of their highest derivatives.*

4.7.2 State-space *IO* system description

The *time* evolution of the extended output vector $\mathbf{Y}^{m-1}(t)$ expresses the internal dynamical behavior of the *IO* system (4.76). Its instantaneous vector value determines the instantaneous internal dynamical situation of the system, which is the instantaneous state vector $\mathbf{S}_{IO}(t)$ of the system. More precisely:

Definition 136 ***The IO system state and state space***

The vector $\mathbf{Y}^{m-1}$ *is* ***the internal dynamics vector of the IO system (4.76),*** *i.e.,* ***the IO system state vector*** $\mathbf{S} = \mathbf{S}_{IO}$:

$$\mathbf{S} = \mathbf{S}_{IO} = \left[\mathbf{S}_1^T \vdots\ \mathbf{S}_2^T \vdots \ldots \vdots\ \mathbf{S}_m^T\right]^T \in \mathfrak{R}^n,\ \mathbf{S}_k \in \mathfrak{R}^N,\ \forall k = 1, 2, ..., m,$$

$$\mathbf{S} = \mathbf{S}_{IO} = \mathbf{Y}^{m-1} = \left[\mathbf{Y}^T \vdots \left(\mathbf{Y}^{(1)}\right)^T \vdots \ldots \vdots \left(\mathbf{Y}^{(m-1)}\right)^T\right]^T. \tag{4.81}$$

The $n = mN$ *dimensional real vector space* $\mathfrak{R}^n = \mathfrak{R}^{mN}$ *is* ***the internal dynamics space,*** *i.e.,* ***the state space of the IO system (4.76).***

The state variables and the state vector $\mathbf{S}_{IO} = \mathbf{Y}^{m-1}$, *Equation (4.81), have the well-known form called* ***the phase form,*** *i.e.,* $\mathbf{S}_{IO}= \mathbf{Y}^{m-1}$ *is* ***the phase state vector,*** *for short:* ***the phase vector*** *or* ***the state vector*** *of the IO system.*

Equations (4.81) and Equation (4.80) furnish:

$$\mathbf{S}_1^{(1)} = \mathbf{S}_2,\ \mathbf{S}_2^{(1)} = \mathbf{S}_3,\ \ldots\ ,\ \mathbf{S}_{m-1}^{(1)} = \mathbf{S}_m,\ \mathbf{S}_m^{(1)} = \mathbf{Y}^{(m)}, \tag{4.82}$$

and for the *IO* system **(4.76):**

$$\mathbf{S}_m^{(1)} = Q^{-1}\left[t, \mathbf{S}(t)\right]\left\{\mathbf{h}\left[t, \mathbf{I}^{\mu}(t)\right] - A^{(m-1)}\mathbf{S}(t) - \mathbf{q}\left[t, \mathbf{S}(t)\right]\right\},$$

This equation and Equations (4.82) yield for the *IO* system **(4.76):**

$$\frac{d\mathbf{S}(t)}{dt} = \mathbf{a}_1\left[t, \mathbf{S}(t)\right] + \mathrm{G}(t)\,\mathbf{g}_1\left[t, \mathbf{I}^{\mu}(t)\right]. \tag{4.83}$$

where $\mathbf{S}_d$ is the system desired state vector induced by the output desired vector $\mathbf{Y}_d$, and

$$\mathbf{a}_1\left[t, \mathbf{S}(t)\right] = \begin{bmatrix} \mathbf{S}_2 \\ \mathbf{S}_3 \\ \ldots \\ \mathbf{S}_m \\ -Q^{-1}\left[t, \mathbf{S}(t)\right]\left\{A^{(m-1)}\mathbf{S}(t) + \mathbf{q}\left[t, \mathbf{S}(t)\right]\right\} \end{bmatrix} \in \mathfrak{R}^n, \tag{4.84}$$

$$\mathrm{G}\left[t, \mathbf{S}(t)\right] = \begin{bmatrix} O_N \\ O_N \\ \ldots \\ O_N \\ Q^{-1}\left[t, \mathbf{S}(t)\right] \end{bmatrix} \in \mathfrak{R}^{nxN}, \tag{4.85}$$

$$\mathbf{g}_1\left[t, \mathbf{S}(t), \mathbf{I}^{\mu}(t)\right] = \mathbf{h}\left[t, \mathbf{I}^{\mu}(t)\right] \in \mathfrak{R}^N, \tag{4.86}$$

so that the system output equation becomes:

$$\mathbf{Y}(t) = \mathrm{C}\mathbf{S}(t),\ \mathrm{C} = \left[I_N \vdots\ O_N \vdots \ldots \vdots\ O_N\right], \tag{4.87}$$

or formally,

$$\mathbf{Y}(t)=\mathbf{a}_2\left[t,\mathbf{S}\left(t\right)\right]+\mathbf{g}_2\left[t,\mathbf{I}\left(t\right)\right],\ \forall t\in\mathfrak{T},$$
$$\textit{for the IO system,} \tag{4.88}$$

where

$$\mathbf{a}_2\left[t,\mathbf{S}\left(t\right)\right]=\mathbf{CS}\left(t\right),\ \ \mathbf{g}_2\left[t,\mathbf{I}\left(t\right)\right]\equiv\mathbf{0}_N. \tag{4.89}$$

Equations (4.83) , (4.88) represent the input-state-output mathematical model of the *IO* system (4.76), in which the matrices and vectors are determined in Equations (4.84)-(4.89).

4.8 *ISO* systems

4.8.1 Original *ISO* system description

The qualitative theory of nonlinear dynamical systems has mostly studied the *Input-State-Output* (*ISO*) nonlinear *systems*. Their mathematical model contains *the first order vector nonlinear differential state equation* (4.90) and *the vector nonlinear algebraic output equation* (4.91),

$$\frac{d\mathbf{X}(t)}{dt}=\mathbf{a}_1\left[t,\mathbf{X}\left(t\right)\right]+\mathbf{g}_1\left[t,\mathbf{I}\left(t\right)\right],\ \forall t\in\mathfrak{T},\ \mathbf{X}\in\mathfrak{R}^n,$$
$$\mathbf{a}_1\left(.\right):\mathfrak{T}\times\mathfrak{R}^n\longrightarrow\mathfrak{R}^n,\ \mathbf{g}_1\left(.\right):\mathfrak{T}\times\mathfrak{R}^M\longrightarrow\mathfrak{R}^n, \tag{4.90}$$

$$\mathbf{Y}(t)=\mathbf{a}_2\left[t,\mathbf{X}\left(t\right)\right]+\mathbf{g}_2\left[t,\mathbf{I}\left(t\right)\right],\ \forall t\in\mathfrak{T},$$
$$\mathbf{a}_2\left(.\right):\mathfrak{T}\times\mathfrak{R}^n\longrightarrow\mathfrak{R}^N,\ \mathbf{g}_2\left(.\right):\mathfrak{T}\times\mathfrak{R}^M\longrightarrow\mathfrak{R}^N, \tag{4.91}$$

where the nonlinear vector functions $\mathbf{a}_1\left(.\right)$, $\mathbf{a}_2\left(.\right)$, $\mathbf{g}_1\left(.\right)$ and $\mathbf{g}_2\left(.\right)$ are defined and continuously differentiable.

4.8.2 State-space *ISO* system description

Definition 137 ***The ISO system state and state space***

The vector $\mathbf{X}$ *is* ***the internal dynamics vector of the*** *ISO* ***system (4.90), (4.91),*** *i.e.,* ***the*** *ISO* ***system state vector*** $\mathbf{S}=\mathbf{S}_{ISO}$***:***

$$\mathbf{S}=\mathbf{S}_{ISO}=\mathbf{X}. \tag{4.92}$$

The n dimensional real vector space $\mathfrak{R}^n$ *is* ***the internal dynamics space,*** *i.e.,* ***the state space, of the*** *ISO* ***system (4.90), (4.91).***

This definition discovers that the *ISO* system (4.90), (4.91) is in its input-state-output form in which the state vector is defined by Equation (4.92), i.e., in which the vector $\mathbf{X}$ is the state vector and to be replaced by $\mathbf{S}$:

$$\frac{d\mathbf{S}(t)}{dt}=\mathbf{a}_1\left[t,\mathbf{S}\left(t\right)\right]+\mathbf{g}_1\left[t,\mathbf{I}\left(t\right)\right],\ \forall t\in\mathfrak{T},\ \mathbf{S}\in\mathfrak{R}^n, \tag{4.93}$$
$$\mathbf{Y}(t)=\mathbf{a}_2\left[t,\mathbf{S}\left(t\right)\right]+\mathbf{g}_2\left[t,\mathbf{I}\left(t\right)\right],\ \forall t\in\mathfrak{T}, \tag{4.94}$$

4.9 *HISO* systems

4.9.1 Original *HISO* system description

The mathematical model of the *Higher order Input-State-Output* (*HISO*) *systems* contains the following nonlinear vector differential state equation:

$$Q\left[t,\mathbf{X}^{m-1}(t)\right]\mathbf{X}^{(m)}(t)+A^{(m-1)}\mathbf{X}^{m-1}(t)+\mathbf{q}_1\left[t,\mathbf{X}^{m-1}(t)\right]=\mathbf{h}_1\left[t,\mathbf{I}^{\mu}(t)\right],\ \forall t\in\mathfrak{T},$$

$$A^{(m-1)}=\left[A_0\ \vdots\ A_1\ \vdots\ ...\ \vdots\ A_{m-1}\right]\in\mathfrak{R}^{\rho\times m\rho},\ \mathbf{X}\in\mathfrak{R}^{\rho},\ \mathbf{X}^{m-1}\in\mathfrak{R}^{m\rho}, \tag{4.95}$$

and the following algebraic vector output equation:

$$\mathbf{Y}(t)=\mathbf{q}_2\left[t,\mathbf{X}^{m-1}(t)\right]+\mathbf{h}_2\left[t,\mathbf{I}(t)\right],\ \forall t\in\mathfrak{T}, \tag{4.96}$$

where the matrix functions and matrices

$$Q(.):\mathfrak{T}\times\mathfrak{R}^{m\rho}\longrightarrow\mathfrak{R}^{\rho\times\rho},\ A_i\in\mathfrak{R}^{\rho\times\rho},\ i=0,1,..,m-1, \tag{4.97}$$

and the nonlinear vector functions

$$\mathbf{q}_1(.):\mathfrak{T}\times\mathfrak{R}^{m\rho}\longrightarrow\mathfrak{R}^{\rho},\ \mathbf{h}_1(.):\mathfrak{T}\times\mathfrak{R}^{(\mu+1)M}\longrightarrow\mathfrak{R}^{\rho}, \tag{4.98}$$

$$\mathbf{q}_2(.):\mathfrak{T}\times\mathfrak{R}^{m\rho}\longrightarrow\mathfrak{R}^{N},\ \mathbf{h}_2(.):\mathfrak{T}\times\mathfrak{R}^{M}\longrightarrow\mathfrak{R}^{N}, \tag{4.99}$$

are defined and continuously differentiable in all their arguments if not stated otherwise.

Property 138 ***Nonsingularity property of the*** $HISO$ ***system***
The matrix $Q(t,\mathbf{X}^{m-1})$ *of the* $HISO$ *system (4.95), (4.96), is nonsingular,*

$$detQ(t,\mathbf{X}^{m-1})\neq 0,\ \forall\left(t,\mathbf{X}^{m-1}\right)\in\mathfrak{T}\mathrm{x}\mathfrak{R}^{m\rho}. \tag{4.100}$$

The $HISO$ system (4.95), (4.96), that possesses Property 138 is called **regular** (i.e., **nonsingular**) $HISO$ system (4.95), (4.96).

The condition (4.100) ensures solvability of the $HISO$ system (4.95), (4.96) in $\mathbf{X}^{(m)}(t)$,

$$\mathbf{X}^{(m)}(t)=Q^{-1}\left[t,\mathbf{X}^{m-1}(t)\right]\left\{\mathbf{h}_1\left[t,\mathbf{I}^{\mu}(t)\right]-A^{(m-1)}\mathbf{X}^{m-1}(t)-\mathbf{q}_1\left[t,\mathbf{X}^{m-1}(t)\right]\right\}. \tag{4.101}$$

4.9.2 State-space *HISO* system description

The instantaneous vector value of the extended vector $\mathbf{X}^{m-1}(t)$ determines the instantaneous internal dynamical situation of the system, which is the instantaneous state vector $\mathbf{S}_{HISO}(t)$ of the system. More precisely:

Definition 139 ***The HISO system state and state space***
The vector $\mathbf{X}^{m-1}$ *is* ***the internal dynamics vector of the*** $HISO$ *system (4.95), (4.96), i.e.,* ***the*** $HISO$ ***system state vector*** $\mathbf{S}=\mathbf{S}_{HISO}$ *in* ***the phase form:***

$$\mathbf{S}=\mathbf{S}_{HISO}=\left[\mathbf{S}_1^T\ \vdots\ \mathbf{S}_2^T\ \vdots\ ...\ \vdots\ \mathbf{S}_m^T\right]^T,\ \mathbf{S}_k=\mathbf{X}^{(k-1)}\in\mathfrak{R}^{\rho},\ \forall k=1,2,...,m,\ \mathbf{X}^{(0)}=\mathbf{X},$$

$$\mathbf{S}=\mathbf{S}_{HISO}=\mathbf{X}^{m-1}=\left[\mathbf{X}^T\ \vdots\ \left(\mathbf{X}^{(1)}\right)^T\ \vdots\ ...\ \vdots\ \left(\mathbf{X}^{(m-1)}\right)^T\right]^T\in\mathfrak{R}^{m\rho}. \tag{4.102}$$

The $n=m\rho$ *dimensional real vector space* $\mathfrak{R}^{m\rho}=\mathfrak{R}^{n}$ *is* ***the internal dynamics space,*** *i.e.,* ***the state space, of the*** $HISO$ *system (4.95) -* ***(4.96).***

Equations (4.102) and (4.101) furnish

$$\mathbf{S}_1^{(1)} = \mathbf{S}_2,\ \mathbf{S}_2^{(1)} = \mathbf{S}_3,\ \ldots\ ,\ \mathbf{S}_{m-1}^{(1)} = \mathbf{S}_m,\ \mathbf{S}_m^{(1)} = \mathbf{X}^{(m)}, \tag{4.103}$$

and

$$\mathbf{S}_m^{(1)} = Q^{-1}\left[t, \mathbf{S}(t)\right]\left\{\mathbf{h}_1\left[t, \mathbf{I}^{\mu}(t)\right] - A^{(m-1)}\mathbf{S}(t) - \mathbf{q}_1\left[t, \mathbf{S}(t)\right]\right\}. \tag{4.104}$$

These equations yield for the *HISO* system (4.95), (4.96):

$$\frac{d\mathbf{S}(t)}{dt} = \mathbf{a}_1\left[t, \mathbf{S}(t)\right] + \mathrm{G}\left[t, \mathbf{S}(t)\right]\mathbf{g}_1\left[t, \mathbf{I}^{\mu}(t)\right], \tag{4.105}$$

where

$$\mathbf{a}_1\left[t, \mathbf{S}(t)\right] = \begin{bmatrix} \mathbf{S}_2 \\ \mathbf{S}_3 \\ \ldots \\ \mathbf{S}_m \\ -Q^{-1}\left[t, \mathbf{S}(t)\right]\left\{A^{(m-1)}\mathbf{S}(t) + \mathbf{q}_1\left[t, \mathbf{S}(t)\right]\right\} \end{bmatrix} \in \mathfrak{R}^n, \tag{4.106}$$

$$\mathrm{G}\left[t, \mathbf{S}(t)\right] = \begin{bmatrix} O_\rho \\ O_\rho \\ \ldots \\ O_\rho \\ Q^{-1}\left[t, \mathbf{S}(t)\right] \end{bmatrix} \in \mathfrak{R}^{n \times \rho}, \tag{4.107}$$

$$\mathbf{g}_1\left[t, \mathbf{I}^{\mu}(t)\right] = \mathbf{h}_1\left[t, \mathbf{I}^{\mu}(t)\right] \in \mathfrak{R}^{\rho}. \tag{4.108}$$

The output equation (4.96) takes the following forms:

$$\mathbf{Y}(t) = \mathbf{a}_2\left[t, \mathbf{S}(t)\right] + \mathbf{g}_2\left[t, \mathbf{I}(t)\right],\ \forall t \in \mathfrak{T}, \tag{4.109}$$

where

$$\mathbf{a}_2\left[t, \mathbf{S}(t)\right] = \mathbf{q}_2\left[t, \mathbf{S}(t)\right],\ \mathbf{g}_2\left[t, \mathbf{I}(t)\right] = \mathbf{h}_2\left[t, \mathbf{I}(t)\right]. \tag{4.110}$$

Equations (4.105), (4.109) represent the input-state-output mathematical model of the *HISO* system (4.95), (4.96), in which the matrices and vectors are determined in Equations (4.106)-(4.110).

4.10 *IIO* systems

4.10.1 Original *IIO* system description

A large class of the nonlinear dynamical systems introduced in [232] differs from other classes of the nonlinear systems for having both internal dynamics, i.e., **the internal state**, and output dynamics, i.e., **the output state**. They are called *Input-Internal and Output Dynamics* (*IIOD*) systems [232] or for short: *Input-Internal and Output state* (*IIO*) systems.

The state equation (4.95) of the *HISO* system is also the state equation of the *Input-Internal and Output state* (*IIO*) *system*:

$$Q\left[t, \mathbf{X}^{m-1}(t)\right]\mathbf{X}^{(m)}(t) + A^{(m-1)}\mathbf{X}^{m-1}(t) + \mathbf{q}_1\left[t, \mathbf{X}^{m-1}(t)\right] = \mathbf{h}_1\left[t, \mathbf{I}^{\mu}(t)\right],\ \forall t \in \mathfrak{T},$$
$$\mathbf{X} \in \mathfrak{R}^{\rho},\ \mathbf{X}^{m-1} \in \mathfrak{R}^{m\rho},\ m \geq 1. \tag{4.111}$$

However, the IIO system output equation is not an algebraic one but a differential vector equation in the output vector $\mathbf{Y}$:

$$E\left[t,\mathbf{Y}^{\nu-1}(t)\right]\mathbf{Y}^{(\nu)}(t)+F^{(\nu-1)}\mathbf{Y}^{\nu-1}(t)+\mathbf{q}_2\left[t,\mathbf{Y}^{\nu-1}(t)\right]=$$
$$=\mathbf{e}\left[t,\mathbf{X}^{m-1}(t)\right]+\mathbf{h}_2\left[t,\mathbf{I}^{\mu}(t)\right],\ \forall t\in\mathfrak{T},$$
$$\mathbf{Y}\in\mathfrak{R}^{N},\ \mathbf{Y}^{\nu-1}\in\mathfrak{R}^{\nu N},\ \nu\geq 1,\ \mu\leq\min\left(m-1,\ \nu-1,\right), \tag{4.112}$$

where the matrix functions $Q(.)$ and $E(.)$, and the matrix $F^{(\nu-1)}$ satisfies the following:

$$Q(.):\mathfrak{T}\times\mathfrak{R}^{m\rho}\longrightarrow\mathfrak{R}^{\rho\times\rho}, \tag{4.113}$$

$$E(.):\mathfrak{T}\times\mathfrak{R}^{\nu N}\longrightarrow\mathfrak{R}^{N\times N},\ F^{(\nu-1)}\in\mathfrak{R}^{N\times\nu N}, \tag{4.114}$$

and the nonlinear vector functions

$$\mathbf{q}_1(.):\mathfrak{T}\times\mathfrak{R}^{m\rho}\longrightarrow\mathfrak{R}^{\rho},\ \mathbf{h}_1(.):\mathfrak{T}\times\mathfrak{R}^{(\mu+1)M}\longrightarrow\mathfrak{R}^{\rho}, \tag{4.115}$$

$$\mathbf{q}_2(.):\mathfrak{T}\times\mathfrak{R}^{\nu N}\longrightarrow\mathfrak{R}^{N}$$
$$\mathbf{e}(.):\mathfrak{T}\times\mathfrak{R}^{m\rho}\longrightarrow\mathfrak{R}^{N},\ \mathbf{h}_2(.):\mathfrak{T}\times\mathfrak{R}^{(\mu+1)M}\longrightarrow\mathfrak{R}^{N}, \tag{4.116}$$

are defined and continuously differentiable in all their arguments if not stated otherwise.

Property 140 ***Nonsingularity properties of the*** *IIO* ***system***

The matrices $Q(t,\mathbf{X}^{m-1})$ *and* $E\left(t,\mathbf{Y}^{\nu-1}\right)$ *of the IIO system (4.111), (4.112) are nonsingular,*

$$detQ(t,\mathbf{X}^{m-1})\neq 0,\ \forall\left(t,\mathbf{X}^{m-1}\right)\in\mathfrak{T}x\mathfrak{R}^{m\rho},$$
$$detE\left(t,\mathbf{Y}^{\nu-1}\right)\neq 0,\ \forall\left(t,\mathbf{Y}^{\nu-1}\right)\in\mathfrak{T}x\mathfrak{R}^{\nu N}. \tag{4.117}$$

The IIO system (4.111), (4.112), which possesses Property 140, is a ***regular*** *(i.e.,* ***nonsingular****) IIO system (4.111), (4.112).*

4.10.2 State-space *IIO* system description

The instantaneous vector value of the extended vector $\mathbf{X}^{m-1}(t)$ determines the instantaneous internal dynamical situation of the system, which is **the instantaneous internal state vector** $\mathbf{S}_{IIOI}(t)$ of the system. More precisely:

Definition 141 ***The*** *IIO* ***system internal state and internal state space***

The vector $\mathbf{X}^{m-1}$ *is* ***the internal dynamics vector of the*** *IIO system (4.111), (4.112), i.e.,* ***the*** *IIO* ***system internal state vector*** $\widehat{\mathbf{S}}=\widehat{\mathbf{S}}_{IIO}$ *in* ***the phase form:***

$$\widehat{\mathbf{S}}=\mathbf{S}_{IIOI}=\left[\widehat{\mathbf{S}}_1^T\vdots\widehat{\mathbf{S}}_2^T\vdots\ldots\vdots\widehat{\mathbf{S}}_m^T\right]^T,\ \widehat{\mathbf{S}}_k\in\mathfrak{R}^{\rho},\ \forall k=1,2,...,m,$$

$$\widehat{\mathbf{S}}=\mathbf{S}_{IIOI}=\mathbf{X}^{m-1}=\left[\mathbf{X}^T\vdots\left(\mathbf{X}^{(1)}\right)^T\vdots\ldots\vdots\left(\mathbf{X}^{(m-1)}\right)^T\right]^T\in\mathfrak{R}^{m\rho}. \tag{4.118}$$

The $\widehat{n}=m\rho$ *dimensional real vector space* $\mathfrak{R}^{m\rho}=\mathfrak{R}^{\widehat{n}}$ ***is the internal dynamics space,*** *i.e.,* ***the internal state space of the*** *IIO system (4.111), (4.112).*

Equations (4.111)-(4.118) furnish:

$$\widehat{\mathbf{S}}_1^{(1)} = \widehat{\mathbf{S}}_2,\ \widehat{\mathbf{S}}_2^{(1)} = \widehat{\mathbf{S}}_3,\ ...\ ,\ \widehat{\mathbf{S}}_{m-1}^{(1)} = \widehat{\mathbf{S}}_m,\ \widehat{\mathbf{S}}_m^{(1)} = \mathbf{X}^{(m)}. \tag{4.119}$$

The condition (4.117) (Property 140) ensures solvability of the *IIO* system (4.111), (4.112), in $\mathbf{X}^{(m)}(t) = \widehat{\mathbf{S}}_m^{(1)}$,

$$\widehat{\mathbf{S}}_m^{(1)} = Q^{-1}\left[t, \widehat{\mathbf{S}}(t)\right]\left\{\mathbf{h}_1\left[t, \mathbf{I}^{\mu}(t)\right] - A^{(m-1)}\widehat{\mathbf{S}}(t) - \mathbf{q}_1\left[t, \widehat{\mathbf{S}}(t)\right]\right\}.$$

This equation and Equations (4.119) yield the internal state space description (4.120)-(4.123) of the *IIO* system (4.111), (4.112):

$$\frac{d\widehat{\mathbf{S}}(t)}{dt} = \widehat{\mathbf{a}}\left[t, \widehat{\mathbf{S}}(t)\right] + \widehat{\mathrm{G}}\left[t, \widehat{\mathbf{S}}(t)\right]\widehat{\mathbf{g}}\left[t, \mathbf{I}^{\mu}(t)\right], \tag{4.120}$$

where

$$\widehat{\mathbf{a}}\left[t, \widehat{\mathbf{S}}(t)\right] = \begin{bmatrix} O_\rho \\ O_\rho \\ \cdots \\ O_\rho \\ -Q^{-1}\left[t, \widehat{\mathbf{S}}(t)\right]\left\{A^{(m-1)}\widehat{\mathbf{S}}(t) + \mathbf{q}_1\left[t, \widehat{\mathbf{S}}(t)\right]\right\} \end{bmatrix} \in \mathfrak{R}^{\widehat{n}},$$

$$\widehat{n} = m\rho, \tag{4.121}$$

$$\widehat{\mathrm{G}}\left[t, \widehat{\mathbf{S}}(t)\right] = \begin{bmatrix} O_\rho \\ O_\rho \\ \cdots \\ O_\rho \\ Q^{-1}\left[t, \widehat{\mathbf{S}}(t)\right] \end{bmatrix} \in \mathfrak{R}^{\widehat{n}\times\rho}, \tag{4.122}$$

$$\widehat{\mathbf{g}}\left[t, \mathbf{I}^{\mu}(t)\right] = \mathbf{h}_1\left[t, \mathbf{I}^{\mu}(t)\right] \in \mathfrak{R}^{\rho}. \tag{4.123}$$

The *IIO* system (4.111), (4.112) has also the output dynamics described by Equation (4.112). It introduces the system output state vector $\mathbf{S}_{IIOO}$ of the system. More precisely (see Definition 115 and Note 116, Section 4.2.1):

Definition 142 ***The IIO system output state and output state space***
The vector $\mathbf{Y}^{\nu-1}$ *is* ***the output dynamics vector of the*** *IIO system (4.111), (4.112), i.e.,* ***the*** *IIO* ***system output state vector*** $\widetilde{\mathbf{S}} = \mathbf{S}_{IIOO}$:

$$\widetilde{\mathbf{S}} = \mathbf{S}_{IIOO} = \left[\widetilde{\mathbf{S}}_1^T \vdots \widetilde{\mathbf{S}}_2^T \vdots \ldots \vdots \widetilde{\mathbf{S}}_\nu^T\right]^T,\ \widetilde{\mathbf{S}}_k \in \mathfrak{R}^N,\ \forall k = 1, 2, ..., \nu,$$

$$\widetilde{\mathbf{S}} = \mathbf{S}_{IIOO} = \mathbf{Y}^{\nu-1} = \left[\mathbf{Y}^T \vdots \left(\mathbf{Y}^{(1)}\right)^T \vdots \ldots \vdots \left(\mathbf{Y}^{(\nu-1)}\right)^T\right]^T \in \mathfrak{R}^{\nu N}. \tag{4.124}$$

The $\widetilde{n} = \nu N$ *dimensional real vector space* $\mathfrak{R}^{\nu N} = \mathfrak{R}^{\widetilde{n}}$ *is* ***the output dynamics vector space****, i.e.,* ***the output state space of the*** *IIO system (4.111), (4.112).*

Comment 143 *The space* $\mathfrak{R}^N$ *is the output space of the IIO system (4.111), (4.112) and* $\mathfrak{R}^{\nu N}$ *is its* ***output state space*** *that is its* ***extended output space****.*

Equations (4.112) - (4.117) and (4.124) furnish:

$$\widetilde{\mathbf{S}}_1^{(1)} = \widetilde{\mathbf{S}}_2,\ \widetilde{\mathbf{S}}_2^{(1)} = \widetilde{\mathbf{S}}_3,\ \dots,\ \widetilde{\mathbf{S}}_{\nu-1}^{(1)} = \widetilde{\mathbf{S}}_\nu,\ \widetilde{\mathbf{S}}_\nu^{(1)} = \mathbf{Y}^{(\nu)}. \tag{4.125}$$

The condition (4.117) ensures solvability of the *IIO* system (4.111), (4.112), in $\mathbf{Y}^{(\nu)}(t) = \widetilde{\mathbf{S}}_\nu^{(1)}$,

$$\widetilde{\mathbf{S}}_\nu^{(1)} = E^{-1}\left[t, \widetilde{\mathbf{S}}(t)\right]\left\{\mathbf{e}\left[t, \widehat{\mathbf{S}}(t)\right] + \mathbf{h}_2\left[t, \mathbf{I}^\mu(t)\right] - F^{(\nu-1)}\widetilde{\mathbf{S}}(t) - \mathbf{q}_2\left[t, \widetilde{\mathbf{S}}(t)\right]\right\}.$$

This equation and Equations (4.125) yield for the *IIO* system (4.111), (4.112):

$$\frac{d\widetilde{\mathbf{S}}(t)}{dt} = \widetilde{\mathbf{a}}\left[t, \widetilde{\mathbf{S}}(t)\right] + \widetilde{\mathrm{E}}\left[t, \widetilde{\mathbf{S}}(t)\right]\widetilde{\mathbf{e}}\left[t, \widehat{\mathbf{S}}(t)\right] + \widetilde{\mathrm{G}}\left[t, \widetilde{\mathbf{S}}(t)\right]\widetilde{\mathbf{g}}\left[t, \mathbf{I}^\mu(t)\right], \tag{4.126}$$

where

$$\widetilde{\mathbf{a}}\left[t, \widetilde{\mathbf{S}}(t)\right] =$$

$$= \begin{bmatrix} O_N \\ O_N \\ \cdots \\ O_N \\ -E^{-1}\left[t, \widetilde{\mathbf{S}}(t)\right]\left\{F^{(\nu-1)}\widetilde{\mathbf{S}}(t) + \mathbf{q}_2\left[t, \widetilde{\mathbf{S}}(t)\right]\right\} \end{bmatrix} \in \mathfrak{R}^{\widetilde{n}}, \tag{4.127}$$

$$\widetilde{\mathrm{G}}\left[t, \widetilde{\mathbf{S}}(t)\right] = \begin{bmatrix} O_N \\ O_N \\ \cdots \\ O_N \\ E^{-1}\left[t, \widetilde{\mathbf{S}}(t)\right] \end{bmatrix} \in \mathfrak{R}^{\widetilde{n}\times N}, \tag{4.128}$$

$$\widetilde{\mathbf{g}}\left[t, \mathbf{I}^\mu(t)\right] = \mathbf{h}_2\left[t, \mathbf{I}^\mu(t)\right] \in \mathfrak{R}^N, \tag{4.129}$$

$$\widetilde{\mathrm{E}}\left[t, \widetilde{\mathbf{S}}(t)\right] = \begin{bmatrix} O_N \\ O_N \\ \cdots \\ O_N \\ E^{-1}\left[t, \widetilde{\mathbf{S}}(t)\right] \end{bmatrix} \in \mathfrak{R}^{\widetilde{n}\times N}, \tag{4.130}$$

$$\widetilde{\mathbf{e}}\left[t, \widehat{\mathbf{S}}(t)\right] = \mathbf{e}\left[t, \widehat{\mathbf{S}}(t)\right] \in \mathfrak{R}^N. \tag{4.131}$$

The internal state vector $\widehat{\mathbf{S}} = \widehat{\mathbf{S}}_{IIOI}$ (4.118) and the output state vector $\widetilde{\mathbf{S}} = \mathbf{S}_{IIOO}$ form the *overall state vector* $\mathbf{S} = \mathbf{S}_{IIO}$ of the *IIO* system (4.111), (4.112):

Definition 144 ***The IIO system state and state space***

The vector $\left[\left(\mathbf{X}^{m-1}\right)^T \vdots \left(\mathbf{Y}^{\nu-1}\right)^T\right]^T$ ***is the dynamics vector of the*** *IIO* ***system*** *(4.111), (4.112), i.e.,* ***the*** *IIO* ***system state vector*** $\mathbf{S} = \mathbf{S}_{IIO}$*:*

$$\mathbf{S} = \mathbf{S}_{IIO} = \left[\left(\mathbf{X}^{m-1}\right)^T \vdots \left(\mathbf{Y}^{\nu-1}\right)^T\right]^T = \left[\widehat{\mathbf{S}}_{IIO}^T \vdots \widetilde{\mathbf{S}}_{IIO}^T\right]^T = \left[\mathbf{S}_1^T \vdots \mathbf{S}_2^T \vdots \dots \vdots \mathbf{S}_{m+\nu}^T\right]^T,$$

$$\mathbf{S}_k \in \mathfrak{R}^\rho,\ \forall k = 1, 2, \dots, m,$$

$$\mathbf{S}_k \in \mathfrak{R}^N,\ \forall k = m+1, m+2, \dots, m+\nu,$$

$$\mathbf{S} = \mathbf{S}_{IIO} = \left[\left(\mathbf{X}^{m-1}\right)^T \vdots \left(\mathbf{Y}^{\nu-1}\right)^T\right]^T \in \mathfrak{R}^{m\rho+\nu N},\ n = m\rho + \nu N. \tag{4.132}$$

The $n = \widehat{n} + \widetilde{n} = m\rho + \nu N$ *dimensional real vector space* $\mathfrak{R}^{m\rho+\nu N} = \mathfrak{R}^n$ *is* ***the dynamics vector space****, i.e.,* ***the state space of the*** *IIO system (4.111), (4.112).*

Equations (4.132) imply

$$\mathbf{S}^{(1)} = \begin{bmatrix} \widehat{\mathbf{S}}_{IIO}^{(1)} \\ \widetilde{\mathbf{S}}_{IIO}^{(1)} \end{bmatrix}. \tag{4.133}$$

This equation permits us to get the input-state-output mathematical description of the *IIO* system (4.111), (4.112) by linking it with Equations (4.120)-(4.123), (4.126)-(4.131) for the *IIO* system (4.111), (4.112):

$$\frac{d\mathbf{S}(t)}{dt} = \mathbf{a}_1[t, \mathbf{S}(t)] + \mathrm{G}[t, \mathbf{S}(t)]\,\mathbf{g}_1[t, \mathbf{I}^{\mu}(t)], \tag{4.134}$$

where

$$\mathbf{a}_1[t, \mathbf{S}(t)] = \begin{bmatrix} \widehat{\mathbf{a}}\left[t, \widehat{\mathbf{S}}(t)\right] \\ \widetilde{\mathbf{a}}\left[t, \widetilde{\mathbf{S}}(t)\right] + \widetilde{\mathrm{E}}\left[t, \widetilde{\mathbf{S}}(t)\right]\widetilde{\mathbf{e}}\left[t, \widehat{\mathbf{S}}(t)\right] \end{bmatrix} \in \mathfrak{R}^{\widehat{n}+\widetilde{n}} = \mathfrak{R}^n, \tag{4.135}$$

$$\mathrm{G}[t, \mathbf{S}(t)] = block\ diaag\left\{\widehat{\mathrm{G}}\left[t, \widehat{\mathbf{S}}(t)\right] \vdots \widetilde{\mathrm{G}}\left[t, \widetilde{\mathbf{S}}(t)\right]\right\} \in \mathfrak{R}^{(\widehat{n}+\widetilde{n})\times(\rho+N)}, \tag{4.136}$$

$$\mathbf{g}_1[t, \mathbf{I}^{\mu}(t)] = \begin{bmatrix} \widehat{\mathbf{g}}[t, \mathbf{I}^{\mu}(t)] \\ \widetilde{\mathbf{g}}[t, \mathbf{I}^{\mu}(t)] \end{bmatrix} \in \mathfrak{R}^{\rho+N}, \tag{4.137}$$

$$\mathbf{Y}(t) = \left[O_{N,\widehat{n}} \vdots I_N \ \vdots O_{N,\widetilde{n}-N}\right]\mathbf{S}(t) = \mathrm{C}\mathbf{S}(t), \tag{4.138}$$

$$\mathrm{C} = \left[O_{N,\widehat{n}} \vdots I_N \ \vdots O_{N,\widetilde{n}-N}\right] \in \mathfrak{R}^{N\times n}. \tag{4.139}$$

The system output equation (4.138) yields the following:

$$\mathbf{Y}(t) = \mathbf{a}_2[t, \mathbf{S}(t)] + \mathbf{g}_2[t, \mathbf{I}(t)], \ \forall t \in \mathfrak{T}, \tag{4.140}$$

where

$$\mathbf{a}_2[t, \mathbf{S}(t)] = \mathrm{C}\mathbf{S}(t) \in \mathfrak{R}^N, \ \mathbf{g}_2[t, \mathbf{I}(t)] \equiv \mathbf{0}_N. \tag{4.141}$$

4.11 *EISO* systems

4.11.1 Original *EISO* system description

The *Extended Input-State-Output* (*EISO*) systems are given in their following original input-state-output form:

$$\begin{gathered}
\frac{d\mathbf{S}(t)}{dt} = \mathbf{a}[t, \mathbf{S}(t)] + B[t, \mathbf{S}(t)]\,\mathbf{b}[t, \mathbf{I}^{\mu}(t)] = \mathbf{f}[t, \mathbf{S}(t), \mathbf{I}^{\mu}(t)], \ \mathbf{S} \in \mathfrak{R}^n, \\
\mathbf{a}(.) : \mathfrak{T} \times \mathfrak{R}^n \longrightarrow \mathfrak{R}^n, \ \mathbf{a}(t, \mathbf{S}) \in \mathfrak{C}(\mathfrak{T}, \mathfrak{R}^n), \\
B(.) : \mathfrak{T} \times \mathfrak{R}^n \longrightarrow \mathfrak{R}^{n\times p}, \ B(t, \mathbf{S}) \in \mathfrak{C}(\mathfrak{T}, \mathfrak{R}^n), \\
\mathbf{b}(.) : \mathfrak{T} \times \mathfrak{R}^{(\mu+1)M} \longrightarrow \mathfrak{R}^p, \ \mathbf{b}(t, \mathbf{I}^{\mu}) \in \mathfrak{C}\left(\mathfrak{T}, \mathfrak{R}^{(\mu+1)M}\right), \\
\mathbf{f}(.) : \mathfrak{T} \times \mathfrak{R}^n \times \mathfrak{R}^{(\mu+1)M} \longrightarrow \mathfrak{R}^n, \ \mathbf{f}(t, \mathbf{S}, \mathbf{I}^{\mu}) \in \mathfrak{C}\left(\mathfrak{T} \times \mathfrak{R}^n \times \mathfrak{R}^{(\mu+1)M}\right),
\end{gathered} \tag{4.142}$$

$$\mathbf{Y}(t) = \mathbf{c}\left[t, \mathbf{S}(t)\right] + \mathbf{d}\left[t, \mathbf{I}(t)\right], \ \forall t \in \mathfrak{T},$$
$$\mathbf{c}(.) : \mathfrak{T} \times \mathfrak{R}^n \longrightarrow \mathfrak{R}^N, \ \mathbf{c}(t, \mathbf{S}) \in \mathfrak{C}\left(\mathfrak{T}, \mathfrak{R}^n\right),$$
$$\mathbf{d}(.) : \mathfrak{T} \times \mathfrak{R}^M \longrightarrow \mathfrak{R}^N, \ \mathbf{d}(t, \mathbf{I}) \in \mathfrak{C}\left(\mathfrak{T}, \mathfrak{R}^M\right). \tag{4.143}$$

The $EISO$ system state-output form (4.142), (4.143) incorporates:

- the input-state-output form (4.83), (4.88) of the IO system (4.76), in which $\mathbf{S} = \mathbf{Y}^{m-1}$, $\mathbf{a}(.) = \mathbf{a}_1(.)$, $B(.) = G(.)$, $\mathbf{b}(.) = \mathbf{g}_1(.)$, $\mathbf{c}(.) = \mathbf{a}_2(.)$, $\mathbf{d}(.) = \mathbf{g}_2(.) \equiv \mathbf{0}_N$,

- the input-state-output form (4.93), (4.94) of the ISO system, which is its original form of the ISO system, in which $\mathbf{S} = \mathbf{X}$, $\mathbf{a}(.) = \mathbf{a}_1(.)$, $B(.) \equiv I_n$, $\mathbf{b}(.) = \mathbf{g}_1(.)$, $\mathbf{a_2}(.) = \mathbf{c}(.)$, $\mathbf{d}(.) \equiv \mathbf{g}_2(.)$,

- the input-state-output form (4.105), (4.109) of the $HISO$ system (4.95), (4.96), in which $\mathbf{S} = \mathbf{X}^{m-1}$, $\mathbf{a}(.) = \mathbf{a}_1(.)$, $B(.) = G(.)$, $\mathbf{b}(.) = \mathbf{g}_1(.)$, $\mathbf{c}(.) = \mathbf{a}_2(.)$, $\mathbf{d}(.) = \mathbf{g}_2(.)$,

- the input-state-output form (4.134), (4.140) of the IIO system (4.134), (4.140), in which $\mathbf{S} = \left[\left(\mathbf{X}^{m-1}\right)^T \vdots \left(\mathbf{Y}^{\nu-1}\right)^T\right]^T$, $\mathbf{a}(.) = \mathbf{a}_1(.)$, $B(.) = G(.)$, $\mathbf{b}(.) = \mathbf{g}_1(.)$, $\mathbf{c}(.) = \mathbf{a}_2(.)$, $\mathbf{d}(.) = \mathbf{g}_2(.) \equiv \mathbf{0}_N$,

The following conclusion is crucial:

Conclusion 145 ***Generality of the EISO system for the stability study***

Studying stability properties of the EISO system (4.142), (4.143) we study simultaneously the same stability properties of the other four systems: of the IO system (4.76), of the ISO system (4.90), (4.91), of the HISO system (4.95), (4.96), and of the IIO system (4.111), (4.112).

4.11.2 Motions of the *EISO* system

What follows applies Definition 117-Definition 123 (Section 4.2) to the $EISO$ system (4.142), (4.143).

"*Almost at every* $t \in \mathfrak{T}_0$" means: "*at every* $t \in \left(\mathfrak{T}_0 \backslash \mathfrak{T}_0^0\right)$", where $\mathfrak{T}_0^0$ is the largest subset of $\mathfrak{T}_0$ with the zero measure.

$D^+\mathbf{S}(t; t_0, \mathbf{S}_0; \mathbf{I})$ denotes the upper right-hand Dini derivative of $\mathbf{S}(t; t_0, \mathbf{S}_0; \mathbf{I})$ at a moment t (Appendix B).

$\mathfrak{L}[\mathbf{S}_0, \mathbf{I}(.)]$ is *the largest constant temporal set of the initial moments* t_0 *for which the given system has a solution* for $\mathbf{I}(.) \in \mathcal{J}$, $\mathfrak{L}[\mathbf{S}_0, \mathbf{I}(.)] \subseteq \mathfrak{T}$.

Let $\mathfrak{L}^c$ and $\mathfrak{T}^c = \mathfrak{T}^c(\mathfrak{L}^c)$ be the largest nonempty connected compact subintervals of a constant set $\mathfrak{L}$ and of $\mathfrak{T}$, respectively, and let $\mathfrak{S}^c$ be a nonempty connected compact subset of a constant set $\mathfrak{S}$,

$$\mathfrak{L}^c \subseteq \mathfrak{L}, \ \mathfrak{T}^c(\mathfrak{L}^c) \subseteq \mathfrak{T}, \ \mathfrak{S}^c(\mathfrak{L}^c) \subseteq \mathfrak{S},$$
$$\mathfrak{L}^c \cap \mathfrak{T}^c(\mathfrak{L}^c) \neq \phi, \ \ \mathfrak{S}^c(\mathfrak{L}^c) \neq \phi. \tag{4.144}$$

Definition 146 ***System motion (solution)***

A vector function $\mathcal{S}(.) : \mathfrak{T}_0 \times \mathfrak{T}_i \times \mathfrak{R}^n \times \mathcal{J} \longrightarrow \mathfrak{R}^n$ *is a* ***solution (motion)*** *of the EISO system (4.142), (4.143) through* $\mathbf{S}_0$ *at* $t = t_0$ *if and only if* $\mathcal{S}(.)$, $\mathbf{S}_0 \in \mathfrak{R}^n$ *and* $t_0 \in \mathfrak{T}_i$ *obey the following for fixed* $\mathbf{I}(.) \in \mathcal{J}$:

- $(E1)$ ***Existence:*** *The vector function* $\mathcal{S}(.)$ *is defined in* $\mathfrak{R}^n$, *on* $\mathfrak{T}_0$, *where* $\mathfrak{T}_0 \subseteq \mathfrak{T}$,

$$\mathbf{S}(t; t_0, \mathbf{S}_0; \mathbf{I}) \equiv \mathbf{S}(t) \in \mathfrak{R}^n, \ \forall t \in \mathfrak{T}_0, \tag{4.145}$$

if and only if $t_0 \in \mathfrak{L}$, $\mathfrak{L} \subseteq \mathfrak{T}_i \subseteq \mathfrak{T}$, *and* $\mathbf{S}_0 \in \mathfrak{S} \subseteq \mathfrak{R}^n$.

- $(E2)$ ***Initial condition:*** *The vector function $\mathcal{S}(.)$ satisfies the initial condition:*

$$\mathcal{S}(t_0; t_0; \mathbf{S}_0; \mathbf{I}) \equiv \mathbf{S}_0. \tag{4.146}$$

- $(E3)$ ***Continuous dependence:*** *The vector function $\mathcal{S}(.)$ is continuous and almost everywhere differentiable on $\mathfrak{T}_0$ and identically satisfies the system (4.142), (4.143),*

$$\mathbf{S}(t; t_0, \mathbf{S}_0; \mathbf{I}) \in \mathfrak{C}(\mathfrak{T}_0) \cup \mathfrak{C}^1(\mathfrak{T}_0 \backslash \mathfrak{T}_0^0), \tag{4.147}$$

$$\frac{d\mathbf{S}(t; t_0, \mathbf{S}_0; \mathbf{I})}{dt} = \left\{ \begin{array}{c} \mathbf{a}[t, \mathbf{S}(t; t_0, \mathbf{S}_0; \mathbf{I})] + \\ +B[t, \mathbf{S}(t; t_0, \mathbf{S}_0; \mathbf{I})]\, \mathbf{b}[t, \mathbf{I}^{\mu}[t, \mathbf{S}(t; t_0, \mathbf{S}_0; \mathbf{I})]] \end{array} \right\},$$
$$\forall [t, t_0, \mathbf{S}_0] \in (\mathfrak{T}_0 \backslash \mathfrak{T}_0^0) \times \mathfrak{L} \times \mathfrak{S},\ \mathbf{I}(.) \in \mathcal{J}, \tag{4.148}$$

and

$$D^+\mathbf{S}(t; t_0, \mathbf{S}_0; \mathbf{I}) = \left\{ \begin{array}{c} \mathbf{a}[t, \mathbf{S}(t; t_0, \mathbf{S}_0; \mathbf{I})] + \\ +B[t, \mathbf{S}(t; t_0, \mathbf{S}_0; \mathbf{I})]\, \mathbf{b}[t, \mathbf{I}^{\mu}[t, \mathbf{S}(t; t_0, \mathbf{S}_0; \mathbf{I})]] \end{array} \right\},$$
$$\forall [t, t_0, \mathbf{S}_0] \in \mathfrak{T}_0^0 \times \mathfrak{L} \times \mathfrak{S},\ \mathbf{I}(.) \in \mathcal{J}, \tag{4.149}$$

- $(E4)$ ***Uniqueness along motions:*** *For every $(t, t_0, \tau, \mathbf{S}_0, \mathbf{I}) \in \mathfrak{T}_0 \times \mathfrak{L}^c \times \mathfrak{L}^c \times \mathfrak{S}^c(t_0) \times \mathcal{J}$ every system motion $\mathcal{S}(.)$ satisfies*

$$\mathbf{S}[t; t_0 + \tau; \mathcal{S}(t_0 + \tau; t_0; \mathbf{S}_0; \mathbf{I}); \mathbf{I}] = \mathcal{S}(t; t_0; \mathbf{S}_0; \mathbf{I}), \tag{4.150}$$

- $(E5)$ ***Motion uniqueness:*** *The vector function $\mathcal{S}(.)$ is **the unique system motion through** $\mathbf{S}_0$ **at** $t = t_0$ if and only if any two vector functions $\mathcal{S}_i(.) : \mathfrak{T}_0 \times \mathfrak{T}_i \times \mathfrak{R}^n \times \mathcal{J} \longrightarrow \mathfrak{R}^n$, $i = 1, 2$, which satisfy (E1)-(E4), are the same system solution (motion) $\mathcal{S}(.)$ through $\mathbf{S}_0$ at $t = t_0$, i.e., if and only if they obey $(E1) - (E4)$, and their vector values $\mathbf{S}_i(t; t_0; \mathbf{S}_0; \mathbf{I})$, $i = 1, 2$, are identical with $\mathbf{S}(t; t_0; \mathbf{S}_0; \mathbf{I})$,*

$$\mathbf{S}_1(t; t_0; \mathbf{S}_0; \mathbf{I}) \equiv \mathbf{S}_2(t; t_0; \mathbf{S}_0; \mathbf{I}) \equiv \mathbf{S}(t; t_0; \mathbf{S}_0; \mathbf{I}). \tag{4.151}$$

4.11.3 System basic properties

Property 147 ***TCUP property of the EISO system***
a) The EISO system (4.142), (4.143) obeys the system form (Definition 28) of TCUP (Principle 27) for any fixed $\mathbf{I}(.) \in \mathcal{J}$.
b) The EISO system (4.142), (4.143) has the motion $\mathcal{S}(.; t_0, \mathbf{S}_0; \mathbf{I})$ for every $(t_0, \mathbf{S}) \in \mathfrak{T} \times \mathfrak{R}^n$ and for any fixed $\mathbf{I}(.) \in \mathcal{J}$.
c) The motion $\mathcal{S}(.; t_0, \mathbf{S}_0; \mathbf{I})$ and the corresponding output response $\mathcal{Y}(.; t_0, \mathbf{Y}_0; \mathbf{I})$ of the EISO system (4.142), (4.143) obey Principle 27 for every $(\mathbf{Y}_0, t_0) \in \mathfrak{R}^N \times \mathfrak{T}$ and for any fixed $\mathbf{I}(.) \in \mathcal{J}$.

Property 148 ***System set properties***
a) The sets $\mathfrak{L}$ and $\mathfrak{S}$ in Definition 153 and $\mathfrak{S}$ in Condition 151 obey $\mathfrak{L} = \mathfrak{T}_i \subseteq \mathfrak{T}$, $\mathfrak{S}(t_0) \equiv \mathfrak{S}(t) \equiv \mathfrak{R}^n$.
b) The motion $S(.; t_0, \mathbf{S}_0; \mathbf{I})$ of the EISO system (4.142), (4.143) is continuous in the initial data $(t_0, \mathbf{S}_0) \in \mathfrak{T} \times \mathfrak{R}^n$ for any fixed input vector function $\mathbf{I}(.) \in \mathcal{J}$.

Note 149 ***On the validity of Property 147 and of Property 148***
Throughout this book we accept the validity of Property 147 and of Property 148 if it is not said otherwise.

4.11.4 Existence and uniqueness of motions

Let the vector function $\mathbf{f}(.) : \mathfrak{T} \times \mathfrak{R}^n \longrightarrow \mathfrak{R}^n$ be defined in (4.142) by:

$$\mathbf{f}(t, \mathbf{S}) = \mathbf{a}[t, \mathbf{S}(t)] + B[t, \mathbf{S}(t)]\, \mathbf{b}[t, \mathbf{I}^{\mu}(t, \mathbf{S})]. \tag{4.152}$$

It is Lipshitz continuous on $\mathfrak{T}^c \times \mathfrak{L}^c$, by definition, if and only if there is Lipshitz constant $L \in \mathfrak{R}^+$ such that:

$$\|\mathbf{f}(t, \mathbf{S}_1) - \mathbf{f}(t, \mathbf{S}_2)\| \leq L \|\mathbf{S}_1 - \mathbf{S}_2\|, \ \forall (t, \mathbf{S}_1, \mathbf{S}_2) \in \mathfrak{T}^c \times \mathfrak{L}^c \times \mathfrak{L}^c. \tag{4.153}$$

Note 150 *The input vector $\mathbf{I}(t, \mathbf{S})$ of a controlled system in a feedback control system depends on a perturbation, i.e.., disturbance, vector $\mathbf{P}$ and the control vector $\mathbf{U}$:*

$$\mathbf{I}(t, \mathbf{S}_k) = P(t)\mathbf{P}(t) + U(t)\mathbf{U}(t, \mathbf{S}_k),\ k = 1, 2. \tag{4.154}$$

Condition 151 ***Lipschitz condition for the system (4.142), (4.143) with $\mathbf{I}^{\mu}(.)$ dependent only on time t***

The vector function $\mathbf{f}(t, \mathbf{S})$ defined by 4.152 with $\mathbf{I}(.)$ dependent only on time t and the state vector $\mathbf{S}$ is Lipschitz continuous on the Cartesian product set $\mathfrak{T} \times \mathfrak{S}$ for fixed $\mathbf{I}(.) \in \mathcal{J}$ and obeys Lipschitz condition (4.153) in $\mathbf{S}$ with Lipschitz constant L if and only if

$$\begin{aligned} &\|[\{\mathbf{a}(t, \mathbf{S}_1) + B(t, \mathbf{S}_1)\, \mathbf{b}[t, \mathbf{I}^{\mu}(t, \mathbf{S}_1)]\}] - \{\mathbf{a}(t, \mathbf{S}_2) + B(t, \mathbf{S}_2)\, \mathbf{b}[t, \mathbf{I}^{\mu}(t, \mathbf{S}_2)]\}\| \leq \\ &\leq L \|\mathbf{S}_1 - \mathbf{S}_2\|,\ \forall [t, \mathbf{S}_1, \mathbf{S}_2] \in \mathfrak{T}^c \times \mathfrak{S}^c \times \mathfrak{S}^c \ \textit{for fixed}\ \mathbf{I}(.) \in \mathcal{J}. \end{aligned} \tag{4.155}$$

Theorem 152 ***Uniqueness of motions of the system***

Let the system (4.142), (4.143) possess Property 147. Let the function $\mathbf{f}(.)$ be defined by (4.152), with $\mathbf{I}(.)$ dependent only on time t and the state vector $\mathbf{S}$, and be Lipschitz continuous on the time-invariant Cartesian product set

$$\mathfrak{T}^c \times \mathfrak{S}^c \ \textit{for fixed}\ \mathbf{I}(.) \in \mathcal{J}.$$

*Then the regular system **(4.142), (4.143)** has the unique solutions $\mathcal{S}(t; t_0; \mathbf{S}_0; \mathbf{I})$ for fixed $\mathbf{I}(.) \in \mathcal{J}$ on a time-invariant Cartesian product set*

$$\mathfrak{T}^c \times \mathfrak{S}^c.$$

Proof. Let the condition of the theorem hold, which implies the validity of (4.155). Then Theorem 77 in [232, pp. 34, 35] becomes Theorem 152. ■

4.11.5 Inherent sets

In order for the mathematical model of the physical system to be physically meaningful, we assume the validity of the following.

Definition 153 ***Inherent sets of the system*** *(4.142), (4.143)*

a) The temporal interval $\mathfrak{L}[\mathbf{S}_0, \mathbf{I}(.)]$ is the largest temporal set of the initial moments t_0 for which the system (4.142), (4.143) has a solution passing through $\mathbf{S}_0$ at $t = t_0$ for any $\mathbf{I}(.) \in \mathcal{J}$. It depends, for the given system (4.142), (4.143), on the initial state vector $\mathbf{S}_0 \in \mathfrak{R}^n$ and on the input vector function $\mathbf{I}(.) \in \mathcal{J}$, $\mathfrak{L}[\mathbf{S}_0, \mathbf{I}(.)] \subseteq \mathfrak{T}$.

b) $\mathfrak{S}[t_0, \mathbf{I}(.)]$ is the largest connected set of the initial vectors $\mathbf{S}_0 \in \mathfrak{R}^n$ at $t_0 \in \mathfrak{T}$, which is a nonempty subset of $\mathfrak{R}^n$, $\mathfrak{S}(t_0) \neq \phi$, $\mathfrak{S}(t_0) \subseteq \mathfrak{R}^n$, such that the system (4.142), (4.143) has a motion passing through $\mathbf{S}_0$ at $t = t_0$ for any $\mathbf{I}(.) \in \mathcal{J}$.

c) The temporal interval $\mathfrak{T}[t_0, \mathbf{S}_0, \mathbf{I}(.)]$ *is the largest temporal set over which the given system (4.142), (4.143) has an extended solution passing through* $\mathbf{S}_0$ *at* $t = t_0$ *for any* $\mathbf{I}(.) \in \mathcal{J}$*. It depends, for the given system (4.142), (4.143), on* $(t_0, \mathbf{S}_0, \mathbf{I}) \in \mathfrak{L} \times \mathfrak{S}(t_0) \times \mathcal{J}$*.*

d) The triplet $(\mathfrak{L}, \mathfrak{S}, \mathfrak{T}_0)$ *determines the largest nonempty connected set product such that for every* $(t_0, \mathbf{S}_0, t) \in \mathfrak{L}\times\mathfrak{S} \times \mathfrak{T}_0$ *the system (4.142), (4.143) has a solution on* $\mathfrak{T}_0$ *passing through* $\mathbf{S}_0$ *at* $t = t_0$ *for any* $\mathbf{I}(.) \in \mathcal{J}$*.*

Comment 154 *The time interval* $\mathfrak{L}$ *of the initial moments* t_0 *depends on* $\mathbf{S}_0$ *and* $\mathbf{I}(.)$ *in general. It can contract as a function of* $\mathbf{S}_0$ *for fixed* $\mathbf{I}(.)$ *to a singleton or to the empty set* ϕ *in general. This is a characteristic of strongly nonuniform systems. The same holds for the set* $\mathfrak{S}(t_0)$*.*

Condition d) of Definition 153 ensures that $\mathfrak{L}$*,* $\mathfrak{S}$ *and* $\mathfrak{T}_0$ *do not contract to singletons or to the empty set.*

Note 155 ***Additional advanced properties***

This book treats stability properties of a positive invariant set of the system. The positive invariant set can be time-invariant, i.e., constant, denoted by $\mathfrak{I}^+$ *or time-varying, i.e., variable, depending on the initial moment* $t_0 \in \mathfrak{T}_i$ *and on the current moment* $t \in \mathfrak{T}_0$ *in general, which is denoted by* $\mathfrak{I}^+\left(t; t_0; \mathfrak{I}_0^+\right)$, $\mathfrak{I}^+\left(t_0; t_0; \mathfrak{I}_0^+\right) \equiv \mathfrak{I}_0^+$, $\forall t_0 \in \mathfrak{T}_i$.

Additional advanced system properties depend on the feature of the system positive invariant set, whether it is constant I^+ *or variable* $\mathfrak{I}^+\left(t; t_0; \mathfrak{I}_0^+\right)$*. In the former case, they are determined in Section 12.2. For the latter case, this book serves as a basis that established the theory of time-varying sets.*

Part IV

STABILITY AND STABILITY DOMAINS: CONCEPT

Chapter 5

Stability, *time* and sets

5.1 On stability concepts in general

Question 156 *What is the meaning, the sense, of* **stability** *in general?*

The general meaning of *stability* is **persistency** of something under certain conditions over some *time* interval. This means that the notion, concept and definition of *stability* should clearly reply to the following questions:

- **Stability of what?** Stability of an equilibrium state or of a set of states or of a motion (of a process)? Stability of a system? Persistency of distances among motions or between motions and a set?

- **How can we define the distance** between two states, or between two motions, or between a motion and a set?

- **How can we specify the closeness** between two states, or between two motions, or between a motion and a set?

- **What is the *time* interval** over which the demanded closeness is to be realized?

- **What are the conditions** under which the required closeness should be achieved? To be more precise this means:

  - **Under which external conditions**, i.e., **under which external actions on the system**, is the requested system behavior defined above to be achieved?

  - **Under which initial internal and external conditions** is the requested system behavior defined above to be realized?

There are many possible replies to these questions. The accepted replies characterize the adopted stability concept and stability properties in the framework of the concept. Various stability concepts exist, for example:

- **The Lyapunov stability concept** is the basic, essentially mathematical, stability concept, the widest and phenomenologically the richest one, and the one the practical and engineering meaning from the system motions accuracy points and energy points of view, although Lyapunov himself originally established it in [335] as a pure mathematical concept by generalizing the concept of continuity of functions to the system motions closeness over the *time* set $\mathfrak{T}$.

DOI: 10.1201/9781003744528-9

- **The practical stability concept** incorporates the **finite-*time* stability concept**, which is aimed at covering stability issues to satisfy practical, engineering needs,
- **The input stability concept** comprises input-state, input-output, and input-state-output stability properties, which is also aim at covering some stability issues to satisfy practical, engineering needs.
- **The L-stability concept** generalizes the input stability concept.
- **The Lagrange stability concept** is a pure mathematical concept.
- **The Poisson stability concept** is also a pure mathematical concept.

Comment 157 ***Time, stability and Lyapunov restriction***
Time is the crucial variable for dynamical systems and dynamical processes. Lyapunov well understood it when he wrote the following, before stating the stability definition [335, p. 13] (see Section 5.2 in the sequel):
"We shall be exclusively interested in those cases in which the solution of the considered question does not depend on a choice of the instant t_0*, when perturbations are acting."*

Property 158 ***Stability properties of time-invariant systems versus time-varying systems***
Time-invariant systems are those systems, the properties, parameters, nonlinearities, interconnections of subsystems and the other characteristics of which, by definition, do not depend on time. Their stability properties are independent of the initial moment $t_0 \in \mathfrak{T}$*. Any moment* $\tau \in \mathfrak{T}$ *can be accepted for its initial moment,* $t_0 = \tau$*. The simplest choice is* $t_0 = \tau = 0$.
The properties and/or characteristics of time-varying systems, by definition, depend on time. Their stability properties are dependent on the initial moment $t_0 \in \mathfrak{T}$ *in general. The initial moment is very important for their properties. The definitions of their stability properties incorporate the influence of the initial moment* $t_0 \in \mathfrak{T}$ *on the system dynamical behavior.*
All definitions given and obtained results for time-varying systems hold simultaneously for the corresponding time-invariant systems, for which $t_0 = 0$ *can be set everywhere in the definitions and in the results.*

Although the book treats Lyapunov stability properties of all five large families of *time-varying* nonlinear systems (see Task 1 in Preface II and Chapter 7) it opens new avenues also for the development of other stability concepts.

5.2 Lyapunov's original definition

In his remarkable dissertation [335, pp. 12-14], Lyapunov established the general stability concept. He presented the stability definition at the beginning of his fundamental memoir written in Russian [335, pp. 12-14]:
"1. Let us consider any material system with k degrees of freedom.
Let

$$q_1,\ q_2,\ \ldots,\ q_k$$

be k independent variables, which we use to determine its position.
We shall assume, that for such quantities are chosen such variables, which rest real for all real positions of the system.

Considering the named variables as functions of time t, we will denote their first derivatives relative to time t as

$$q'_1,\ q'_2,\ \ldots,\ q'_k.$$

In every dynamical task, in which forces are prespecified in a certain way, such functions will satisfy some k second order differential equations.

Let for such equations a particular solution has been found

$$q_1 = f_1(t),\ q_2 = f_2(t),\ \ldots,\ q_k = f_k(t),$$

in which the quantities q_j are real functions of t, which at every t give only possible values to them.

To that particular solution will correspond some motion of the system. Comparing it in a known sense with others, which are possible under the same forces, we shall call that motion *unperturbed*, and all others, with which it is compared, *perturbed.*

Understanding by t_0 some given moment of time, let us denote the values of the quantities q_j, q'_j that correspond to it along any motion by q_{j0}, q'_{j0}.

Let

$$\begin{aligned} q_{10} &= f_1(t_0) + \varepsilon_1,\ q_{20} = f_2(t_0) + \varepsilon_2,\ \ldots,\ q_{k0} = f_k(t_0) + \varepsilon_k, \\ q'_{10} &= f_1'(t_0) + \varepsilon'_1,\ q'_{20} = f_2'(t_0) + \varepsilon'_2,\ \ldots,\ q'_{k0} = f_k'(t_0) + \varepsilon'_k, \end{aligned}$$

where $\varepsilon_j, \varepsilon'_j$ are real-valued constants.

Prespecifying the constants, which will be called *perturbations*, the perturbed motion is determined. We shall suppose that we may specify arbitrary small values to them.

Speaking about perturbed motions, *close* to the unperturbed one, we shall comprehend motions, for which the perturbations are numerically sufficiently small.

Let $Q_1,\ Q_2,\ \ldots,\ Q_n$ be any given continuous real-valued functions of quantities

$$q_1,\ q_2,\ \ldots,\ q_k,\ q'_1,\ q'_2,\ \ldots,\ q'_k.$$

Along the unperturbed motion, they become known functions of t, which we denote by $F_1,\ F_2,\ \ldots,\ F_n$. Along a perturbed motion they will be functions of

$$t, \varepsilon_1,\ \varepsilon_2,\ \ldots,\ \varepsilon_k,\ \varepsilon'_1,\ \varepsilon'_2,\ \ldots,\ \varepsilon'_k.$$

When all $\varepsilon_j, \varepsilon'_j$ are equal to zero, then the quantities

$$Q_1 - F_1,\ Q_2 - F_2,\ \ldots\ ,\ Q_n - F_n$$

are all equal to zero for every t. However, if the constants $\varepsilon_j, \varepsilon'_j$ are not equal to zero, but all are infinitely small, then a question appears: Is it possible to specify such infinitely small bounds on the quantities $Q_s - F_s$, which the latter would never exceed numerically?

An answer to that question, which is the topic of our investigations, depends on both a characteristic of the considered unperturbed motion and a choice of the functions Q_1, $Q_2,\ \ldots,\ Q_n$ as well as on the instant t_0. Under a specific choice of the latter, the reply to the question will, respectively, characterize in some sense the unperturbed motion, by determining a feature of the latter, which we will call *stability*, or one contrary to it, which we will call *instability.*

We shall be exclusively interested in those cases in which the solution of the posed question does not depend on the choice of the moment t_0, when the perturbations act. Thus we accept the following definition:

Let L_1, L_2, ..., L_{n} be arbitrary given positive numbers. If for all L_s, regardless of how small they are, there can be selected positive numbers E_1, E_2, ..., E_k, E_1‘, E_2‘, ..., E_k‘ such that for all real ε_j, ε'_j, which satisfy conditions

$$|\varepsilon_j| \leq E_j, \ \ |\varepsilon_j`| \leq E_j`, \ \ (j = 1, 2, ..., k)\,,^1$$

and for all t greater than t_0, the inequalities

$$|Q_1 - F_1| < L_1, \ |Q_2 - F_2| < L_2, \ ... \ , \ |Q_{\mathrm{n}} - F_{\mathrm{n}}| < L_{\mathrm{n}},$$

are satisfied, then the unperturbed motion is stable with respect to the quantities Q_1, Q_2, ..., Q_{n}; otherwise unstable relative to the same quantities** [335, pp. 12-14].

[1] In general $|x|$ means the absolute value of a real-, or modulus of a complex variable x."

Lyapunov continued in [335, p. 15]:

"2. The solution of our question depends on the investigation of differential equations of the perturbed motion or, if appropriate, on the investigation of differential equations, which satisfy the functions

$$Q_1 - F_1 = x_1, \ Q_2 - F_2 = x_2, \ ... \ , \ Q_{\mathrm{n}} - F_{\mathrm{n}} = x_{\mathrm{n}}.$$

The order of these last equations in general will be the same, i.e., $2k$; but in some cases can be lower.

We will assume that the number n and the functions Q_s are such that the order of the system will be n and that it will take the normal form

$$\frac{dx_1}{dt} = X_1, \ \frac{dx_2}{dt} = X_2, \ ... \ , \ \frac{dx_{\mathrm{n}}}{dt} = X_{\mathrm{n}}, \qquad (1)$$

and everywhere in the sequel we will think of the last equations, calling them the differential equations of the perturbed motion.

All X_s in equations (1) are known functions of the quantities

$$x_1, \ x_2, \ ... \ , \ x_{\mathrm{n}}, \ t,$$

vanishing for

$$x_1 = x_2 = \ ... \ = x_{\mathrm{n}} = 0\text{ " } \ [335, \text{p.}15].$$

5.3 Comments on Lyapunov's original definition

Comment 159 ***Unperturbed, desired (nominal) motion***

Lyapunov considered stability of a fixed, single, unperturbed motion (solution).

Lyapunov's term "unperturbed" variables means "desired" or "nominal valued" variables, denoted by the subscript d or N, respectively.

Lyapunov's term "perturbed" variables means "real-valued" variables denoted without the subscript d or N.

Comment 160 ***Initial value of the unperturbed motion***

Lyapunov's unperturbed variables $q_1 = f_1(t)$, $q_2 = f_2(t)$, ..., $q_k = f_k(t)$ and their derivatives $q_1^{(1)} = f_1^{(1)}(t)$, $q_2^{(1)} = f_2^{(1)}(t)$, ..., $q_k^{(1)} = f_k^{(1)}(t)$ determine the system unperturbed motion $\mathcal{S}_{Up}(.; t_0; \mathbf{S}_{Up0})$ since they are the system unperturbed (desired: subscript d, nominal: subscript N) state variables $S_{Upi} = S_{id} = S_{iN}$, $i = 1, 2, ..., n = 2k$:

$$S_{Upi}(t) = S_{id}(t) = S_{iN}(t) = f_i(t) \ \ for \ i = 1, 2, ..., k \ and$$
$$S_{Upi}(t) = S_{id}(t) = S_{iN}(t) = f_i^{(1)}(t) \ \ for \ i = k+1, k+2, ..., 2k. \qquad (5.1)$$

Their initial values are fixed constants, independent of the initial moment t_0*:*

$$S_{Upi}(t_0) = S_{id}(t_0) = S_{iN}(t_0) = f_i(t_0) = const.\ for\ i = 1, 2, ..., k\ and$$
$$S_{Upi}(t_0) = S_{id}(t_0) = S_{iN}(t_0) = f_i^{(1)}(t_0) = const.\ for\ i = k+1, k+2, ..., 2k. \quad (5.2)$$

This means that the unperturbed (desired, nominal) initial state vector $\mathbf{S}_{Up0}$ *(*$\mathbf{S}_{d0}$*,* $\mathbf{S}_{N0}$*), respectively, is fixed constant vector. It is independent of the initial moment* t_0*.*

Note 161 ***Time, Lyapunov's stability definition and stability concepts and definitions beyond Lyapunov's***

Lyapunov established his theory for the systems whose stability properties do not depend on the choice of the initial moment t_0 *and for a fixed constant initial desired state vector* $\mathbf{S}_{d0}$*. All properties of every time-invariant dynamical system are independent of time. This book treats the systems whose stability properties can depend on the initial moment* t_0 *in the framework of the general time-varying nonlinear systems. All definitions and results valid for the latter become simpler for the former by accepting the initial time* $t_0 = 0$*. The moment* t_o *can be any constant moment; thus,* $t_0 = 0$ *is acceptable for the former without loss of generality.*

There are classes of time-varying dynamical systems that satisfy Lyapunov's restriction relative to the initial moment. Such their stability properties are uniform relative to the choice of the initial moment t_0 *from a time set* $\mathfrak{T}_i$*, or even from the whole time set* $\mathfrak{T}$*. This means, if such a system possesses a stability property at some initial instant* t_{01} *from the time set* $\mathfrak{T}_i$ *(or* $\mathfrak{T}$*), then the system possesses the same stability property at every initial moment* t_0 *from the time set* $\mathfrak{T}_i$ *(i.e.,* $\mathfrak{T}$*), respectively. However, even for such systems, the position in the space, outer diameter, volume and the shape of the domain of the corresponding stability property can essentially vary with a variation of the initial moment* t_0 *over the time set* $\mathfrak{T}_i$ *(*$\mathfrak{T}$*), respectively. Lyapunov did not mention stability domains.*

This book also treats the systems whose stability properties depend on the initial moment t_0*.*

The concept of stability domains and definitions of new stability properties in Lyapunov's sense of the system's desired motion and/or of sets, which can depend on time, characterize the next chapters of the book. In order to read the book easily with full comprehension, the reader should be familiar with the development of set theory in time-varying sets, which occupies Chapter 6.

Lyapunov permitted $\mathrm{n} \neq 2k$. Let

$$\mathbf{Q} = \begin{bmatrix} Q_1 & Q_2 & \cdots & Q_{\mathrm{n}-1} & Q_{\mathrm{n}} \end{bmatrix}^T \in \mathfrak{R}^{\mathrm{n}}, \quad (5.3)$$

and

$$S_j = \left\{ \begin{array}{c} q_j,\ j = 1, 2, ..., k, \\ q_{j-k}^{'},\ j = k+1, k+2, ..., 2k = n, \end{array} \right\},\ \forall j = 1, 2, ..., 2k = n. \quad (5.4)$$

Comment 162 *The following theorem justifies the usage of Lyapunov functions* $Q_j(.)$*, which compose the vector function* $\mathbf{Q}(.)$*, in the simplest form,*

$$Q_j\left(q_1,\ q_2,\ ...,\ q_k,\ q_1^{'},\ q_2^{'},\ ...,\ q_k^{'}\right) =$$
$$= Q_j(S_1, S_2, ..., S_n) \equiv S_j,\ \forall j = 1, 2, ..., \mathrm{n} = n. \quad (5.5)$$

i.e., to accept the vector function $\mathbf{Q}(.)$ *in the simplest form,*

$$\mathbf{Q}(\mathbf{S}) = \mathbf{S} = \begin{bmatrix} S_1 & S_2 & \cdots & S_{n-1} & S_n \end{bmatrix}^T \in R^n. \quad (5.6)$$

The theorem (discovered, stated, and proved in [195, Theorem 1, p.7], [247, Theorem 1.1, pp. 2-4] by the author of this book) explains why we can use the original Lyapunov stability definition in its simplest form.

Theorem 163 *[195, Theorem 1, p.7], [247, Theorem 1.1, pp. 2-4] Stability of the zero equilibrium state of the Lyapunov system (1) with respect to the vector* $\mathbf{Q}(\mathbf{S}) = \mathbf{S}$ *is necessary and sufficient for its stability with respect to any continuous real-valued vector function* $\mathbf{Q}(.)$.

Comment 164 *This theorem clarifies why the use of* $\mathbf{Q} = \mathbf{S}$ *is possible without loss of generality. The Lyapunov differences* $Q_j - F_j$ *form the state vector deviation:*

$$\left[Q_1 - F_1 \vdots\, Q_2 - F_2 \vdots\, ... \vdots\, Q_n - F_n\right]^T = \mathbf{S} - \mathbf{S}_d = \mathbf{s}.$$

Instead of using Lyapunov's numbers $L_1, L_2, ..., L_n$ *we use a positive real number* δ*, and instead of using Lyapunov's numbers* $E_1, E_2, ..., E_k, E_1{}', E_2{}', ..., E_k{}'$ *we use a positive real number* ε*. Furthermore, instead of using Lyapunov's scalar absolute values of the variables differences*

$$|\varepsilon_j| \leq E_j, \;\; |\varepsilon_j{}'| \leq E_j{}', \;\; (j = 1, 2, ..., k),$$
$$|Q_1 - F_1| < L_1, \;\; |Q_2 - F_2| < L_2, \;\; ... \;, \;\; |Q_n - F_n| < L_n,$$

which induce vector norms and vector inequality

$$\left\| \left[\varepsilon_1 \vdots\, \varepsilon_2 \vdots\, ... \vdots\, \varepsilon_k \vdots\, \varepsilon_1{}' \vdots\, \varepsilon_2{}', \, ... \vdots\, \varepsilon_k{}'\right] \right\|^T <$$
$$< \left\| \left[Q_1 - F_1 \vdots\, Q_2 - F_2 \vdots\, ... \vdots\, Q_k - F_k \vdots\, ... \vdots\, Q_n - F_n\right] \right\|^T, \;\; 2k = \mathrm{n} = n,$$

we use the norms of the deviation vectors, respectively:

$$\|\mathbf{S} - \mathbf{S}_d\| = \|\mathbf{s}\| \Longrightarrow \|\mathbf{S}_0 - \mathbf{S}_{d0}\| = \|\mathbf{s}_0\|.$$

The result is that instead of using Lyapunov's scalar mutually inconsistent inequalities (because there is the weak inequality for the initial conditions while the inequality for instantaneous differences $Q_j - F_j$ *is strong):*

$$|\varepsilon_j| \leq E_j, \;\; |\varepsilon_j{}'| \leq E_j{}', \;\; (j = 1, 2, ..., k),$$
$$|Q_1 - F_1| < L_1, \;\; |Q_2 - F_2| < L_2, \;\; ... \;, \;\; |Q_n - F_n| < L_n, \;\; \forall t > t_0 \tag{5.7}$$

we consistently use the inequalities ("strictly less than"), respectively,

$$\|\mathbf{S}_0 - \mathbf{S}_{d0}\| = \|\mathbf{S}_{d0} - \mathbf{S}_0\| = \|\mathbf{s}_0\| < \delta,$$
$$\|\mathbf{S}(t) - \mathbf{S}_d(t)\| = \|\mathbf{S}_d(t) - \mathbf{S}(t)\| = \|\mathbf{s}(t)\| < \varepsilon, \;\; \forall t \geq t_0. \tag{5.8}$$

Lyapunov's scalar inequalities are inconsistent because they are weak ("less than" or "equal to" $:\leq$*) relative to the initial conditions, but strong ("less than":* $<$*) with respect to the differences* $|Q_j - F_j|$*. Besides, we replace Lyapunov's weak condition that the temporal inequalities should be satisfied only for every* $t > t_0$ *by the stricter condition that they should hold for every* $t \geq t_0$*.*

The numbers E_j *and* $E_j{}'$ *are replaced by a positive real number* δ*, and the numbers* L_i *are replaced by a positive real number* ε *in the Lyapunov stability theory developed after Lyapunov.*

These changes do not influence definitions of stability properties, but influence those of the stability domains that Lyapunov did not treat.

Conclusion 165 *We have accepted Equations (5.5) and (5.6). The Lyapunov variables*

$$q_1,\ q_2,\ ...,\ q_k,\ q_1\text{‘},\ q_2\text{‘},\ ...,\ q_k\text{‘}$$

are the state variables denoted herein by

$$S_1,\ S_2,\ ...,\ S_n,\ \mathrm{n} = n = 2k.$$

They compose the state vector $\mathbf{S}$,

$$\mathbf{S} = \begin{bmatrix} S_1 \vdots\ S_2 \vdots\ ... \vdots\ S_n \end{bmatrix}^T \in \mathfrak{R}^n.$$

The number $n = 2k$ *is the number of linearly independent state variables* S_1, S_2, ..., S_n. *The consequence is that the number* n *of the state variables is equal to Lyapunov's number "*n*" of the functions* Q_i, $i = 1, 2, ..., \mathrm{n} = n$.

Conclusion 166 *The particular solution defined by Lyapunov:*

$$q_1 = f_1(t),\ q_2 = f_2(t),\ ...,\ q_k = f_k(t),$$

is the first part

$$S_{1d},\ S_{2d},\ ...,\ S_{k,d}$$

of the nominal (i.e., desired) state vector $\mathbf{S}_d$,

$$\mathbf{S}_d = \begin{bmatrix} S_{1d} \vdots\ S_{2d} \vdots\ ... \vdots\ S_{n,d} \end{bmatrix}^T,\ n = 2k.$$

The Lyapunov quantities $\varepsilon_j, \varepsilon_j\text{‘}$ *represent initial deviations* s_j *of the state variables,*

$$s_i = S_j - S_{jd},\ \forall j = 1, 2, ..., n.$$

They form the initial state deviation vector $\mathbf{s}_0$,

$$\mathbf{s}_0 = \mathbf{S}_0 - \mathbf{S}_{d0}. \tag{5.9}$$

Comment 167 *Lyapunov's equations (1) are the scalar state differential equations of the ISO system (4.90), (4.90).*

This book establishes the stability theory for the general nonlinear dynamical EISO systems (4.142), (4.143), which incorporate IO, ISO, HISO and IIO systems.

5.4 Lyapunov's original definition and sets

Definition 168 ***Lyapunov's original stability definition***

"If for all L_s*, nevertheless how small they are, there can be selected positive numbers* E_1, E_2, ..., E_k, $E_1\text{‘}$, $E_2\text{‘}$, ..., $E_k\text{‘}$ *such that for all real* ε_j, $\varepsilon_j\text{‘}$*, which satisfy conditions (5.10):*

$$|\varepsilon_j| \leq E_j,\ |\varepsilon_j\text{‘}| \leq E_j\text{‘},\ (j = 1, 2, ..., k),^1 \tag{5.10}$$

and for all t*, greater than* t_0*, the inequalities*

$$|Q_1 - F_1| < L_1,\ |Q_2 - F_2| < L_2,\ ...,\ |Q_n - F_n| < L_n, \tag{5.11}$$

are satisfied then the unperturbed motion is stable with respect to the quantities Q_1, Q_2, ..., Q_n*; otherwise, unstable".*

Note 169 *The above consideration and Theorem 163 explain and prove the equivalence between Lyapunov's original definition 168 and the usually accepted following definition:*

Definition 170 ***Equivalent definition to Lyapunov's original stability Definition 168***

If for every positive real number ε, regardless of how small it is, there can be selected a positive real number δ, $\delta = \delta(\varepsilon)$, such that for all real initial state deviations $\mathbf{S}_0 - \mathbf{S}_{d0} = \mathbf{s}_0$, which satisfy the condition

$$\|\mathbf{S}_0 - \mathbf{S}_{d0}\| = \|\mathbf{s}_0\| < \delta, \tag{5.12}$$

and for all $t \geq t_0$, the inequality

$$\|\mathbf{S}(t; \mathbf{S}_0) - \mathbf{S}_d(t; \mathbf{S}_{d0})\| = \|\mathbf{s}(t; \mathbf{s}_0)\| < \varepsilon, \tag{5.13}$$

is satisfied, then the unperturbed motion is stable; otherwise, unstable.

Note 171 *From the stability property point of view, the Lyapunov original conditions (5.10), (5.11) and the modern stability conditions (5.12), (5.13) are mutually equivalent, respectively. While this holds also for the Lyapunov original condition (5.11) and the modern stability condition (5.13), it is not valid for the Lyapunov original condition (5.10) and the modern stability condition (5.12) from the stability domain point of view.*

The Lyapunov original condition (5.10) strictly represented in the modern manner should read:

$$\|\mathbf{S}_0 - \mathbf{S}_{d0}\| = \|\mathbf{s}_0\| \leq \delta \tag{5.14}$$

instead of the modern stability condition (5.12). Interpreting these conditions from the set point of view, the condition (5.14) demands that $\mathbf{S}_0$ belongs to the closed hyperball $Cl\mathfrak{B}_\delta(\mathbf{S}_{d0})$ centered at $\mathbf{S}_{d0}$ with radius δ, i.e., that $\mathbf{s}_0$ belongs to the closed hyperball $Cl\mathfrak{B}_\delta$ centered at $\mathbf{s} = \mathbf{0}_n$,

$$\mathbf{S}_0 \in Cl\mathfrak{B}_\delta(\mathbf{S}_{d0}) = \{\mathbf{S} : \|\mathbf{S} - \mathbf{S}_{d0}\| \leq \delta\}, \ \mathbf{s}_0 \in Cl\mathfrak{B}_\delta = \{\mathbf{s} : \|\mathbf{s}\| \leq \delta\}. \tag{5.15}$$

However, the set representation of the modern stability condition (5.12) requires that $\mathbf{S}_0$ belongs to the open hyperball $\mathfrak{B}_\delta(\mathbf{S}_{d0})$ centered at $\mathbf{S}_{d0}$ with radius δ, i.e., that $\mathbf{s}_0$ belongs to the open hyperball $\mathfrak{B}_\delta$ centered at $\mathbf{s} = \mathbf{0}_n$,

$$\mathbf{S}_0 \in \mathfrak{B}_\delta(\mathbf{S}_{d0}) = \{\mathbf{S} : \|\mathbf{S} - \mathbf{S}_{d0}\| < \delta\}, \ \mathbf{s}_0 \in \mathfrak{B}_\delta = \{\mathbf{s} : \|\mathbf{s}\| < \delta\}. \tag{5.16}$$

The stability domain can be a closed set [247, Example 2.1, pp. 22-25, and Example 2.2, pp. 25-27]. We should permit the system properties to determine whether the system stability domain is open or closed or neither of both.

Note 172 *The sets of all $\mathbf{s}$ obeying $\|\mathbf{s}\| < \delta$ and of all $\mathbf{s}$ satisfying $\|\mathbf{s}\| < \varepsilon$ are the hyperballs $\mathfrak{B}_\delta$ and $\mathfrak{B}_\varepsilon$ centered at the origin $\mathbf{s} = \mathbf{0}_n$ with the radius δ, i.e., ε, respectively. They are (open)* ***connected neighborhoods*** *$\mathfrak{N}_\delta$ and $\mathfrak{N}_\varepsilon$ of the origin $\mathbf{s} = \mathbf{0}_n$. They have the special form of the hyperball.*

The stability domains, which Lyapunov did not consider, are ***connected neighborhoods*** *of the reference state, but need not be hyperballs. For that reason it is more adequate to use a connected neighborhood of a form that is determined by the system itself rather than to be fixed, a priori, by the definition. This explains why all stability definitions herein are expressed in terms of sets in general, without any restriction on their forms because they are determined exclusively by the system properties and characteristics.*

Note 173 *Notice that:*
- the ε-neighborhood $\mathfrak{N}_{\varepsilon}(t;\mathbf{S}_d)$ of well defined uniquely $\mathbf{S}_d(t)$ for every $t \in \mathfrak{T}_0$ is connected and obeys the following:

$$0 < \varepsilon_1 < \varepsilon_2 \Longrightarrow \mathfrak{N}_{\varepsilon_1}(t;\mathbf{S}_d) \subset \mathfrak{N}_{\varepsilon_2}(t;\mathbf{S}_d),\ \forall t \in \mathfrak{T}_0,$$
$$\partial\mathfrak{N}_{\varepsilon_1}(t;\mathbf{S}_d) \cap \partial\mathfrak{N}_{\varepsilon_2}(t;\mathbf{S}_d) = \phi,\ \forall t \in \mathfrak{T}_0, \tag{5.17}$$

- $\mathfrak{N}(\varepsilon;t_0;\mathbf{S}_{d0};\mathbf{I}_N)$ denotes a connected neighborhood of $\mathbf{S}_{d0}$ at $t_0 \in \mathfrak{T}$, which is induced by the connected neighborhood $\mathfrak{N}_{\varepsilon}(t;\mathbf{S}_d)$ at any $t \in \mathfrak{T}_0$.

Comment 174 ***Crucial lack of Lyapunov's original stability definition***

Lyapunov's original stability Definition 168 and its equivalent Definition 170 emphasize the condition on L_s: "nevertheless how small they are," i.e., on ε: "nevertheless how small it is," for the following reason: if the condition (5.11), i.e., (5.13), is satisfied for all $t \geq t_0$ for arbitrary small L_s, i.e., for arbitrary small ε, then the condition (5.11), i.e., (5.13), is simultaneously satisfied for arbitrary big L_s, i.e., for arbitrary big ε, for all $t \geq t_0$, respectively.

From the mathematical point of view, this result can be satisfactory and sufficient. Lyapunov's original definition is a generalization of the definition of the function continuity as its extension to system motions. Notice that Lyapunov was a mathematician, not an engineer.

However, it is difficult to satisfy engineering demands and be sufficient from the practical, engineering point of view. It can happen that the maximal $\delta = \delta(\varepsilon)$ denoted by $\delta_M(\varepsilon)$ can be very small for every $\varepsilon > \varepsilon^* > 0$ where ε^* can be small, e.g., $\delta_M(\varepsilon) = 10^{-10^{10}}$ for every $\varepsilon > \varepsilon^* = 10^{-10^{10}}$. This poses the following question:

What is the maximal $\delta_M(\varepsilon)$ for all $\varepsilon > \varepsilon^* > 0$ and what is the value of $\varepsilon^* > 0$?

Even more, in view of Note 172 and Note 173, it is essential to find the largest neighborhood $\mathfrak{N}(\varepsilon;t_0;\mathbf{S}_{d0};\mathbf{I}_N)$ for every $\varepsilon \in \mathfrak{R}^+$ and the largest $\mathfrak{N}_L(\mathfrak{R}^+;t_0;\mathbf{S}_{d0};\mathbf{I}_N)$ valid for all $\varepsilon \in \mathfrak{R}^+$, which is $\mathfrak{N}(\varepsilon^*;t_0;\mathbf{S}_{d0};\mathbf{I}_N)$,

$$\mathfrak{N}_L(\mathfrak{R}^+;t_0;\mathbf{S}_{d0};\mathbf{I}_N) = \mathfrak{N}(\varepsilon^*;t_0;\mathbf{S}_{d0};\mathbf{I}_N), \tag{5.18}$$

such that

$$\mathfrak{N}(\varepsilon;t_0;\mathbf{S}_{d0};\mathbf{I}_N) \subseteq \mathfrak{N}_L(\mathfrak{R}^+;t_0;\mathbf{S}_{d0};\mathbf{I}_N) = \mathfrak{N}(\varepsilon^*;t_0;\mathbf{S}_{d0};\mathbf{I}_N),\ \forall \varepsilon \in \mathfrak{R}^+. \tag{5.19}$$

This implies that the largest neighborhood $\mathfrak{N}_L(\mathfrak{R}^+;t_0;\mathbf{S}_{d0};\mathbf{I}_N)$ of $\mathbf{S}_{d0}$ is the union of all neighborhoods $\mathfrak{N}(\varepsilon;t_0;\mathbf{S}_{d0};\mathbf{I}_N)$ in ε over $\mathfrak{R}^+$,

$$\mathfrak{N}_L(\mathfrak{R}^+;t_0;\mathbf{S}_{d0};\mathbf{I}_N) = \mathfrak{N}(\varepsilon^*;t_0;\mathbf{S}_{d0};\mathbf{I}_N) = \cup\left[\mathfrak{N}(\varepsilon;t_0;\mathbf{S}_{d0};\mathbf{I}_N) : \varepsilon \in \mathfrak{R}^+\right] \tag{5.20}$$

and explains the reason for accepting the neighborhood $\mathfrak{N}_L(\mathfrak{R}^+;t_0;\mathbf{S}_{d0};\mathbf{I}_N)$ of $\mathbf{S}_{d0}$ to be ***the domain of stability of the desired motion $\mathcal{S}_d(.;\mathbf{S}_{d0})$ at $t_0 \in \mathfrak{T}$.***

5.5 Initial data, motion and stability

Note 175 *Lyapunov restricted his stability study to systems with properties independent of the initial moment t_0 [335]. Stability properties of the systems treated in this book can depend on the initial moment $t_0 \in \mathfrak{T}$. The system can possess a stability property at some initial moment $t_{01} \in \mathfrak{T}_i$ but not at $t_{02} \in \mathfrak{T}_i$ if $t_{02} \neq t_{01}$.*

The desired initial vector $\mathbf{S}_{d0}$ *is a fixed constant vector (Comment 160). It can be a stationary vector* $\mathbf{S}_s$ *or an equilibrium vector* $\mathbf{S}_e$*, relative to the nominal input vector* $\mathbf{I}_N(t)$*. Then the system desired motion*

$$\mathcal{S}_d(.;t_0;\mathbf{S}_{d0};\mathbf{I}_N)$$

through $\mathbf{S}_{d0}$ *is constant and always equal to* $\mathbf{S}_{d0}$*,*

$$\mathbf{S}_d(t;t_0;\mathbf{S}_{d0};\mathbf{I}_N) \equiv \mathbf{S}_{d0}.$$

When we study a stability property of the system stationary vector $\mathbf{S}_s$ *or of the equilibrium state* $\mathbf{S}_e$*, the system desired initial state vector* $\mathbf{S}_{d0}$ *is fixed to be the stationary vector* $\mathbf{S}_s$ *or the equilibrium vector* $\mathbf{S}_e$*,*

$$\mathbf{S}_{d0} \in \{\mathbf{S}_s, \mathbf{S}_e\},$$

and the system desired motion $\mathcal{S}_d(.;t_0;\mathbf{S}_{d0};\mathbf{I}_N)$ *is constant stationary or equilibrium motion:*

$$\mathbf{S}_d(t;t_0;\mathbf{S}_{d0};\mathbf{I}_N) \equiv \mathbf{const} \equiv \mathbf{S}_{d0} \in \{\mathbf{S}_s, \mathbf{S}_e\}.$$

The desired initial state vector $\mathbf{S}_{d0}$ *contains and expresses the influence of the whole past, of the whole history, of the system on its future desired behavior on* $\mathfrak{T}_0$ *since the initial moment* $t_0 \in \mathfrak{T}_i$*, in general. Two different initial moments* $t_{01} \in \mathfrak{T}_i$ *and* $t_{02} \in \mathfrak{T}_i$*,* $t_{01} \neq t_{02}$*, determine different durations of the system past:* $(-\infty, t_{01}] \neq (-\infty, t_{021}]$*, respectively. The desired initial state vector* $\mathbf{S}_{d0}$ *can, but need not, depend on the initial moment* $t_0 \in \mathfrak{T}_i$*,* $\mathbf{S}_{d0} = \mathbf{S}_d(t_0)$*. This characterizes the desired motions of the planes, of the aircraft. At different initial instants, their desired initial states are different in general. They determine their different desired motions.*

The single desired vector value $\mathbf{S}_d(t;t_0;\mathbf{S}_{d0};\mathbf{I}_N)$ *of the desired motion* $\mathcal{S}_d(.;t_0;\mathbf{S}_{d0};\mathbf{I}_N)$ *at a moment* $t \in \mathfrak{T}_0$ *is the system instantaneous desired state vector* $\mathbf{S}_d(t)$ *at the moment* $t \in \mathfrak{T}_0$*,* $\mathbf{S}_d(t;t_0;\mathbf{S}_{d0};\mathbf{I}_N) \equiv \mathbf{S}_d(t)$*. This implies* $\mathbf{S}_{d0} = \mathbf{S}_d(t_0) = \mathbf{S}_d(t_0;t_0;\mathbf{S}_{d0};\mathbf{I}_N)$*, which shows that along the system desired motion, the initial desired state vector* $\mathbf{S}_{d0}$ *depends on the initial instant* $t_0 \in \mathfrak{T}_i$*, i.e., the desired initial state vector* $\mathbf{S}_{d0k} = \mathbf{S}_d(t_{0k})$ *along the system motion at another initial moment* $t_{0k} \neq t_0$ *is not* $\mathbf{S}_{d0} = \mathbf{S}_d(t_0) = \mathbf{S}_d(t_0;t_0;\mathbf{S}_{d0};\mathbf{I}_N)$ *but in general*

$$\begin{aligned}\mathbf{S}_{d0k} &= \mathbf{S}_d(t_{0k}) = \mathbf{S}_d\left[t_{0k};t_{0k};\mathbf{S}_d\left(t_{0k}\right);\mathbf{I}_N\right] \equiv \\ &\equiv \mathbf{S}_d\left[t_{0k};t_{0k};\mathbf{S}_d\left(t_{0k};t_0;\mathbf{S}_{d0};\mathbf{I}_N\right);\mathbf{I}_N\right] \neq \mathbf{S}_{d0}.\end{aligned}$$

This is due to the uniqueness property P4 of the system motions (Definition 123), which is expressed in general by Equation (4.38) in Section 4.11:

$$\mathbf{S}(t;t_{01};\mathbf{S}_{01};\mathbf{I}_N) \equiv \mathbf{S}\left[t;t_{02};\mathbf{S}(t_{02};t_{01};\mathbf{S}_{01};\mathbf{I}_N);\mathbf{I}_N\right]).$$

If the desired motion is fixed but the initial moment changes from $t_{01} \in \mathfrak{T}_i$ *to* $t_{02} \in \mathfrak{T}_i$*,* $t_{02} \neq t_{01}$*, then the desired initial state vector* $\mathbf{S}_{d02} = \mathbf{S}_d(t_{02})$ *at the initial moment* t_{02} *is* $\mathbf{S}_d(t_{02}) = \mathbf{S}_d(t_{02};t_{02};\mathbf{S}_{d02};\mathbf{I}_N)$*, which is, in general, different from*

$$\mathbf{S}_{d01} = \mathbf{S}_d(t_{01}) = \mathbf{S}_d(t_{01};t_{01};\mathbf{S}_{d01};\mathbf{I}_N).$$

If we treat a stability property of a particular single fixed system desired motion, i.e., of a system desired motion $\mathcal{S}_d(.;t_0;\mathbf{S}_{d0};\mathbf{I}_N)$*, and if we change the initial conditions from* $(t_{01};\mathbf{S}_{d01})$ *to* $(t_{02};\mathbf{S}_{d02})$*, then must be* $\mathbf{S}_{d02} = \mathbf{S}_d(t_{02};t_{01};\mathbf{S}_{d01};\mathbf{I}_N)$*. This holds when the desired initial state at any moment after* t_0 *taken as initial should belong to that desired motion. This is* *beyond Lyapunov's stability theory. This is treated in the nonlinear systems tracking theory [232].*

This explains that the situation essentially changes if the desired initial state is not an equilibrium vector [i.e., is not a stationary vector relative to the nominal input $\mathbf{I}_N$*] or if we do not consider the initial state vector to be determined along the system motion at a moment taken for the initial moment.*

If the initial state vector $\mathbf{S}_0^*$ *is fixed, i.e., if it is independent of the initial moment* t_0*, and if it is not an equilibrium vector,* $\mathbf{S}_0^* \neq \mathbf{S}_e$*, i.e., if it is the same whatever the initial moment* $t_0 \in \mathfrak{T}_i$*, then the motions* $\mathcal{S}_1(.;t_{01};\mathbf{S}_0^*;\mathbf{I}_N)$ *and* $\mathcal{S}_2(.;t_{02};\mathbf{S}_0^*;\mathbf{I}_N)$*, for which the initial state vector is the same* $\mathbf{S}_0^*$*, but they start at two different initial moments* t_{01} *and* $t_{02} \neq t_{01}$*, are different due to* $\mathbf{S}(t_{02};t_{01};\mathbf{S}_0^*;\mathbf{I}_N) \neq \mathbf{S}_0^*$ *in general:*

$$\mathbf{S}(t;t_{02};\mathbf{S}_0^*;\mathbf{I}_N) \neq \mathbf{S}\left[t;t_{02};\mathbf{S}(t_{02};t_{01};\mathbf{S}_0^*;\mathbf{I}_N);\mathbf{I}_N\right] = \mathcal{S}(t;t_{01};\mathbf{S}_0^*;\mathbf{I}_N).$$

The fixed initial state vector $\mathbf{S}_0^*$ *and* $\mathbf{S}_0 = \mathbf{S}(t_0;t_0;\mathbf{S}_0;\mathbf{I}_N)$ *can be arbitrarily chosen. The system motion* $\mathcal{S}(.;t_0;\mathbf{S}_0;\mathbf{I}_N)$ *passing through, or starting from,* $\mathbf{S}_0$ *at* t_0 *can converge to* $\mathbf{S}(t;t_0;\mathbf{S}_0^*;\mathbf{I}_N)$ *as* $t \longrightarrow \infty$*. However, the system motion* $\mathcal{S}_1(.;t_{01};\mathbf{S}_0;\mathbf{I}_N)$ *passing also through, or starting also from,* $\mathbf{S}_0$ *but at another initial moment* t_{01} *different from* t_0*,* $t_{01} \neq t_0$*, can diverge from* $\mathbf{S}(t;t_0;\mathbf{S}_0^*;\mathbf{I}_N)$ *as* $t \longrightarrow \infty$*.*

Notice also that $\mathcal{S}(.;t_0;\mathbf{S}_0;\mathbf{I}_N)$ *and* $\mathcal{S}_1(.;t_{01};\mathbf{S}_0;\mathbf{I}_N)$ *are two different motions in general:* $\mathcal{S}(.;t_0;\mathbf{S}_0;\mathbf{I}_N) \neq \mathcal{S}_1(.;t_{01};\mathbf{S}_0;\mathbf{I}_N)$*.*

The system desired motion $\mathcal{S}_d(.;t_0;\mathbf{S}_{d0};\mathbf{I}_N)$ *with fixed initial desired state* $\mathbf{S}_{d0}$ *can be stable if the initial moment is* t_0 *but the system desired motion* $\mathcal{S}_{d1}(.;t_{01};\mathbf{S}_{d0};\mathbf{I}_N)$ *with the same fixed initial desired state* $\mathbf{S}_{d0}$ *can be unstable if the initial moment* $t_{01} \neq t_0$*. This already happens in the framework of linear time-varying systems, as the following example shows.*

Example 176 *[115] Let* $s \in \mathfrak{R}$ *and*

$$\frac{ds}{dt} = \frac{1}{a-t}s, \quad (a \neq 0) \in \mathfrak{R}.$$

The motions of this system are found to be

$$\mathcal{S}(.;t_0;s_0;0) = \frac{t_0 - a}{t-a}s_0, \ t_0 \neq a, \ t \neq a.$$

For $t_0 = a$ *and* $t = t_0 = a$ *the motion is not defined. For* $t_0 = a$ *the following holds for every fixed* $s_0 \neq 0 \in \mathfrak{R}$ *and for any* $t_{0k} \neq a$*,* $k = 1, 2, ...$*:*

$$|s(t;t_{0k};s_0;0)| = \left|\frac{t_{0k}-a}{t-a}s_0\right| \left\{ \begin{array}{c} \longrightarrow \infty \ \ as \ t \longrightarrow a, \ t < a \ if \ t_{0k} < a, \\ \longrightarrow 0 \ as \ t \longrightarrow \infty \ if \ t_{0k} > a, \end{array} \right\}, \ \forall (s_0 \neq 0) \in \mathfrak{R}.$$

This shows the crucial influence of the initial moment $t_{0k} \neq a$ *on the system behavior without changing the initial state* $s_0 \neq 0$*. The zero equilibrium state* $s_e = 0$ *is:*

- *unstable if* $t_{0k} < a$*,* $k = 1, 2, ...$
- *globally: stable, attractive and asymptotically stable if* $t_{0k} > a$*,* $k = 1, 2, ...$

Notice that this is a stability investigation of the constant desired state, of the zero equilibrium state.

Comment 177 *Different desired initial conditions at the same, fixed, initial moment induce different desired motions starting at that initial moment.*

Conclusion 178 ***Case 1.*** *At different initial moments* $t_{0k} \in \mathfrak{T}$, $k = 1, 2, ...$, *the system desired initial vector* $\mathbf{S}_d(t_{0k})$ *can be different. It can vary with the variation of the initial moment. Then the system initial desired state vector* $\mathbf{S}_d(t_{0k})$ *is not fixed but depends on the initial moment* $t_{0k} \in \mathfrak{T}_i$, $\mathbf{S}_d(t_{0i}) \neq \mathbf{S}_d(t_{0k})$, $i, k = 1, 2, ...$, $i \neq k$. $\mathbf{S}_d(t_{0i})$ *and* $\mathbf{S}_d(t_{0k})$ *are constant but different vectors. In this sense, for short,* $\mathbf{S}_d(t_0) = \mathbf{S}_{d0}$, $t_0 \in \mathfrak{T}_i$ *for any* $t_0 \in \{t_{0k}\colon k = 1, 2, ...\} \subset \mathfrak{T}_i$.

Analogously, the reference or desired initial set value $\mathfrak{S}(t_0)$ *of a time-varying set* $\mathfrak{S}(t)$ *at* $t_0 \in \mathfrak{T}_i$ *need not be fixed but can depend on the initial moment* $t_{0k} \in \mathfrak{T}_i$, $\mathfrak{S}(t_{01}) \neq \mathfrak{S}(t_{02})$ *for* $t_{01} \neq t_{02}$, t_{01} *and* t_{02} *belong to* $\mathfrak{T}_i$. *The sets* $\mathfrak{S}(t_{01})$ *and* $\mathfrak{S}(t_{02})$ *are constant, i.e., time-invariant, but different. In this sense, for short,* $\mathfrak{S}(t_0) = \mathfrak{S}_0$, *for any* $t_0 \in \{t_{0k}\colon k = 1, 2, ...\} \subset \mathfrak{T}_i$.

The Lyapunov stability concept does not treat this case.

Case 2. *The system initial desired state vector* $\mathbf{S}_d(t_0)$ *is constant and fixed, i.e., it is independent of the initial moment* $t_0 \in \mathfrak{T}_i$. *For short,* $\mathbf{S}_d(t_0) \equiv \mathbf{S}_0$.

Analogously, the initial set value $\mathfrak{S}(t_0)$ *of a time-varying set* $\mathfrak{S}(t)$ *at* $t_0 \in \mathfrak{T}_i$ *is fixed, i.e., it is independent of the initial moment* $t_0 \in \mathfrak{T}_i$. *For short,* $\mathfrak{S}(t_0) \equiv \mathfrak{S}_0$; *for all* $t_0 \in \mathfrak{T}_i$ *the same* $\mathfrak{S}_0$.

The Lyapunov stability concept concerns this case, which is valid throughout the book.

Condition 179 ***On the desired initial data***

The Lyapunov stability theory assumes the following:

The initial desired state vector $\mathbf{S}_d(t_0)$ *is a known, single, and fixed constant at every initial moment* $t_0 \in \mathfrak{T}_i$. *It is independent of the initial moment* $t_0 \in \mathfrak{T}_i$, $\mathbf{S}_d(t_0) = \mathbf{S}_{d0}$, $\forall t_0 \in \mathfrak{T}_i$.

The initial set $\mathfrak{S}(t_0)$ *of a time-varying set* $\mathfrak{S}(t)$ *is a known and fixed constant set* $\mathfrak{S}_0$; *i.e., it is independent of the initial moment* $t_0 \in \mathfrak{T}_i$. *For short then,* $\mathfrak{S}(t_0) = \mathfrak{S}_0$, $\forall t_0 \in \mathfrak{T}_i$.

5.6 Input vector, motion, and stability

System motion depends on the input vector $\mathbf{I}$ and on the initial state $\mathbf{S}_0$. Lyapunov stability theory concerns stability properties of a single (desired, nominal) motion $\mathcal{S}_d(.; t_0; \mathbf{S}_{d0}; \mathbf{I}_N)$ under the action of the nominal input vector $\mathbf{I}_N$ and under the influence of arbitrary, unknown, unpredictable initial conditions $\mathbf{S}_0$ on its nonnominal motion $\mathcal{S}(.; t_0; \mathbf{S}_0; \mathbf{I}_N)$.

Lyapunov stability properties reflect the influence of the non-nominal, i.e., non-desired, initial state conditions under the action of a known, accepted and fixed nominal input vector function $\mathbf{I}_N(.)$. This permits us to simplify the notation of motion by omitting the symbol $\mathbf{I}_N$ of the nominal input as follows:

$$\mathcal{S}(.; t_0; \mathbf{S}_0; \mathbf{I}_N) \equiv \mathcal{S}(.; t_0; \mathbf{S}_0) \text{ and } \mathbf{S}(t; t_0; \mathbf{S}_0; \mathbf{I}_N) \equiv \mathbf{S}(t; t_0; \mathbf{S}_0). \tag{5.21}$$

With Identities (5.21) in mind, the following condition is valid throughout the book (see also Note 125, Section 4.4):

Condition 180 *The symbol* $\mathbf{I}_N$ *is omitted everywhere in the sequel, meaning that the input vector* $\mathbf{I}(t)$ *is a known nominal* $\mathbf{I}(t) \equiv \mathbf{I}_N(t)$, *accepted and fixed on* $\mathfrak{T}_0$ *for every* $t_0 \in \mathfrak{T}$.

Corollary 181 ***Final form of the system (4.142), (4.143)*** *reads:*

$$\frac{d\mathbf{S}(t)}{dt} = \mathbf{a}_1[t, \mathbf{S}(t)] + \mathrm{G}[t, \mathbf{S}(t)]\,\mathbf{g}_1(t, \mathbf{S}) = \mathbf{f}(t, \mathbf{S}),$$
$$\mathbf{g}_1(t, \mathbf{S}) \equiv \mathbf{h}_1[t, \mathbf{I}_N^{\mu}(t, \mathbf{S})], \tag{5.22}$$

$$\mathbf{Y}(t) = \mathbf{a}_2 [t, \mathbf{S}(t)] + \mathbf{g}_2 (t, \mathbf{S}), \ \forall t \in \mathfrak{T},$$
$$\mathbf{g}_2 (t, \mathbf{S}) \equiv \mathbf{h}_2 [t, \mathbf{I}_N (t, \mathbf{S})]. \tag{5.23}$$

Note 182 ***They hold beginning now and throughout the book***

Condition 179, Condition 180, and Corollary 181 hold beginning now and throughout the book.

Comment 183 *The restrictions of the Lyapunov stability theory are overcome in the tracking theory [232]. It treats dynamical and control systems subjected to the simultaneous actions of unknown, unpredictable both input vector and initial conditions for any desired system motion from a given motion family. Its goal is to assure high performance of the real system output behavior relative to its every desired motion from a functional family of desired output behaviors. The Lyapunov stability theory is a very useful prerequisite for the tracking theory [232].*

Chapter 6

Stability domains concept

6.1 Introduction

The essence of the stability domains concept is the topic of this chapter.

Note 184 *Condition 179 (Section 5.5, Chapter 9) and Condition 180 (Section 5.6, Chapter 9) are valid throughout the book.*

6.2 "stability domain" or "stability region"?

In the stability literature, e.g., [71, p. 36], [323], [541], the term *region* is mainly used instead of the term *domain*. *Region* is an open set, but the largest neighborhood that obeys the stability definition for any $\varepsilon > 0$ need not be an open set, which corresponds to the notion *domain [247, pp. 21,22]*. This explains the reason for which the notion **domain** is used in this book rather than the term *region*. In order to avoid any ambiguity, the following is accepted:

Definition 185 ***Region and domain***

*i) A set $\mathfrak{B}$ is **a region of a constant set** $\mathfrak{A}$, which is denoted by $\mathcal{R}(\mathfrak{A})$, if and only if it is an open neighborhood of $\mathfrak{A}$:*

$$\mathfrak{B} = \mathcal{R}(\mathfrak{A}) \Longleftrightarrow \mathfrak{A} \subset \mathfrak{B},\ \partial\mathfrak{A} \cap \partial\mathfrak{B} = \phi \text{ and } \mathfrak{B} = In\mathfrak{B}. \tag{6.1}$$

*ii) A set $\mathfrak{B}$ is **a domain of the constant set** $\mathfrak{A}$, which is denoted by $\mathcal{D}(\mathfrak{A})$, if and only if it is a connected neighborhood of $\mathfrak{A}$:*

$$\mathfrak{B} = \mathcal{D}(\mathfrak{A}) \Longleftrightarrow \mathfrak{A} \subset In\mathfrak{B} \text{ and } \partial\mathfrak{A} \cap \partial\mathfrak{B} = \phi. \tag{6.2}$$

iii) The domain $\mathcal{D}(\mathfrak{A})$ of the set $\mathfrak{A}$ can be either an open set or closed set, but need not be either one of them.

See Note 171 in Section 5.4, also Example 187 together with Example 189, with regard to condition iii) of the above definition.

The Lyapunov stability concept demands *the existence* of the neighborhoods of an equilibrium state, or of a desired motion, or of a set, which are determined by their stability property. From the practical and engineering point of view, such information on *the existence* of the requested neighborhoods can be useless. Moreover, an equilibrium state, or a desired

DOI: 10.1201/9781003744528-10

motion or a set can possess a Lyapunov stability property, but from the practical, engineering point of view, the system dynamical behavior can be very unsatisfactory, unacceptable. This is due to the fact that even though ε is very big, say $\varepsilon = 10^{10^{10}}$, the corresponding maximal $\delta_{\max} = \delta_{\max}(\varepsilon)$ can be very small, say $\delta_{\max}(\varepsilon) = \delta_{\max}\left(10^{10^{10}}\right)^{-10} = 10^{-10}$. According to the Lyapunov stability definition, the equilibrium state, or the desired motion or the set, is, respectively, stable. However, if the norm $\|\mathbf{x}_0\|$ of the initial state deviation is equal to $10^{10^{10}}$, the maximal corresponding δ can be very small: $\delta_{\max}\left(10^{10^{10}}\right) = 10^{-10}$, then the system state deviation will escape the boundary of $\varepsilon = 10^{10^{10}}$ at some moment!

In conclusion, we need to enrich definitions of Lyapunov stability properties so they permit us to determine the appropriate maximal neighborhood of the equilibrium state or of the desired motion or of a set, so that for all initial state vectors from it, the motions are bounded and converge to the equilibrium vector or to the desired motion or to the set.

This poses the following question and opens the following problem:

Problem 186 ***Lyapunov stability problem***

What is the largest connected neighborhood of an equilibrium state, or of a desired motion or of a set for any $\varepsilon > 0$ at an initial moment t_0, such that for every state deviation vector or the state vector from that neighborhood, the instantaneous state deviation vector or the instantaneous state vector at every moment, since the initial moment, is in the specified ε-neighborhood of the equilibrium state, or of the desired motion or of the set, respectively?

6.3 Introductory example

The following example illustrates an essential weakness of Lyapunov's original concept and definitions of stability properties:

Example 187 *Let the system be defined by*

$$Y^{(1)} + 2tY(\alpha - Y) = 0, \ a \in \mathfrak{R}^+, \ let\ S = Y. \tag{6.3}$$

The system has two equilibrium states denoted by $S_{e1} = Y_{e1}$ and $S_{e2} = Y_{e2}$,

$$S_{e1} = Y_{e1} = 0, \ S_{e2} = Y_{e2} = a.$$

Its output response, which is simultaneously its motion, reads, respectively,

$$Y(t;t_0;Y_0) = \frac{\alpha Y_0}{Y_0 - (Y_0 - \alpha)e^{\alpha(t^2 - t_0^2)}} = S(t;t_0;S_0) = \frac{\alpha S_0}{S_0 - (S_0 - \alpha)e^{\alpha(t^2 - t_0^2)}}. \tag{6.4}$$

The motion blows up at $t = t_\infty$ iff $S_0 = Y_0 > \alpha$,

$$S_0 > \alpha \Longrightarrow t_\infty = \sqrt{t_0^2 + \frac{1}{\alpha}\ln\frac{S_0}{(S_0 - \alpha)}}.$$

$$S_0 = -\alpha \Longrightarrow S(t;t_0;S_0) = \frac{S_0}{-1 + 2e^{\alpha(t^2 - t_0^2)}}$$

We can summarize the main characteristics of the system motions:

$$S(t;t_0;S_0)\left\{\begin{array}{c} \longrightarrow \infty \ for\ t \longrightarrow t_\infty^-, \ S_0 > \alpha, \\ \longrightarrow 0^- \ for\ t \longrightarrow \infty, \ S_0 > \alpha, \\ = \alpha \ for\ every\ t \in \mathfrak{T}, \ S_0 = \alpha, \\ \longrightarrow 0^+ \ for\ t \longrightarrow t_\infty, \ 0 < S_0 < \alpha, \\ = 0 \ for\ every\ t \in \mathfrak{T}, \ S_0 = 0, \\ \longrightarrow 0^- \ for\ t \longrightarrow \infty, \ S_0 < 0 \end{array}\right\}; \ see\ Figure\ 6.1.$$

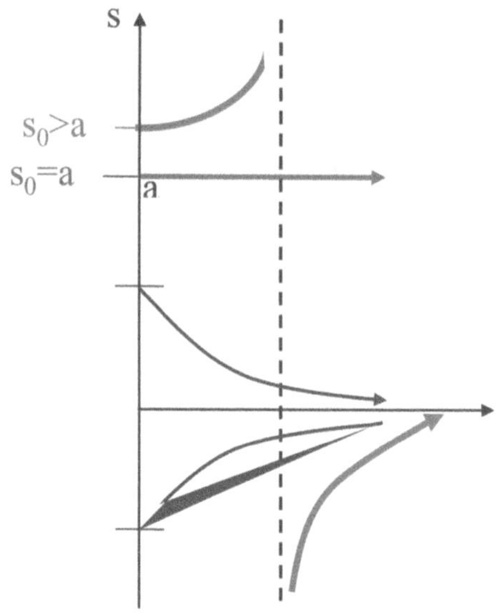

Figure 6.1: Diagram of the system solutions (6.4). In this figure $t_8 = t_\infty$.

The first equilibrium state (in the origin) $S_{e1} = 0$ *is:*
- stable (but not globally stable) because for every $\varepsilon > 0$ *there is* $\delta(t_0, \varepsilon) > 0$ *such that*

$$|S_0| < \delta(t_0, \varepsilon) \Longrightarrow |S(t; t_0; S_0)| < \varepsilon, \ \forall t \in \mathfrak{T}_0, \ \forall t_0 \in \mathfrak{T},$$

where the maximal $\delta(t_0, \varepsilon)$ *for a fixed pair* (t_0, ε), *which is denoted by* $\delta_{\max}(t_0, \varepsilon)$, *reads*

$$\delta_{\max}(t_0, \varepsilon) = \left\{ \begin{array}{c} \varepsilon, \ \varepsilon < \alpha \\ \alpha, \ \varepsilon \geq \alpha \end{array} \right\},$$

and $\delta_{\max}(t_0, \varepsilon)$ *does not diverge to* ∞ *as* $\varepsilon \to \infty$,
- attractive but not globally because $S_0 = \alpha$ *implies* $S(t; t_0; a) \equiv a$ *so that*

$$\Delta_{\max}(t_0) = \Delta_{\max} = a.$$

Hence, $|S_0| < \Delta_{\max} = a$ *yields*

$$|S_0| < \Delta_{\max}(t_0) \Longrightarrow |S(t; t_0; S_0)| \to 0 \ as \ t \to \infty.$$

If $S_0 > \alpha$, *then the motions diverge to infinity as time goes to* t_∞, *but for* $t \longrightarrow \infty$ *they converge to the equilibrium state* $S_{e1} = 0$; *see Figure 6.1.*

The equilibrium vector $S_{e1} = 0$ *is stable, attractive for every* $S_0 \neq a$, *and therefore, asymptotically stable but not globally.*

Since $\alpha > 0$ *can be very small, i.e., it can be arbitrarily small, asymptotic stability of* $S_{e1} = 0$ *does not have a practical value because for very small initial* $S_0 > \alpha$ *the motions become unbounded as* $t \to t_\infty$.

This example explains the importance of the knowledge of the maximal $\delta_{\max}(t_0, \varepsilon)$ for every $\varepsilon \in \mathfrak{R}^+$ in the definition of the stability and $\Delta_{\max}(t_0)$ in the definition of the attraction. This means that we should know the maximal hyperballs $\mathfrak{B}_{\delta_{\max}(t_0,\varepsilon)}$ for every $\varepsilon \in \mathfrak{R}^+$

and $\mathfrak{B}_{\Delta_{\max}(t_0)}$. These hyperballs are open connected sets in the state space of the system. They represent a special case of connected neighborhoods $\mathcal{N}$ of the origin. This explains why we should look for the largest open connected neighborhood of the origin, so that all motions starting from it satisfy the corresponding stability requirements. The neighborhood need not be in the form of a hyperball.

6.4 Domains of stability properties

The concept of *the stability domain* and the concept of *the asymptotic stability domain* were introduced in [115] by developing the ideas of [110]. The references [124], [140], [119], [127]-[129], [139], [142]-[134], [146], [154]-[160], [166]-[169], [181], [195], [198], [207], [223], [232], [247] further developed and used these concepts.

The book [247] synthesizes the concept of the domains of Lyapunov stability properties, establishes and develops the concepts of practical stability properties, of their domains, and contains illustrative examples in the framework of the *time*-invariant nonlinear dynamical systems in the Cauchy (i.e., normal) form. The reader familiar with [247] will find that this book broadens and generalizes the theory of the stability domains covered in [247] for *ISO* *time*-invariant nonlinear systems.

Comment 174 (Section 5.4) explains the need for the introduction and use of **the stability domain**.

Property 188 ***Meaning and property of the stability domain of an equilibrium vector*** $\mathbf{S}_e$

The stability domain $\mathcal{D}_S(\mathfrak{T}_0;\mathbf{S}_e)$ *of an equilibrium vector* $\mathbf{S}_e$ *means that* ***for every*** $\varepsilon \in \mathfrak{R}^+$ ***there is a connected subset*** $\mathcal{N}(\varepsilon;\mathfrak{T}_0;\mathbf{S}_e)$ ***of the*** $\varepsilon-$***neighborhood*** $\mathcal{N}_\varepsilon(\mathfrak{T}_0;\mathbf{S}_e)$ ***of the equilibrium point*** $\mathbf{S}_e$ ***and of*** $\mathcal{D}_S(\mathfrak{T}_0;\mathbf{S}_e)$,

$$\mathcal{N}(\varepsilon;\mathfrak{T}_0;\mathbf{S}_e) \subseteq \mathcal{D}_S(\mathfrak{T}_0;\mathbf{S}_e) \cap \mathcal{N}_\varepsilon(\mathfrak{T}_0;\mathbf{S}_e), \tag{6.5}$$

which is a connected neighborhood of the equilibrium point $\mathbf{S}_e$, ***with the property that for every*** $\mathbf{S}_0 \in \mathcal{N}(\varepsilon;\mathfrak{T}_0;\mathbf{S}_e)$ ***the system state*** $\mathbf{S}(t;t_0;\mathbf{S}_0)$ ***is always in*** $\mathcal{N}_\varepsilon(\mathfrak{T}_0;\mathbf{S}_e)$,

$$\begin{gathered}\forall\varepsilon \in \mathfrak{R}^+,\ \exists\mathcal{N}(\varepsilon;\mathfrak{T}_0;\mathbf{S}_e),\ \mathcal{N}(\varepsilon;\mathfrak{T}_0;\mathbf{S}_e) \subseteq \mathcal{D}_S(\mathfrak{T}_0;\mathbf{S}_e) \cap \mathcal{N}_\varepsilon(\mathfrak{T}_0;\mathbf{S}_e): \\ \mathbf{S}_0 \in \mathcal{N}(\varepsilon;\mathfrak{T}_0;\mathbf{S}_e) \Longrightarrow \mathbf{S}(t;t_0;\mathbf{S}_0) \in \mathcal{N}_\varepsilon(\mathfrak{T}_0;\mathbf{S}_e),\ \forall t \in \mathfrak{T}_0.\end{gathered} \tag{6.6}$$

The stability domain $\mathcal{D}_S(\mathfrak{T}_0;\mathbf{S}_e)$ *of the equilibrium state* $\mathbf{S}_e$ *is the largest neighborhood* $\mathcal{N}(\varepsilon;\mathfrak{T}_0;\mathbf{S}_e)$ *of* $\mathbf{S}_e$ *with this property.*

The equilibrium singleton set $\mathfrak{S}_E = \{\mathbf{S}_e\}$ is a trivially connected set. It is natural for its stability domain to be its connected neighborhood.

Zubov [541] defined *the attraction region*, which is called *the attraction domain* in this book. It was also well treated by Hahn [257]. Its meaning is the following:

The attraction domain of the zero equilibrium state $\mathbf{S}_e = \mathbf{0}_n$, i.e., of the singleton $\mathfrak{O} = \{\mathbf{0}_n\}$, is the largest connected neighborhood $\mathcal{N}(\mathbf{0}_n) = \mathcal{N}(\mathfrak{O})$ of $\mathbf{S}_e = \mathbf{0}_n$, i.e., of $\mathfrak{O} = \{\mathbf{0}_n\}$, such that for every initial state $\mathbf{S}_0$ from the neighborhood $\mathcal{N}(\mathbf{0}_n)$ at the initial moment t_0 the system motion $\mathcal{S}(.;t_0;\mathbf{S}_0)$ approaches asymptotically $\mathbf{S}_e = \mathbf{0}_n$ as time escapes to infinity,

$$\mathbf{S}(t;t_0;\mathbf{S}_0) \longrightarrow \mathbf{0}_n \ as\ t \to \infty \Longleftrightarrow \mathbf{S}_0 \in \mathcal{N}(\mathbf{0}_n).$$

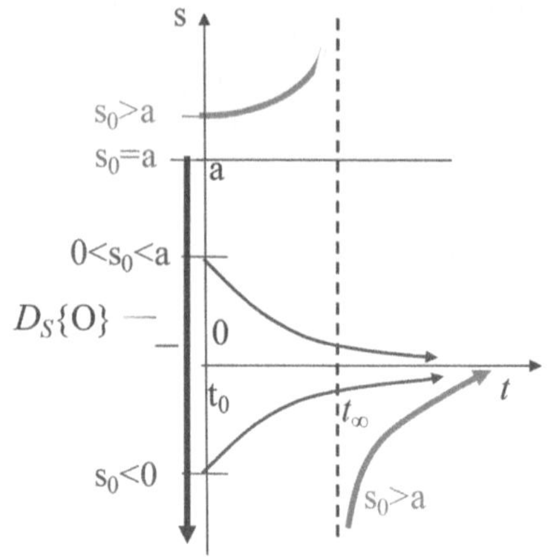

Figure 6.2: The stability domain $\mathcal{D}_S(\mathfrak{O})$ of the set $\mathfrak{O} = \{0\}$.

Example 189 *In Example 187:*
1) The connected set $\{S: \; S \leq a\}$ *is the stability domain* $\mathcal{D}_S(\mathfrak{O})$ *of the singleton* $\mathfrak{O}$,

$$\mathfrak{O} = \{S: \; S = 0\},$$

which contains the equilibrium state $S_{e1} = 0$. *The stability domain* $\mathcal{D}_S(\mathfrak{O}) = \{S : S \leq a\}$ *is, in this example, connected neighborhood of the set* $\mathfrak{O}$. *It is neither an open nor a closed set. It is not the stability region of the set* $\mathfrak{O}$ *and verifies the statement iii) of Definition 185. It has the nonempty boundary* $\partial\mathcal{D}_S(\mathfrak{O}) = \{S: \; S = a\}$. *For every* $\varepsilon > 0$ *there is a subset* $\mathcal{N}(\varepsilon;\mathfrak{O})$ *of* $\mathcal{D}_S(\mathfrak{O})$, *which is a connected neighborhood of the set* $\mathfrak{O}$, *such that for every* $S_0 \in \mathcal{N}(\varepsilon;\mathfrak{O})$ *the system state* $S(t;t_0;S_0)$ *is in the* ε*-neighborhood* $\mathcal{N}_\varepsilon(\mathfrak{O})$ *of* $\mathfrak{O}$, *i.e.,*

$$|S(t;0;S_0)| < \varepsilon, \; \forall (t, S_0) \in \mathfrak{T}_0 \times \mathcal{N}(\varepsilon;\mathfrak{O}),$$

where

$$\mathcal{N}(\varepsilon;\mathfrak{O}) = \left\{S: \; \left\{ \begin{array}{c} |S| < \varepsilon \; if \; \varepsilon < a \\ -\varepsilon < S \leq a \; if \; \varepsilon \geq a \end{array} \right\}\right\},$$

so that

$$\mathcal{D}_S(\mathfrak{O}) = \{S: \; S \leq a\}; \; see \; Figure \; 6.2.$$

The stability domain is neither an open nor a closed set. It is the domain but not the region.
2) The set $\{S: \; 0 \leq S < a\}$ *is the attraction domain* $\mathcal{D}_A(\mathfrak{O})$ *of the singleton* $\mathfrak{O}$. *It is the open connected neighborhood* $\mathcal{N}_A(\mathfrak{O}) = \{S : 0 \leq |S| < a\}$ *of the set* $\mathfrak{O}$. *For every* S_0 *from the set* $\mathcal{N}_A(\mathfrak{O})$, *the system motion converges asymptotically and continuously to the set* $\mathfrak{O}$ *as* $t \longrightarrow \infty$. *It is the largest set with such properties. The set* $\mathcal{N}_A(\mathfrak{O})$ *is the domain* $\mathcal{D}_A(\mathfrak{O})$ *of attraction of the set* $\mathfrak{O}$.*The attraction domain* $\mathcal{D}_A(\mathfrak{O})$ *is an open set. It is both a region and a domain.*

For every $S_0 \in \mathcal{D}_A(\mathfrak{O})$, *the system motion* $S(t;t_0;S_0)$ *asymptotically converges to the set* $\mathfrak{O}$ *as* $t \longrightarrow \infty$,

$$S(t;0;S_0) \longrightarrow \mathfrak{O}, \; \forall S_0 \in \mathcal{D}_A(\mathfrak{O}).$$

The set $\mathfrak{O}$ *attracts all system motions that start from the attraction domain* $\mathcal{D}_A(\mathfrak{O})$; *see Figure 6.3. This illustrates a difference between the stability domain and the attraction domain. They are not equal. The interior of the former is equal to the latter,*

$$\mathcal{D}_A(\mathfrak{O}) \neq \mathcal{D}_S(\mathfrak{O}), \; \mathcal{D}_A(\mathfrak{O}) \subset \mathcal{D}_S(\mathfrak{O}), \; In\mathcal{D}_S(\mathfrak{O}) = \mathcal{D}_A(\mathfrak{O}).$$

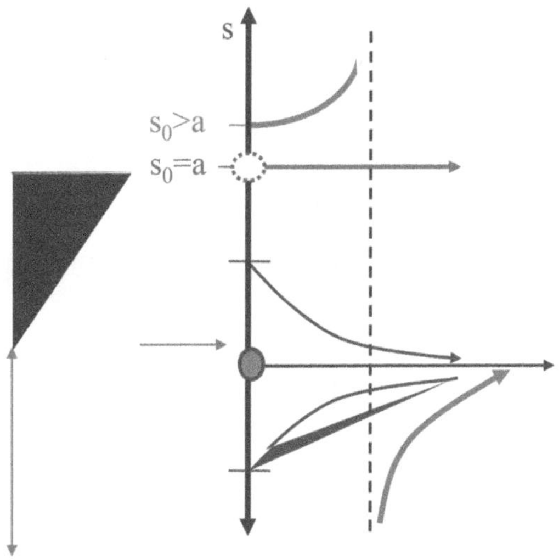

Figure 6.3: The attractiom domain $\mathcal{D}_A(\mathfrak{D})$ of the set $\mathfrak{D}$.

This illustrates the difference between the stability domain and the attraction domain. They do not coincide. They need not be equal, which holds in this example.

3) The set $\mathcal{D}_{AS}(\mathfrak{D}) = \mathcal{D}_S(\mathfrak{D}) \cap \mathcal{D}_A(\mathfrak{D})$,

$$\mathcal{D}_{AS}(\mathfrak{D}) = \{S : S < a\} = \mathcal{D}_A(\mathfrak{D}) = In\mathcal{D}_S(\mathfrak{D}) \subset \mathcal{D}_S(\mathfrak{D}),$$

is the asymptotic stability domain of the set $\mathfrak{D}$*; see Figure 6.4. It is open connected neighborhood of the set* $\mathfrak{D}$*. It is also the strict asymptotic stability domain* $\mathcal{D}_{SAS}(\mathfrak{D})$ *of the set* $\mathfrak{D}$ *in this example,*

$$\mathcal{D}_{SAS}(\mathfrak{D}) = \mathcal{D}_{SS}(\mathfrak{D}) \cap \mathcal{D}_A(\mathfrak{D}) = \mathcal{D}_{AS}(\mathfrak{D}).$$

Example 190 *[115, Example 4, p. 531] The zero equilibrium state* $\mathbf{S}_e = \mathbf{0}_2$ *of the first order, two dimensional system,* $n = 2$, $\mathbf{S} = [S_1\ S_2]^T$:

$$\frac{d\mathbf{S}}{dt} = \begin{bmatrix} \frac{dS_1}{dt} \\ \frac{dS_2}{dt} \end{bmatrix} = 2\left(1+t^2\right)\left(-1+4|S_1|+2|S_2|\right)\begin{bmatrix} S_1 \\ 2S_2 \end{bmatrix}$$

is stable with the domain of stability

$$\mathcal{D}_{US}(\mathfrak{T}) = \mathcal{D}_{US} = \left\{S\colon \mathbf{S} \in \mathfrak{R}^2,\ 2|S_1| + |S_2| \le \frac{1}{2}\right\}.$$

The proof is in [115, Example 4, p. 531].

Example 191 *[115, Example 5, p. 531] The instantaneous domain of stability* $\mathcal{D}_S(t_0)$ *of the zero equilibrium state* $\mathbf{S}_e = \mathbf{0}_2$ *at any* $t_0 \in \mathfrak{T}$ *of the first order, two dimensional system,* $n = 2$, $\mathbf{S} = [S_1\ \ S_2]^T$:

$$\frac{d\mathbf{S}}{dt} = \begin{bmatrix} \frac{dS_1}{dt} \\ \frac{dS_2}{dt} \end{bmatrix} = 2\left(\frac{t+4}{1+t^2} + 8|S_1| + 4|S_2|\right)\begin{bmatrix} S_1 \\ S_2 \end{bmatrix}$$

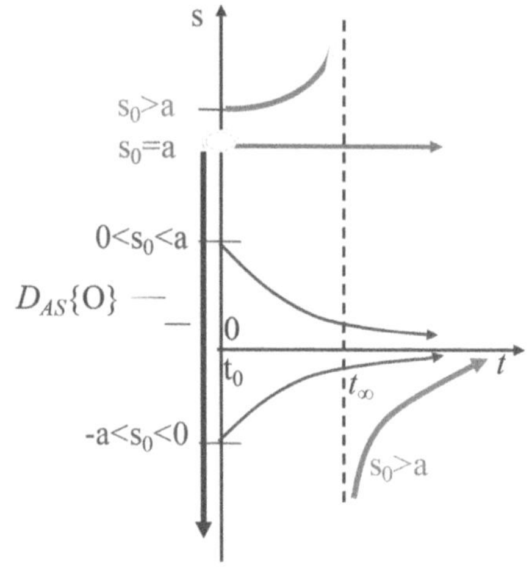

Figure 6.4: The asymptotic stability domain $\mathcal{D}_{AS}(\mathfrak{O})$ of the set $\mathfrak{O}$.

reads

$$\mathcal{D}_S(t_0) = \left\{ S: \mathbf{S} \in \mathfrak{R}^2,\ 2|S_1| + |S_2| \leq \frac{1}{1+t_0^2} \right\}.$$

For the proof, see [115, Example 5, p. 531].

The instantaneous domain of stability $\mathcal{D}_S(t_0)$ of the zero equilibrium state $\mathbf{S}_e = \mathbf{0}_2$ at $t_0 \in \mathfrak{T}$ is time-varying, compact at every $t_0 \in \mathfrak{T}$, and asymptotically contractive:

$$t_0 \longrightarrow \infty \Longrightarrow \mathcal{D}_S(t_0) \longrightarrow \mathfrak{O} = \{\mathbf{0}_2\},$$

so that

$$\cap [\mathcal{D}_S(t_0) : t_0 \in \mathfrak{T}] = \{\mathbf{0}_2\} \Longrightarrow \mathcal{D}_S(\mathfrak{T}) = \phi.$$

The zero equilibrium state $\mathbf{S}_e = \mathbf{0}_2$ is stable.

6.5 Crucial counterexamples

Example 189 and the following counterexamples reject the claim that the attraction domain is also the stability domain (e.g., [71]) or asymptotic stability domain (e.g., [323]). Lyapunov [335] did not define either *attraction* or *asymptotic stability* or *their domains.* Hahn [257] explained well that the stability and the attraction are mutually independent qualitative dynamical properties (see [247, Example 2.5, pp. 30-32], Example 187 and Example 189), which is sufficient reason to understand that their domains are different in general. In order to illustrate it convincingly once more, let us consider the following examples.

Example 192 ***Counterexample: Unstable but globally attractive equilibrium state of a time-invariant system***

The solution of the second order N-dimensional time-invariant IO system described by

$$\|\mathbf{Y}(t)\|^2 \mathbf{Y}^{(2)}(t) - 2\left[\mathbf{Y}^T(t)\mathbf{Y}^{(1)}(t)\right]\mathbf{Y}^{(1)}(t) = \mathbf{0}_N,\ \forall t \in \mathfrak{T}. \tag{6.7}$$

has the following form:

$$\mathbf{Y}(t;t_0;\mathbf{Y}_0) = \mathbf{Y}(t) = \frac{1}{1 - \|\mathbf{Y}_0\|^2 (t-t_0)} \mathbf{Y}_0,\ \forall (t,\mathbf{Y}_0) \in \mathfrak{T} \times \mathfrak{R}^n. \tag{6.8}$$

The solution obeys the differential equation, which is easy to verify, satisfies the initial condition at the initial moment t_0, $\mathbf{Y}(t_0) = \mathbf{Y}_0$, *and blows to infinite at* $t_\infty = t_0 + \|\mathbf{Y}_0\|^{-2}$ *for every* $\mathbf{Y}_0 \neq \mathbf{0}_N$,

$$\mathbf{Y}(t_\infty; t_0; \mathbf{Y}_0) = \infty \mathbf{Y}_0 \Longrightarrow \|\mathbf{Y}(t_\infty; t_0; \mathbf{Y}_0)\| = \infty, \ \forall (\mathbf{Y}_0 \neq \mathbf{0}_N) \in \mathfrak{R}^N.$$

The zero equilibrium state $\mathbf{S}_e = \left[\mathbf{Y}_e^T \ \ \mathbf{Y}_e^{(1)T}\right]^T = \left[\mathbf{0}_N^T \ \ \mathbf{0}_N^T\right]^T$ *is unstable. Such a physical system would explode at* $t = t_\infty$. *Its domain of stability is the empty set,* $\mathcal{D}_S(\mathbf{S}_e) = \phi$, *i.e., it does not exist, because for any* $\varepsilon \in \mathfrak{R}^+$ ***there is not*** $\delta(\varepsilon) \in \mathfrak{R}^+$ *such that* $\|\mathbf{S}_0\| < \delta(\varepsilon)$ *implies*

$$\|\mathbf{S}(t; t_0; \mathbf{S}_0)\| = \left\| \left[\mathbf{Y}^T(t) \ \ \mathbf{Y}^{(1)T}(t)\right]^T \right\| < \varepsilon, \ \forall t \in \mathfrak{T}_0.$$

However, for every $\mathbf{S}_0 \neq \mathbf{0}_{2N}$, $\mathbf{S}_0 \in \mathfrak{R}^{2N}$, *the system state* $\mathbf{S}(t; t_0; \mathbf{S}_0)$ *approaches asymptotically* $\mathbf{S}_e = \mathbf{0}_{2N}$ *as* $t \longrightarrow \infty$; *Figure 6.5. The zero equilibrium state* $\mathbf{S}_e = \mathbf{0}_{2N}$ *is globally attractive. Its domain of attraction is the whole state space,* $\mathcal{D}_A(\mathbf{S}_e) = \mathfrak{R}^{2N}$. *The result is that the stability domain is different from the attraction domain,*

$$\mathcal{D}_S(\mathbf{S}_e) = \phi \neq \mathfrak{R}^{2N} = \mathcal{D}_A(\mathbf{S}_e).$$

In the framework of time-invariant nonlinear systems, this already disproves the claim that the attraction domain is also the stability domain.

Example 193 ***Counterexample: Unstable but globally attractive equilibrium state of a time-varying system***
The second order N*-dimensional time-varying IO system is determined by*

$$t\|\mathbf{Y}(t)\|^4 \mathbf{Y}^{(2)}(t) - \mathbf{Y}^T(t)\mathbf{Y}^{(1)}(t)\left[\|\mathbf{Y}(t)\|^2 + 2t\mathbf{Y}^T(t)\mathbf{Y}^{(1)}(t)\right]\mathbf{Y}(t) = \mathbf{0}_N, \ \forall t \in \mathfrak{T}. \quad (6.9)$$

Its solution reads

$$\mathbf{Y}(t; t_0; \mathbf{Y}_0) = \mathbf{Y}(t) = \frac{1}{1 - \|\mathbf{Y}_0\|^2 (t^2 - t_0^2)} \mathbf{Y}_0, \ \forall (t, \mathbf{Y}_0) \in \mathfrak{T} \times \mathfrak{R}^n. \quad (6.10)$$

In this case $t_\infty = \sqrt{t_0^2 + \|\mathbf{Y}_0\|^{-2}} \rightarrow .t_0$ *as* $\|\mathbf{Y}_0\| \rightarrow \infty$. *The system solution escapes to infinity as time* t *approaches* t_∞, *but asymptotically converges to the zero vector as time* t *goes to infinity. As in Example 192, the zero equilibrium state* $\mathbf{S}_e = \mathbf{0}_{2N}$ *of the system (6.9) is unstable, but attractive in the whole. This proves the cardinal mistake if somebody claims that the attraction domain is also the stability domain, or even the asymptotic stability domain.*

6.6 Stability domain versus attraction domain

Some authors equalized the stability region (i.e., the stability domain) with the attraction region (i.e., with the attraction domain), or equalized the asymptotic stability region (i.e., the asymptotic stability domain) with the attraction region (i.e., with the attraction domain). It is the consequence of the ignorance of the stability domain concept and of the definitions of the stability domains. They will be given for *time*-varying nonlinear *EISO* systems in the sequel. They hold simultaneously for *IO*, *ISO*, *HISO* and *IIO* *time*-varying nonlinear systems.

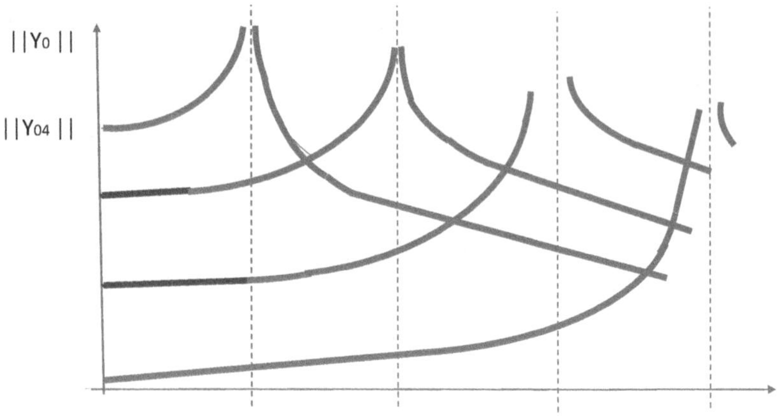

Figure 6.5: Unstable equilibrium state, but globally attractive. In the figure, $t_0 = 0$. The figure shows the motions (SolCntExl2) of the system (CntExl2).

Example [247, Example 2.5, pp. 30-32] and Example 189 show that the attraction domain $\mathcal{D}_A(0)$ of the zero equilibrium state is the proper subset of its stability domain $\mathcal{D}_S(0)$, $\mathcal{D}_A(0) \subset \mathcal{D}_S(0)$.

Since the Lyapunov stability theory has been established so far directly in the framework of the ISO dynamical systems only, the same holds for the stability domains theory. We will broaden them herein directly to the $EISO$ dynamical nonlinear systems because they incorporate IO, ISO, $HISO$ and IIO systems.

Remark 194 ***Substantial mistake, catastrophic consequence***
The acceptance of the fundamentally and conceptually wrong claim that the attraction domain is also the (asymptotic) stability domain is a substantial mistake with possible catastrophic consequences (Example 189, Example 192 and Example 193).

Note 195 *Example [247, Example 2.5, pp. 30-32], Example 187, Example 189, Example 192, and Example 193 clarify the necessity to define rigorously, exactly, and clearly the stability domain, the attraction domain, and the asymptotic stability domain.*

Problem 186, Remark 194, and Note 195 emphasize the importance of the following fundamental stability task:

Task 196 ***Fundamental stability task***
Determine the necessary and sufficient conditions:
1. for asymptotic stability of an equilibrium state, or of a desired motion, or of a set, and
2. for a set to be the asymptotic stability domain of an equilibrium state, or of a desired motion, or of a set.

A goal of this book is to present solutions to Problem 186 and to achieve Task 196 in the framework of the very large class of the $EISO$ *time*-varying nonlinear systems, which incorporate the other four classes of *time*-varying nonlinear systems.

Part V

STABILITY AND ITS DOMAINS: DEFINITIONS

Chapter 7

Stability and domains

7.1 Introduction

The chapter presents the extension and the generalization of the Lyapunov stability concept of the zero equilibrium state to stability of any equilibrium state and to stability of a *time*-invariant set all in the framework of general nonlinear time-varying systems.

Stability of the zero equilibrium state is a special case of stability of any equilibrium state, which is further a special case of stability of a *time*-invariant set. The study of stability of the zero equilibrium state and of any equilibrium state will be imbedded in the study of a *time*-invariant set stability.

We recall Note 182, Section 5.6.

7.2 Characteristics of the set $\mathcal{I}^+$

A special system constant set $\mathfrak{S}$ denoted by $\mathcal{I}^+$, (see Definition 130 in Section 4.5),

$$\mathfrak{S} = \mathcal{I}^+, \tag{7.1}$$

has the following characteristics:

Property 197 ***Properties of the time-invariant set*** $\mathcal{I}^+$

The time-invariant, i.e., constant, set $\mathcal{I}^+ \subset \mathfrak{R}^n$*, the stability properties of which will be studied, satisfies the following:*

1) It is a positively invariant set (Definition 130, Section 4.5) relative to the motions of the system (5.22), (5.23).

2) It is a connected compact nonempty set with a nonempty boundary $\partial\mathcal{I}^+$*,*

$$Cl\mathcal{I}^+ = \mathcal{I}^+ \neq \phi, \ \partial\mathcal{I}^+ \neq \phi. \tag{7.2}$$

In this regard see Note 155 in Section 4.11.

7.3 Stability of a constant set

Let us remember (2.5) (Section 2.2) that the *time* set $\mathfrak{T}_0$ of any current instant $t \geq t_0$ is a subset of the set $\mathfrak{T}_i$ of all initial moments t_0,

$$\mathfrak{T}_0 \subseteq \mathfrak{T}_i. \tag{7.3}$$

DOI: 10.1201/9781003744528-12

Let $\mathcal{D}(t_0)$ be a connected nonempty domain at every $t_0 \in \mathfrak{T}_i$. Let $\mathcal{D}(\mathfrak{T}_i)$ and $\mathcal{D}(t;\mathfrak{T}_i)$ be defined as follows:

$$\mathcal{D}(\mathfrak{T}_i) = \mathcal{D}_m(\mathfrak{T}_i) = \cap [\mathcal{D}(t_0) : t_0 \in \mathfrak{T}_i] \Longleftrightarrow \cap [\mathcal{D}(t_0) : t_0 \in \mathfrak{T}_i] \neq \phi,$$
$$\mathcal{D}(t;\mathfrak{T}_i) = \{\mathbf{S} = \mathbf{S}(t) = \mathbf{S}(t;t_0;\mathbf{S}_0) : \ (t_0,\mathbf{S}_0) \in \mathfrak{T}_i \times \mathcal{D}(\mathfrak{T}_i)\}, \ t \in \mathfrak{T}_0. \tag{7.4}$$

$\mathcal{D}(t;\mathfrak{T}_i)$ is essentially different from the *time evolution* $Ev_{\mathfrak{T}_i}\mathcal{D}(t)$ of the set $\mathcal{D}(t)$ in t over the *time* set $\mathfrak{T}_i$ (3.46), Section 3.2,

$$Ev_{\mathfrak{T}_i}\mathcal{D}(t) = \{(t,\mathbf{S}) : \ (t,\mathbf{S}) \in \mathfrak{T}_i \times \mathcal{D}(t)\} \subseteq \mathfrak{T}_i \times \mathfrak{R}^n. \tag{7.5}$$

Note 198 ***Initial set value, time-invariant set and time-varying system***
In view of Definition 62, Equations (3.23), Section 3.2, and Definition 130 in Section 4.5, the notation $\mathcal{I}^+ = \mathcal{I}^+(t_0)$ denotes the initial set value of the time-invariant set $\mathcal{I}^+$, which is the set $\mathcal{I}^+$ itself, $\mathcal{I}^+(t_0) \equiv \mathcal{I}^+$. The initial set value of any time-invariant, i.e., constant, set is independent of the initial instant t_0. However, this does not hold, due to the time variations of the system itself, for stability domains $\mathcal{D}_{..}(t_0;\mathcal{I}^+)$ of the set $\mathcal{I}^+$ of a time-varying system.

The letter $\mathcal{I}^+$ in the domains notation $\mathcal{D}_{..}(t;\mathcal{I}^+)$ specifies symbolically that the domains are related to the set $\mathcal{I}^+$ of a time-varying system.

Note 199 ***Stability domain, attraction domain and asymptotic stability domain***
*Zubov [541] defined the **attraction region**. It has been well-known and widely accepted. In what follows, the study of the **attraction domain** occupies our interest. For more details on it, see Section 6.2 and Section 6.6.*

*Neither Lyapunov [335] nor Zubov [541] defined either the **stability domain** or the **asymptotic stability domain**.*

Some authors wrongly call the attraction domain also the stability domain (stability region) or the asymptotic stability domain (asymptotic stability region). It is severely wrong. Stability and attraction are qualitatively different dynamical properties. The attraction domain does not reflect the stability property and the stability domain does not express the attraction property (Chapter 10).

For example, let the system equilibrium state be asymptotically stable. Let the system have an unstable limit cycle around the equilibrium state. Its stability domain is its neighborhood composed of all system states in the interior of the limit cycle and on the limit cycle. The attraction domain of the equilibrium state is only the interior of the limit cycle. The attraction domain is a proper subset of the stability domain and equals the asymptotic stability domain. The stability domain is a compact set. The attraction domain is an open set in this example.

Without precisely and adequately defining the stability domain, it is impossible to define the asymptotic stability domain. The stability domain should reflect the essence of the stability property. Then, in view of the sense of the asymptotic stability, its domain is to be the intersection of the stability domain and the attraction domain. The system possesses the asymptotic stability property if and only if the upper intersection 7.4 is a nonempty connected set that is a neighborhood of the corresponding set or state. At the author's knowledge, it was first publicly recognized in 1975 in [115].

The attraction domain itself is not the asymptotic stability domain in general.

Let us generalize the meaning and the property of the stability domain explained briefly in Comment 174, Section 5.4, and in Property 188, Section 6.4:

Property 200 ***Meaning and property of the stability domain of the set $\mathcal{I}^+$***

The stability domain $\mathcal{D}_S(t_0;\mathcal{I}^+)$ ***of the set*** $\mathcal{I}^+$ ***at*** $t_0 \in \mathfrak{T}_i$ *is the largest connected neighborhood of the set* $\mathcal{I}^+$ *at* $t_0 \in \mathfrak{T}_i$ *such that for every* $\varepsilon \in \mathfrak{R}^+$*, which determines the connected* $\varepsilon-$*neighborhood* $\mathcal{N}_\varepsilon(\mathcal{I}^+)$ *of the set* $\mathcal{I}^+$ *at* $t_0 \in \mathfrak{T}_i$*, there is a connected subset* $\mathcal{N}(\varepsilon;t_0;\mathcal{I}^+)$ *of both* $\mathcal{D}_S(t_0;\mathcal{I}^+)$ *and* $\mathcal{N}_\varepsilon(\mathcal{I}^+)$ *at* $t_0 \in \mathfrak{T}_i$*,*

$$\mathcal{N}(\varepsilon;t_0;\mathcal{I}^+) \subseteq \left[\mathcal{D}_S(t_0;\mathcal{I}^+) \cap \mathcal{N}_\varepsilon(\mathcal{I}^+)\right],\ t_0 \in \mathfrak{T}_i,$$

which is a neighborhood of the set $\mathcal{I}^+$*, with the property that for every* $\mathbf{S}_0 \in \mathcal{N}(\varepsilon;t_0;\mathcal{I}^+)$ *the system motion* $\mathcal{S}(.;t_0;\mathbf{S}_0)$ *is always in* $\mathcal{N}_\varepsilon(\mathcal{I}^+)$ *since* t_0*,*

$$\begin{gathered} t_0 \in \mathfrak{T}_i,\ \forall\varepsilon \in \mathfrak{R}^+,\ \exists\mathcal{N}(\varepsilon;t_0;\mathcal{I}^+) \subseteq \mathfrak{R}^n,\ Cl\mathcal{I}^+ \subset In\mathcal{N}(\varepsilon;t_0;\mathcal{I}^+) : \\ \mathbf{S}_0 \in \mathcal{N}(\varepsilon;t_0;\mathcal{I}^+) \Longrightarrow \mathbf{S}(t;t_0;\mathbf{S}_0) \in \mathcal{N}_\varepsilon(\mathcal{I}^+),\ \forall t \in \mathfrak{T}_0 \Longrightarrow \\ \mathcal{N}(\varepsilon;t_0;\mathcal{I}^+) \subseteq \mathcal{D}_S(t_0;\mathcal{I}^+) \cap \mathcal{N}_\varepsilon(\mathcal{I}^+). \end{gathered}$$

The stability domain $\mathcal{D}_S(t_0;\mathcal{I}^+)$ ***of the set*** $\mathcal{I}^+$ ***at*** $t_0 \in \mathfrak{T}_i$ *is the largest connected neighborhood* $\mathcal{N}(\varepsilon;t_0;\mathcal{I}^+)$ *of* $\mathcal{I}^+$ *in* $\varepsilon \in \mathfrak{R}^+$ *at* $t_0 \in \mathfrak{T}_i$ *with such property, or, more precisely:*

$$\mathcal{D}_S(t_0;\mathcal{I}^+) = \mathcal{N}_M(\varepsilon;t_0;\mathcal{I}^+) = \cup\left[\mathcal{N}(\varepsilon;t_0;\mathcal{I}^+) :\ \varepsilon \in \mathfrak{R}^+.\right]$$

This holds also in the case when the set $\mathcal{I}^+$ *is the constant singleton set* $\mathcal{I}^+$*, which can be a singleton* $\mathcal{I}^+ = \{\mathbf{S}_E\}$ *containing any equilibrium state* $\mathbf{S}_E$ *or the singleton* $\mathcal{I}^+ = \{\mathbf{0}_n\}$*.*

A *time*-invariant compact connected nonempty set $\mathcal{I}^+$ can be the singleton $\mathcal{I}^+$, the single element of which is an equilibrium state $\mathbf{S}_E \in \mathfrak{R}^n$,

$$\mathcal{I}^+ = \{\mathbf{S}_E\} \subset \mathfrak{R}^n.$$

The stability properties of the set $\mathcal{I}^+$ are then the stability properties of the equilibrium state $\mathbf{S}_E$. The equilibrium state can be at the origin: $\mathbf{S}_E = \mathbf{0}_n$. Then $\mathcal{I}^+ = \{\mathbf{0}_n\} = \mathfrak{O}$. The set $\mathcal{I}^+$ can be, for example, an infinite set. All its elements can be equilibrium states, or it can be a limit cycle. In the former case it is an equilibrium set containing an infinite number of equilibrium states. Its stability properties are then the stability properties of the equilibrium set. In the latter case, its stability properties are those of the limit cycle.

Property 200 is valid in a generalized form for all these cases and for both *time*-varying systems and sets $\mathcal{I}^+(t;t_0)$.

We embed stability studies of an equilibrium state, of an equilibrium set and of the limit cycle in the study of the stability properties of the set $\mathcal{I}^+ \subset \mathfrak{R}^n$. Its $\varepsilon-$neighborhood $\mathcal{N}_\varepsilon(\mathcal{I}^+)$ is constant, connected and nonempty (Comment 56, Section 3.1.2).

Definition 201 ***Stability of the set*** $\mathcal{I}^+$ ***of the system (5.22), (5.23)***

a) ***The set*** $\mathcal{I}^+$ ***of the system (5.22), (5.23) is stable at*** $t_0 \in \mathfrak{T}_i$ *if and only if 1) to 3) are satisfied:*

1) for every $\varepsilon \in \mathfrak{R}^+$*, which determines the constant connected* $\varepsilon-$*neighborhood* $\mathcal{N}_\varepsilon(\mathcal{I}^+) \subseteq \mathfrak{R}^n$ *of the set* $\mathcal{I}^+$*, there is a connected nonempty neighborhood* $\mathcal{N}(\varepsilon;t_0;\mathcal{I}^+)$ *of the set* $\mathcal{I}^+$*, such that:*

2) it is nondecreasing in ε *and its closure* $Cl\mathcal{N}(\varepsilon;t_0;\mathcal{I}^+)$ *converges to* $\mathcal{I}^+$ *as* $\varepsilon \longrightarrow 0^+$*:*

$$\begin{gathered} t_0 \in \mathfrak{T}_i,\ \forall\varepsilon \in \mathfrak{R}^+,\ \mathcal{N}_\varepsilon(\mathcal{I}^+) \subseteq \mathfrak{R}^n \Longrightarrow \exists\mathcal{N}(\varepsilon;t_0;\mathcal{I}^+) \subseteq \mathfrak{R}^n, \\ Cl\mathcal{I}^+ \subset In\mathcal{N}(\varepsilon;t_0;\mathcal{I}^+),\ \mathcal{N}(\varepsilon;t_0;\mathcal{I}^+) \subseteq \mathcal{N}_\varepsilon(\mathcal{I}^+), \\ 0 < \varepsilon_1 < \varepsilon_2 \Longrightarrow \mathcal{N}(\varepsilon_1;t_0;\mathcal{I}^+) \subseteq \mathcal{N}(\varepsilon_2;t_0;\mathcal{I}^+), \\ \varepsilon \longrightarrow 0^+ \Longrightarrow Cl\mathcal{N}(\varepsilon;t_0;\mathcal{I}^+) \longrightarrow \mathcal{I}^+, \end{gathered} \tag{7.6}$$

and

3) for every initial state vector $\mathbf{S}_0$ from $\mathcal{N}(\varepsilon; t_0; \mathcal{I}^+)$, (7.6), the state vector $\mathbf{S}(t; t_0; \mathbf{S}_0)$ stays in the $\varepsilon-$neighborhood $\mathcal{N}_\varepsilon(\mathcal{I}^+)$ of $\mathcal{I}^+$ at every moment $t \in \mathfrak{T}_0$, i.e.,

$$\forall\varepsilon \in \mathfrak{R}^+,\ \mathcal{N}_\varepsilon(\mathcal{I}^+) \subseteq \mathfrak{R}^n,\ t_0 \in \mathfrak{T}_i,\ \exists\mathcal{N}(\varepsilon; t_0; \mathcal{I}^+) \subseteq \mathfrak{R}^n,\ (7.6),$$
$$\forall\mathbf{S}_0 \in \mathcal{N}(\varepsilon; t_0; \mathcal{I}^+) \Longrightarrow \mathbf{S}(t; t_0; \mathbf{S}_0) \in \mathcal{N}_\varepsilon(\mathcal{I}^+),\ \forall t \in \mathfrak{T}_0. \tag{7.7}$$

The set $\mathcal{I}^+$ of the system (5.22), (5.23) is globally (in the whole) stable at *$t_0 \in \mathfrak{T}_i$ if and only if it is stable at $t_0 \in \mathfrak{T}_i$ and $\mathcal{N}(\varepsilon; t_0; \mathcal{I}^+) \longrightarrow \mathfrak{R}^n$ as $\varepsilon \longrightarrow \infty$.*

The set $\mathcal{I}^+$ of the system (5.22), (5.23) is stable on *$\mathfrak{T}_i$ if and only if it is stable at every $t_0 \in \mathfrak{T}_i$.*

The set $\mathcal{I}^+$ of the system (5.22), (5.23) is globally (in the whole) stable on *$\mathfrak{T}_i$ if and only if it is globally stable at every $t_0 \in \mathfrak{T}_i$.*

If and only if this holds when $\mathfrak{T}_i$ is replaced by $\mathfrak{T}$, then the set $\mathcal{I}^+$ of the system (5.22), (5.23) is ***stable.***

This definition holds for the invariant set $\mathcal{I}$ of the system ***(5.22), (5.23)*** *if and only if the set $\mathcal{I}$ replaces the set $\mathcal{I}^+$ everywhere.*

Note 202 *Notice that the third condition and the fourth condition in 7.6 imply the following:*

$$\varepsilon \in \mathfrak{R}^+,\ \cup\left\{\mathcal{N}(\eta; t_0; \mathcal{I}^+) : \eta \in [0, \varepsilon]\right\} = \mathcal{N}(\varepsilon; t_0; \mathcal{I}^+), \tag{7.8}$$
$$\varepsilon \in \mathfrak{R}^+,\ \cap\left\{\mathcal{N}(\eta; t_0; \mathcal{I}^+) : \eta \in]0, \varepsilon]\right\} = \mathcal{I}^+. \tag{7.9}$$

This explains the use of the operator $\cup$ rather than the operator $\cap$ in the next Definition 204 of the stability domain: Equation 7.12.

Comment 203 *Let the set $\mathcal{I}^+$ be an equilibrium set $\mathcal{I}_E^+$, $\mathcal{I}^+ = \mathcal{I}_E^+ = \{\mathbf{S}_E\}$. The neighborhood $\mathcal{N}_\varepsilon(\mathcal{I}_E^+)$ may be accepted to be the "hyperball" $\mathfrak{B}_\varepsilon(\mathcal{I}_E^+) = \left\{\mathbf{S} : d(\mathbf{S}, \mathcal{I}_E^+) < \boldsymbol{\varepsilon}\right\}$. Then the neighborhood $\mathcal{N}(\varepsilon; t_0; \mathcal{I}_E^+)$ is also accepted in the form of a "hyperball" $\mathfrak{B}_\delta(\mathcal{I}_E^+) = \left\{\mathbf{S} :\ d(\mathbf{S}, \mathcal{I}_E^+) < \delta\right\}$ with $\boldsymbol{\delta} \in \mathfrak{R}^+$ dependent in general on t_0 and ε, $\boldsymbol{\delta} = \boldsymbol{\delta}(t_0, \varepsilon)$. Such restricted choice of the form, of the shape, of the neighborhoods $\mathcal{N}_\varepsilon(\mathcal{I}_E^+)$ and $\mathcal{N}(\varepsilon; t_0; \mathcal{I}_E^+)$ prevents us from defining the stability domains. They need not be in the form of a "hyperball" in general.*

7.4 Stability domains

Definition of stability domains of the set $\mathcal{I}^+$ results from Definition 201.

Definition 204 ***Stability domains of the set $\mathcal{I}^+$ of the system (5.22), (5.23)***

The largest connected neighborhood $\mathcal{N}(\varepsilon; t_0; \mathcal{I}^+)$ satisfying Definition (201) at $t_0 \in \mathfrak{T}_i$ of the set $\mathcal{I}^+$ denoted by $\mathcal{N}_L(\varepsilon; t_0; \mathcal{I}^+)$ for any chosen $\varepsilon \in \mathfrak{R}^+$ so that

$$\left[\mathbf{S}(t; t_0; \mathbf{S}_0) \in \mathcal{N}_\varepsilon(\mathcal{I}^+),\ \forall t \in \mathfrak{T}_0\right] \Longleftrightarrow \left[\mathbf{S}_0 \in \mathcal{N}_L(\varepsilon; t_0; \mathcal{I}^+)\right],\ \forall\varepsilon \in \mathfrak{R}^+, \tag{7.10}$$

is ***the $\varepsilon-$stability domain $\mathcal{D}_S(\varepsilon; t_0; \mathcal{I}^+)$ of the set $\mathcal{I}^+$ at the initial moment*** *$t_0 \in \mathfrak{T}_i$,*

$$\mathcal{D}_S(\varepsilon; t_0; \mathcal{I}^+) = \mathcal{N}_L(\varepsilon; t_0; \mathcal{I}^+),\ \forall\varepsilon \in \mathfrak{R}^+,\ t_0 \in \mathfrak{T}_i. \tag{7.11}$$

The stability domain $\mathcal{D}_S(t_0; \mathcal{I}^+)$ of the set $\mathcal{I}^+$ at the initial moment *t_0 is the union of all $\varepsilon-$stability domains $\mathcal{D}_S(\varepsilon; t_0; \mathcal{I}^+)$ in $\varepsilon \in \mathfrak{R}^+$ at the initial moment t_0,*

$$\mathcal{D}_S(t_0; \mathcal{I}^+) = \cup\left[\mathcal{D}_S(\varepsilon; t_0; \mathcal{I}^+) : \varepsilon \in \mathfrak{R}^+\right],\ t_0 \in \mathfrak{T}_i, \tag{7.12}$$

see Figure 7.1.

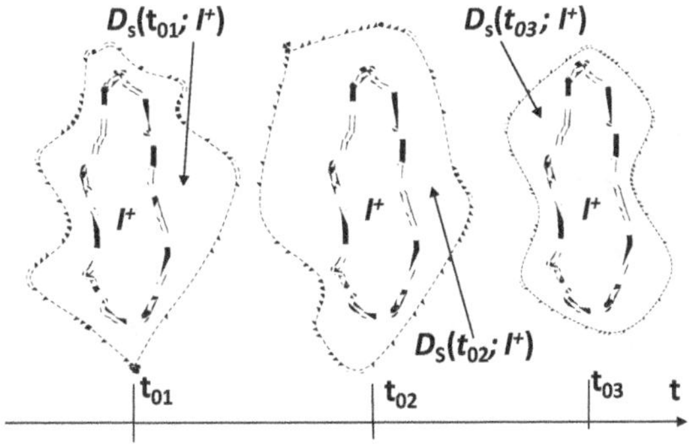

Figure 7.1: Stability domain $\mathcal{D}_S\left(t_0; \mathcal{I}^+\right)$ of the set $\mathcal{I}^+$ at different initial moments t_0: t_{01}, t_{02} and t_{03}. D stands for $\mathcal{D}$ and I^+ for $\mathcal{I}^+$ in this figure.

*$\mathcal{D}_S\left(t; t_0; \mathcal{I}^+\right)$ is **the time development of the stability domain** $\mathcal{D}_S\left(t_0; \mathcal{I}^+\right)$ **along the system motions** $\mathcal{S}(.; t_0; \mathbf{S}_0)$ **at any moment** $t \in \mathfrak{T}_0$,*

$$\mathcal{D}_S\left(t; t_0; \mathcal{I}^+\right) = \left\{\mathbf{S} = \mathbf{S}(t) = \mathbf{S}(t; t_0; \mathbf{S}_0) : \ \mathbf{S}_0 \in \mathcal{D}_S\left(t_0; \mathcal{I}^+\right)\right\}, \ t \in \mathfrak{T}_0 \Longrightarrow \tag{7.13}$$

$$\mathcal{D}_S\left(t_0; t_0; \mathcal{I}^+\right) = \mathcal{D}_S\left(t_0; \mathcal{I}^+\right), \ t_0 \in \mathfrak{T}_i. \tag{7.14}$$

***The stability domain** $\mathcal{D}_S\left(t; \mathfrak{T}_i; \mathcal{I}^+\right)$ **of the set** $\mathcal{I}^+$ **on** $\mathfrak{T}_i$ exists if and only if the following conditions hold:*

1) The set $\mathcal{I}^+$ is stable on $\mathfrak{T}_i$.

2) $\mathcal{D}_S\left(t; \mathfrak{T}_i; \mathcal{I}^+\right)$ is a continuous connected neighborhood of the set $\mathcal{I}^+$ on $\mathfrak{T}_0$ for all $t_0 \in \mathfrak{T}_i$ such that it is equal to the stability domain $\mathcal{D}_S\left(t_0; \mathcal{I}^+\right)$ at every $t = t_0 \in \mathfrak{T}_i$,

$$\mathcal{D}_S\left(t; \mathfrak{T}_i; \mathcal{I}^+\right) \in \mathfrak{C}\left(\mathfrak{T}_0 \times \mathfrak{T}_i\right), \ Cl\mathcal{I}^+ \subset In\mathcal{D}_S\left(t; \mathfrak{T}_i; \mathcal{I}^+\right), \ \forall (t, t_0) \in \mathfrak{T}_0 \times \mathfrak{T}_i,$$

$$\mathcal{D}_S\left(t_0; \mathfrak{T}_i; \mathcal{I}^+\right) = \mathcal{D}_S\left(t_0; \mathcal{I}^+\right), \ \forall t_0 \in \mathfrak{T}_i. \tag{7.15}$$

*This is **the state space stability domain** $\mathcal{D}_S\left(t; \mathfrak{T}_i; \mathcal{I}^+\right)$ **of the set** $\mathcal{I}^+$ **on** $\mathfrak{T}_0 \times \mathfrak{T}_i$; see Figure 7.2.*

***The integral space stability domain** $\mathbb{I}_S\left(\mathfrak{T}_i; \mathcal{I}^+\right)$ **of the set** $\mathcal{I}^+$ **on** $\mathfrak{T}_i$ is the time evolution $Ev_{\mathfrak{T}_i}\mathcal{D}_S\left(t_0; \mathcal{I}^+\right)$ of $\mathcal{D}_S\left(t_0; \mathcal{I}^+\right)$ in t_0 over $\mathfrak{T}_i$,*

$$\mathbb{I}_S\left(\mathfrak{T}_i; \mathcal{I}^+\right) = Ev_{\mathfrak{T}_i}\mathcal{D}_S\left(t_0; \mathcal{I}^+\right) = \left\{(t_0, \mathbf{S}) : (t_0, \mathbf{S}) \in \mathfrak{T}_i \times \mathcal{D}_S\left(t_0; \mathcal{I}^+\right)\right\} \subseteq \mathfrak{T}_i \times \mathfrak{R}^n, \tag{7.16}$$

see Figure 7.3.

***The stability of** $\mathcal{I}^+$ **at** $t_0 \in \mathfrak{T}_i$ **is global [in the whole]** if and only if the set $\mathcal{I}^+$ is stable and*

$$\varepsilon \to \infty \Longrightarrow \mathcal{D}_S\left(\varepsilon; t_0; \mathcal{I}^+\right) \longrightarrow \mathfrak{R}^n.$$

*Equivalently, **the stability of** $\mathcal{I}^+$ **at** $t_0 \in \mathfrak{T}_i$ **is global [in the whole]** if and only if the set $\mathcal{I}^+$ is stable and the stability domain $\mathcal{D}_S\left(t_0; \mathcal{I}^+\right)$ of $\mathcal{I}^+$ at $t_0 \in \mathfrak{T}_i$ is the whole state space $\mathfrak{R}^n$:*

$$\mathcal{D}_S\left(t_0; \mathcal{I}^+\right) = \mathfrak{R}^n, \ t_0 \in \mathfrak{T}_i.$$

***Stability of the set** $\mathcal{I}^+$ **is global [in the whole] on** $\mathfrak{T}_i$ if and only if the set $\mathcal{I}^+$ is stable on $\mathfrak{T}_i$ and the stability domain $\mathcal{D}_S\left(t_0; \mathcal{I}^+\right)$ of the set $\mathcal{I}^+$ is the whole state space $\mathfrak{R}^n$ for all $t_0 \in \mathfrak{T}_i$,*

$$\mathcal{D}_S\left(t_0; \mathcal{I}^+\right) = \mathfrak{R}^n, \ \forall t_0 \in \mathfrak{T}_i.$$

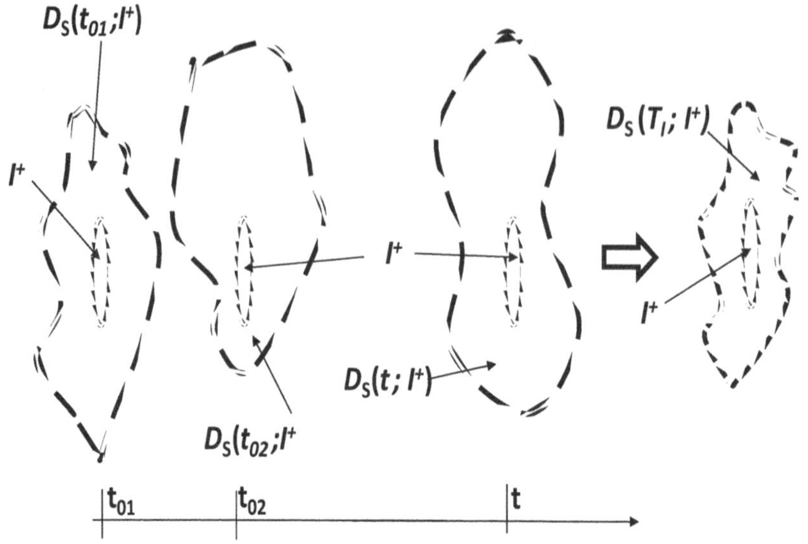

Figure 7.2: The state space stability domain $\mathcal{D}_S\left(t;\mathfrak{T}_i;\mathcal{I}^+\right)$ of the set $\mathcal{I}^+$ on $\mathfrak{T}_i$. D, I^+ and T stand for $\mathcal{D}$, $\mathcal{I}^+$ and $\mathfrak{T}$.

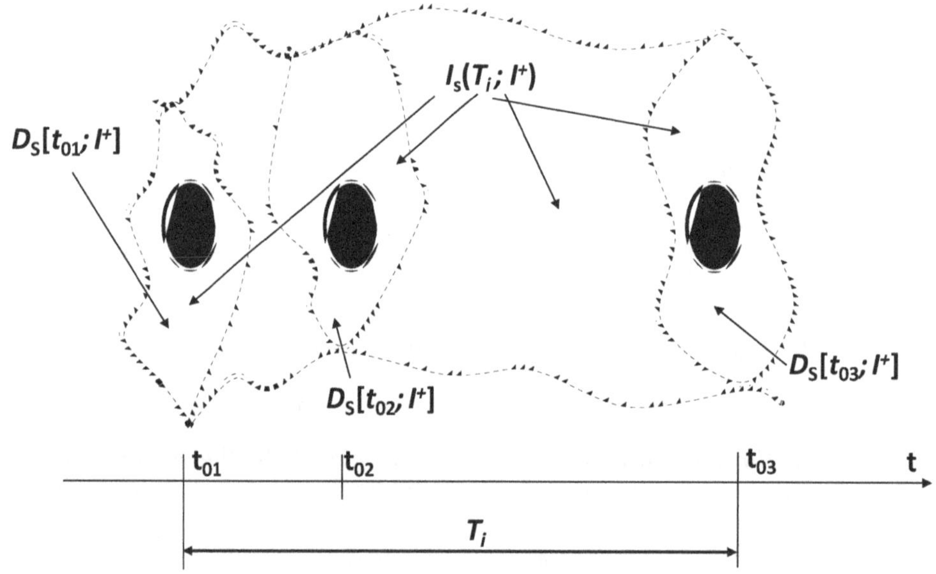

Figure 7.3: The integral space stability domain $\mathbb{I}_S\left(\mathfrak{T}_i;\mathfrak{I}^+\right)$ of the set $\mathcal{I}^+$ on $\mathfrak{T}_i$.

If and only if the above conditions hold on $\mathfrak{T}$ *instead of only on* $\mathfrak{T}_i$*, then the term* ***"on*** $\mathfrak{T}_i$*" is to be omitted.*

This definition holds for the invariant set $\mathcal{I}$ *of the system* ***(5.22), (5.23)*** *if and only if the set* $\mathcal{I}$ *replaces* $\mathcal{I}^+$ *everywhere.*

Comment 205 ***On the largest connected neighborhood*** $\mathcal{N}_L(\varepsilon; t_0; \mathcal{I}^+)$ *and the* $\varepsilon-$***stability domain*** $\mathcal{D}_S(\varepsilon; t_0; \mathcal{I}^+)$

The meaning of the largest connected neighborhood (7.10) of the set $\mathcal{I}^+$*, equivalently of the* $\varepsilon-$***stability domain*** $\mathcal{D}_S(\varepsilon; t_0; \mathcal{I}^+)$*, (7.11), is that for the chosen and fixed* $\varepsilon \in \mathfrak{R}^+$ *the stability condition* $\mathbf{S}(t; t_0; \mathbf{S}_0) \in \mathcal{N}_\varepsilon(\mathcal{I}^+)$, $\forall t \in \mathfrak{T}_0$*, is satisfied if and only if* $\mathbf{S}_0 \in \mathcal{N}_L(\varepsilon; t_0; \mathcal{I}^+)$*, i.e., if and only if* $\mathbf{S}_0 \in \mathcal{D}_S(\varepsilon; t_0; \mathcal{I}^+)$ *due to Equation (7.11).*

7.5 Strict stability domains

Definition 206 ***Strict stability domain of the set*** $\mathcal{I}^+$ ***of the system (5.22), (5.23)***

Let $[0, \varepsilon_M)$ *be the maximal interval over which* $\mathcal{D}_S(\varepsilon; t_0; \mathcal{I}^+)$ *is continuous in* $\varepsilon \in \mathfrak{R}_+$,

$$\mathcal{D}_S(\varepsilon; t_0; \mathcal{I}^+) \in \mathfrak{C}\langle [0, \varepsilon_M) \rangle. \tag{7.17}$$

The strict stability domain $\mathcal{D}_{SS}(t_0; \mathcal{I}^+)$ ***of*** $\mathcal{I}^+$ ***at the initial moment*** t_0 *is the union over* $\varepsilon \in [0, \varepsilon_M)$ *of all* $\mathcal{D}_S(\varepsilon; t_0; \mathcal{I}^+)$,

$$\mathcal{D}_{SS}(t_0; \mathcal{I}^+) = \cup \{\mathcal{D}_S(\varepsilon; t_0; \mathcal{I}^+) : \varepsilon \in [0, \varepsilon_M)\} = \mathcal{D}_{SS}(\varepsilon_M; t_0; \mathcal{I}^+). \tag{7.18}$$

$\mathcal{D}_{SS}(t; t_0; \mathcal{I}^+)$ *is* ***the time development of the strict stability domain*** $\mathcal{D}_{SS}(t_0; \mathcal{I}^+)$ ***along the system motions*** $\mathcal{S}(.; t_0; \mathbf{S}_0)$ ***at any moment*** $t \in \mathfrak{T}_0$,

$$\mathcal{D}_{SS}(t; t_0; \mathcal{I}^+) = \{\mathbf{S} : \mathbf{S} = \mathbf{S}(t) = \mathbf{S}(t; t_0; \mathbf{S}_0),\ \mathbf{S}_0 \in \mathcal{D}_{SS}(t_0; \mathcal{I}^+)\},\ \forall t \in \mathfrak{T}_0,$$
$$\mathcal{D}_{SS}(t_0; t_0; \mathcal{I}^+) = \mathcal{D}_{SS}(t_0; \mathcal{I}^+) \tag{7.19}$$

The strict stability domain $\mathcal{D}_{SS}(t; \mathfrak{T}_i; \mathcal{I}^+)$ ***of the set*** $\mathcal{I}^+$ ***on*** $\mathfrak{T}_i$ *exists if and only if the following conditions hold:*

1) The set $\mathcal{I}^+$ *is strictly stable on* $\mathfrak{T}_i$.

2) $\mathcal{D}_{SS}(t; \mathfrak{T}_i; \mathcal{I}^+)$ *is a continuous connected neighborhood of the set* $\mathcal{I}^+$ *on* $\mathfrak{T}_0$ *for all* $t_0 \in \mathfrak{T}_i$ *such that it is equal to the strict stability domain* $\mathcal{D}_{SS}(t_0; \mathcal{I}^+)$ *at every* $t_0 \in \mathfrak{T}_i$,

$$\mathcal{D}_{SS}(t; \mathfrak{T}_i; \mathcal{I}^+) \in \mathfrak{C}(\mathfrak{T}_i),\ Cl\mathcal{I}^+ \subset In\mathcal{D}_{SS}(t; \mathfrak{T}_i; \mathcal{I}^+),\ \forall (t, t_0) \in \mathfrak{T}_0 \times \mathfrak{T}_i,$$
$$\mathcal{D}_{SS}(t_0; \mathfrak{T}_i; \mathcal{I}^+) = \mathcal{D}_{SS}(t_0; \mathcal{I}^+),\ \forall t_0 \in \mathfrak{T}_i. \tag{7.20}$$

This is ***the state space strict stability domain*** $\mathcal{D}_{SS}(t; \mathfrak{T}_i; \mathcal{I}^+)$ ***of the set*** $\mathcal{I}^+$ ***on*** $\mathfrak{T}_0 \times \mathfrak{T}_i$.

The integral space strict stability domain $\mathbb{I}_{SS}(\mathfrak{T}_i; \mathcal{I}^+)$ ***of the set*** $\mathcal{I}^+$ ***on*** $\mathfrak{T}_0 \times \mathfrak{T}_i$ *is the time evolution* $Ev_{\mathfrak{T}_i}\mathcal{D}_{SS}(t_0; \mathcal{I}^+)$ *of* $\mathcal{D}_{SS}(t_0; \mathcal{I}^+)$ *in* t_0 *over* $\mathfrak{T}_i$,

$$\mathbb{I}_{SS}(\mathfrak{T}_i; \mathcal{I}^+) = Ev_{\mathfrak{T}_i}\mathcal{D}_{SS}(t_0; \mathcal{I}^+) = \{(t_0, \mathbf{S}) : (t_0, \mathbf{S}) \in \mathfrak{T}_i \times \mathcal{D}_{SS}(t_0; \mathcal{I}^+)\} \subseteq \mathfrak{T}_i \times \mathfrak{R}^n. \tag{7.21}$$

The strict stability of $\mathcal{I}^+$ ***at*** $t_0 \in \mathfrak{T}_i$ ***is global [in the whole]*** *if and only if* $\mathcal{I}^+$ *is strictly stable and*

$$\varepsilon \to \infty \Longrightarrow \mathcal{D}_{SS}(\varepsilon; t_0; \mathcal{I}^+) \longrightarrow \mathfrak{R}^n.$$

Equivalently, ***the strict stability of*** $\mathcal{I}^+$ ***at*** $t_0 \in \mathfrak{T}_i$ ***is global [in the whole]*** *if and only if the strict stability domain* $\mathcal{D}_{SS}(t_0; \mathcal{I}^+)$ *of the set* $\mathcal{I}^+$ *is the whole state space,*

$$\mathcal{D}_{SS}(t_0; \mathcal{I}^+) = \mathfrak{R}^n,\ t_0 \in \mathfrak{T}_i.$$

*The set $\mathcal{I}^+$ is **globally [in the whole] strictly stable on** $\mathfrak{T}_i$ if and only if the strict stability domain $\mathcal{D}_{SS}(t_0;\mathcal{I}^+)$ of the set $\mathcal{I}^+$ is the whole state space $\mathfrak{R}^n$ for all $t_0 \in \mathfrak{T}_i$,*

$$\mathcal{D}_{SS}\left(t_0;\mathcal{I}^+\right) = \mathfrak{R}^n, \ \forall t_0 \in \mathfrak{T}_i.$$

*If and only if the above conditions hold on $\mathfrak{T}$ instead of only on $\mathfrak{T}_i$, then the term "**on** $\mathfrak{T}_i$" is to be omitted.*

*This definition holds for the invariant set $\mathcal{I}$ of the system **(5.22), (5.23)** if and only if the set $\mathcal{I}$ replaces the set $\mathcal{I}^+$ everywhere.*

Note 207 *Definition 204 and Definition 206 refine Definition 2.1 and Definition 2.2 of [247, pp. 22, 23].*

7.6 Properties of the stability domains

Lemma 208 ***Positive invariance and continuity of the stability domains of the set $\mathcal{I}^+$ of the system (5.22), (5.23)***

a) If the set $\mathcal{I}^+$ of the system (5.22), (5.23) is stable at $t_0 \in \mathfrak{T}$ then it is also positively invariant on $\mathfrak{T}_0$ with respect to system motions, that is that

$$\mathbf{S}_0 \in \mathcal{I}^+, \ t_0 \in \mathfrak{T}_i \Longrightarrow \mathbf{S}(t;t_0;\mathbf{S}_0) \in \mathcal{I}^+, \ \forall t \in \mathfrak{T}_0. \tag{7.22}$$

If the set $\mathcal{I}^+$ of the system (5.22), (5.23) is stable on $\mathfrak{T}_i$ then it is also positively invariant on $\mathfrak{T}_0 \times \mathfrak{T}_i$.

b) If the set $\mathcal{I}^+$ of the system (5.22), (5.23) is stable on $\mathfrak{T}_i$ then its domain of stability $\mathcal{D}_S(t;\mathcal{I}^+)$ is a positively invariant set on $\mathfrak{T}_i$.

c) Let the set $\mathcal{I}^+$ of the system (5.22), (5.23) be stable at $t_0 \in \mathfrak{T}_i$. The time development $\mathcal{D}_S(t;t_0;\mathcal{I}^+)$ $/\mathcal{D}_{SS}(t;t_0;\mathcal{I}^+)/$ of the stability (strict stability) domain $\mathcal{D}_S(t_0;\mathcal{I}^+)$ $/\mathcal{D}_{SS}(t_0;\mathcal{I}^+)/$ of $\mathcal{I}^+$ of the system (5.22), (5.23) is positively invariant on $\mathfrak{T}_0$ with respect to system motions, that is that

$$t_0 \in \mathfrak{T}_i, \ \mathbf{S}_0 \in \mathcal{D}_S\left(t_0;\mathcal{I}^+\right) \ /t_0 \in \mathfrak{T}_i, \ \mathbf{S}_0 \in \mathcal{D}_{SS}\left(t_0;\mathcal{I}^+\right)/ \Longrightarrow$$
$$\mathbf{S}(t;t_0;\mathbf{S}_0) \in \mathcal{D}_S\left(t;t_0;\mathcal{I}^+\right) \ /\mathbf{S}(t;t_0;\mathbf{S}_0) \in \mathcal{D}_{SS}\left(t;t_0;\mathcal{I}^+\right)/, \ \forall t \in \mathfrak{T}_0, \tag{7.23}$$

respectively.

d) Let the set $\mathcal{I}^+$ of the system (5.22), (5.23) be stable at $t_0 \in \mathfrak{T}_i$. The time development $\mathcal{D}_S(t;t_0;\mathcal{I}^+)$ $/\mathcal{D}_{SS}(t;t_0;\mathcal{I}^+)/$ of the stability (strict stability) domain $\mathcal{D}_S(t_0;\mathcal{I}^+)$ $/\mathcal{D}_{SS}(t_0;\mathcal{I}^+)/$ is continuous in $t \in \mathfrak{T}_0$, respectively.

Proof. a) Let a nonempty *time*-invariant set $\mathcal{I}^+$ of the system (5.22), (5.23) be stable at $t_0 \in \mathfrak{T}$. Definition 201 holds. Let $\mathbf{S}_0 \in \mathcal{I}^+$, which implies $d(\mathbf{S}_0,\mathcal{I}^+) = 0$. Let it be assumed that the set $\mathcal{I}^+$ is not a positively invariant set of the system on $\mathfrak{T}_0$, i.e., that there is an instant $\tau \in In\mathfrak{T}_0$ at which

$$\mathbf{S}(\tau) = \mathbf{S}(\tau;t_0;\mathbf{S}_0) \notin \mathcal{I}^+. \tag{7.24}$$

This means that there is $\varepsilon \in \mathfrak{R}^+$ such that $d[\mathbf{S}(\tau),\mathcal{I}^+] = \varepsilon > 0$. Let $(\eta < \varepsilon) \in \mathfrak{R}^+$ be arbitrarily small and let $\mathcal{N}(\eta;t_0;\mathcal{I}^+)$ obey (7.7). Since $d(\mathbf{S}_0,\mathcal{I}^+) = 0$, then $\mathbf{S}_0 \in \mathcal{N}(\eta;t_0;\mathcal{I}^+)$, which guarantees $\mathbf{S}(t;t_0;\mathbf{S}_0) \in \mathcal{N}_\eta(\mathcal{I}^+)$, $\forall t \in \mathfrak{T}_0$, i.e., $d[\mathbf{S}(\tau),\mathcal{I}^+] < \eta < \varepsilon$, which contradicts $d[\mathbf{S}(\tau),\mathcal{I}^+] = \varepsilon$. The contradiction is the consequence of the assumption expressed by (7.24). The result is that the assumed $\tau \in In\mathfrak{T}_0$ does not exist, i.e., that $\mathbf{S}(t) = \mathbf{S}(t;t_0;\mathbf{S}_0) \in \mathcal{I}^+$ for every $t \in \mathfrak{T}_0$. The statement under a) of Definition 130,

(Section 4.5), is satisfied. The set $\mathcal{I}^+$ is a positively invariant set of the system on $\mathfrak{T}_0$. If the set $\mathcal{I}^+$ of the system (5.22), (5.23) is stable on $\mathfrak{T}_i$, then the preceding result is valid for every $t_0 \in \mathfrak{T}_i$, which proves the positive invariance of the set $\mathcal{I}^+$ of the system on $\mathfrak{T}_0 \times \mathfrak{T}_i$.

b) Let the set $\mathcal{I}^+$ of the system (5.22), (5.23) be stable on $\mathfrak{T}_i$. Let t_0 and τ belong to $\mathfrak{T}_i$, be arbitrary and $t_0 < \tau$. Let $\varepsilon \in \mathfrak{R}^+$ be arbitrary and $\mathbf{S}_0 \in \mathcal{D}_S(t_0;\varepsilon;\mathcal{I}^+)$ be also arbitrarily chosen. At the moment τ, the system state along the system motion is $\mathbf{S}_\tau = \mathbf{S}(\tau;\tau;\mathbf{S}_\tau)$ and $\mathbf{S}_\tau = \mathbf{S}(\tau;t_0;\mathbf{S}_0) = \mathbf{S}[\tau;\tau;\mathbf{S}(\tau;t_0;\mathbf{S}_0)]$. The choice $\mathbf{S}_0 \in \mathcal{D}_S(t_0;\varepsilon;\mathcal{I}^+)$ guarantees that $\mathbf{S}(\sigma;t_0;\mathbf{S}_0) \in \mathcal{N}_\varepsilon(\mathcal{I}^+)$ for every $(\sigma \geq t_0) \in \mathfrak{T}_i$ hence also for $\sigma \geq \tau$, which yields $\mathbf{S}_\sigma = \mathbf{S}(\sigma;t_0;\mathbf{S}_0) \in \mathcal{N}_\varepsilon(\mathcal{I}^+)$. This and $\mathbf{S}(\sigma;t_0;\mathbf{S}_0) = \mathbf{S}(\sigma;\tau;\mathbf{S}_\tau)$ imply $\mathbf{S}(\sigma;\tau;\mathbf{S}_\tau) \in \mathcal{N}_\varepsilon(\mathcal{I}^+)$ for every $(\sigma \geq t_0) \in \mathfrak{T}_i$, The system state $\mathbf{S}_\tau = \mathbf{S}(\tau;t_0;\mathbf{S}_0) \in \mathcal{D}_S(\tau;\varepsilon;\mathcal{I}^+)$ due to Definition 204, which together with $\mathcal{D}_S(\tau;\varepsilon;\mathcal{I}^+) \subset \mathcal{D}_S(\tau;\mathcal{I}^+)$ verifies $\mathbf{S}(\tau;t_0;\mathbf{S}_0) \in \mathcal{D}_S(\tau;\mathcal{I}^+)$, which proves the positive invariance of the stability domain $\mathcal{D}_S(t;\mathcal{I}^+)$ on $\mathfrak{T}_i$.

c) Let the set $\mathcal{I}^+$ of the system (5.22), (5.23) be stable at $t_0 \in \mathfrak{T}_i$. The set $\mathcal{I}^+$ has the stability /strict stability/ domain $\mathcal{D}_S(t_0;\mathcal{I}^+)$ $/\mathcal{D}_{SS}(t_0;\mathcal{I}^+)/$ at $t_0 \in \mathfrak{T}_i$. Let $\mathbf{S}_0 \in \mathcal{D}_S(t_0;\mathcal{I}^+)$ $/\mathbf{S}_0 \in \mathcal{D}_{SS}(t_0;\mathcal{I}^+)/$. This, (7.10)-(7.12) and (7.13) /(7.10), (7.18), (7.19)/ prove

$$t_0 \in \mathfrak{T}_i,\ \forall \mathbf{S}_0 \in \mathcal{D}_S(t_0;\mathcal{I}^+) \Longrightarrow \mathbf{S}(t;t_0;\mathbf{S}_0) \in \mathcal{D}_S(t;t_0;\mathcal{I}^+),\ \forall t \in \mathfrak{T}_0,$$
$$/t_0 \in \mathfrak{T}_i,\ \forall \mathbf{S}_0 \in \mathcal{D}_{SS}(t_0;\mathcal{I}^+) \Longrightarrow \mathbf{S}(t;t_0;\mathbf{S}_0) \in \mathcal{D}_{SS}(t;t_0;\mathcal{I}^+),\ \forall t \in \mathfrak{T}_0/,$$

respectively. This proves the statement under c) i.e., it proves (7.23).

d) Let the set $\mathcal{I}^+$ of the system (5.22), (5.23) be stable at $t_0 \in \mathfrak{T}_i$. Continuity of the system's motions (Property 147 and Property 148, Note 149, Section 4.11), and positive invariance of $\mathcal{D}_S(t;t_0;\mathcal{I}^+)$ $/\mathcal{D}_{SS}(t;t_0;\mathcal{I}^+)/$ on $\mathfrak{T}_0$ prove the statement under d), respectively. Let us show it in detail. Let us assume that $\mathcal{D}_S(t;t_0;\mathcal{I}^+)$ $/\mathcal{D}_{SS}(t;t_0;\mathcal{I}^+)/$ is not continuous in $t \in \mathfrak{T}_0$. Then, in view of Definition 104 (Section 3.2.3), there are $\mathbf{S}_0^* \in \mathcal{D}_S(t_0;\mathcal{I}^+)$, $\epsilon \in \mathfrak{R}^+$, $t^* \in \mathfrak{T}_0$, and the *time* set $\mathfrak{T}_{t^*}$ composed of a *time*-sequence t_k converging to t^* as $k \longrightarrow \infty$,

$$\mathfrak{T}_{t^*} = \{t_k : t_k \in \mathfrak{T},\ k = 0,1,2,...,\ k \longrightarrow \infty \Longrightarrow\ t_k \longrightarrow t^*\},$$
$$t^* \in \mathfrak{T},$$

which imply the following:

$$\left[\mathcal{D}_S(t_k;t_0;\mathcal{I}^+)\,\Delta\mathcal{D}_S(t^*;t_0;\mathcal{I}^+)\right] \nsubseteq \mathcal{N}_\epsilon\left[\partial\mathcal{D}_S(t^*;t_0;\mathcal{I}^+)\right] \Longrightarrow$$
$$\|\mathbf{S}(t^*;t_0;\mathbf{S}_0^*) - \mathbf{S}(t_k;t_0;\mathbf{S}_0^*)\| > \epsilon \Longrightarrow \mathbf{S}(t;t_0;\mathbf{S}_0) \notin \mathfrak{C}(\mathfrak{T}_0),$$

which contradicts Property P4, i.e., (4.15) of Definition 117 (Section 4.2). Since the contradiction is the consequence of the assumption that the claim d) is invalid, it follows that the claim d) is valid for $\mathcal{D}_S(t;t_0;\mathcal{I}^+)$. The proof for $\mathcal{D}_{SS}(t;t_0;\mathcal{I}^+)$ is the literal repetition of the proof for $\mathcal{D}_S(t;t_0;\mathcal{I}^+)$ if $\mathcal{D}_S(t;t_0;\mathcal{I}^+)$ is replaced everywhere by $\mathcal{D}_{SS}(t;t_0;\mathcal{I}^+)$ ∎

Comment 209 *Stability of the set $\mathcal{I}^+$ of the time-varying nonlinear system (5.22), (5.23) depends on the initial moment t_0. It can be stable at t_{01} but not at another initial moment t_{02}, $t_{01} \neq t_{02}$. The dependence of $\mathcal{N}_L(\varepsilon;t_0;\mathcal{I}^+)$, hence of $\mathcal{D}_S(\varepsilon;t_0;\mathcal{I}^+)$, of $\mathcal{D}_S(t_0;\mathcal{I}^+)$ and of $\mathcal{D}_{SS}(t_0;\mathcal{I}^+)$ on the initial instant t_0 expresses this phenomenon.*

This phenomenon does not exist among time-invariant systems. For them, all these sets are independent of the initial moment t_0 because the stability of the set $\mathcal{I}^+$ at some $t_0 \in \mathfrak{T}$ implies its stability at every initial instant $t_0 \in \mathfrak{T}$. This results from the independence of the system motions of the initial instant $t_0 \in \mathfrak{T}$,

$$\mathbf{S}(.;t_0;\mathbf{S}_0) \equiv \mathbf{S}(.;\mathbf{S}_0).$$

There exist time-varying systems with the analogous property. It means that their stability is uniform in $t_0 \in \mathfrak{T}_i$ *or even in* $t_0 \in \mathfrak{T}$.

Note 210 *If the initial moment* $t_0 \in \mathfrak{T}_i$ *is fixed, then the stability domain* $\mathcal{D}_S(t_0;\mathcal{I}^+)$ *is also fixed and it can be, but need not be, positively invariant set of the time-varying system (5.22), (5.23).*

However, if the system is time-invariant, then its stability domain $\mathcal{D}_S(t_0;\mathcal{I}^+)$ *is independent of the initial moment,* $\mathcal{D}_S(t_0;\mathcal{I}^+) \equiv \mathcal{D}_S(\mathcal{I}^+)$ *and it is the system positively invariant set [247].*

7.7 Uniform stability

In what follows, $\mathcal{D}_S(\varepsilon;\mathfrak{T}_i;\mathcal{I}^+)$ and $\mathcal{D}_S(\mathfrak{T}_i;\mathcal{I}^+)$ represent the lower set limits, i.e., the intersections, $\mathcal{D}_{Sm}(\varepsilon;\mathfrak{T}_i;\mathcal{I}^+)$ and $\mathcal{D}_{Sm}(\mathfrak{T}_i;\mathcal{I}^+)$ of the sets $\mathcal{D}_S(\varepsilon;t_0;\mathcal{I}^+)$ and $\mathcal{D}_S(t_0;\mathcal{I}^+)$ on $\mathfrak{T}_i$ (see Equation 3.41 - Equation 3.44 in Section 3.2).

Definition 211 ***Uniform stability of the set*** $\mathcal{I}^+$ ***of the system*** *(5.22), (5.23)*

The set $\mathcal{I}^+$ ***of the system*** *(5.22), (5.23)* ***is uniformly /strictly/ stable on*** $\mathfrak{T}_i$ *if and only if the conditions 1) through 3) are valid:*

1) The set $\mathcal{I}^+$ *is stable /strictly stable/ on* $\mathfrak{T}_i$.

2) For every $\varepsilon \in \mathfrak{R}^+$, *the intersection* $\mathcal{D}_S(\varepsilon;\mathfrak{T}_i;\mathcal{I}^+)$ */*$\mathcal{D}_{SS}(\varepsilon;\mathfrak{T}_i;\mathcal{I}^+)$*/ of the* ε*-stability /*ε*-strict stability/ domain* $\mathcal{D}_S(\varepsilon;t_0;\mathcal{I}^+)$ */*$\mathcal{D}_{SS}(\varepsilon;t_0;\mathcal{I}^+)$*/ in* $t_0 \in \mathfrak{T}_i$ *is a connected neighborhood of the set* $\mathcal{I}^+$,

$$\begin{gathered}\forall\varepsilon \in \mathfrak{R}^+,\ \mathcal{D}_S(\varepsilon;\mathfrak{T}_i;\mathcal{I}^+) = \cap\left[\mathcal{D}_S(\varepsilon;t_0;\mathcal{I}^+) : t_0 \in \mathfrak{T}_i\right] \Longrightarrow \\ Cl\mathcal{I}^+ \subset In\mathcal{D}_S(\varepsilon;\mathfrak{T}_i;\mathcal{I}^+), \\ /\mathcal{D}_{SS}(\varepsilon;\mathfrak{T}_i;\mathcal{I}^+) = \cap\left[\mathcal{D}_{SS}(\varepsilon;t_0;\mathcal{I}^+) : t_0 \in \mathfrak{T}_i\right] \Longrightarrow \\ Cl\mathcal{I}^+ \subset In\mathcal{D}_{SS}(\varepsilon;\mathfrak{T}_i;\mathcal{I}^+)/, \end{gathered} \tag{7.25}$$

respectively.

3) The intersection $\mathcal{D}_S(\mathfrak{T}_i;\mathcal{I}^+)$ */*$\mathcal{D}_{SS}(\mathfrak{T}_i;\mathcal{I}^+)$*/ of the stability /strict stability/ domain* $\mathcal{D}_S(t_0;\mathcal{I}^+)$ */*$\mathcal{D}_{SS}(t_0;\mathcal{I}^+)$*/ in* $t_0 \in \mathfrak{T}_i$ *is a connected neighborhood of the set* $\mathcal{I}^+$,

$$\begin{gathered}\mathcal{D}_S(\mathfrak{T}_i;\mathcal{I}^+) = \cap\left[\mathcal{D}_S(t_0;\mathcal{I}^+) : t_0 \in \mathfrak{T}_i\right] \Longleftrightarrow \\ Cl\mathcal{I}^+ \subset In\mathcal{D}_S(\mathfrak{T}_i;\mathcal{I}^+), \\ /\mathcal{D}_{SS}(\mathfrak{T}_i;\mathcal{I}^+) = \cap\left[\mathcal{D}_{SS}(t_0;\mathcal{I}^+) : t_0 \in \mathfrak{T}_i\right] \Longleftrightarrow \\ Cl\mathcal{I}^+ \subset In\mathcal{D}_{SS}(\mathfrak{T}_i;\mathcal{I}^+)/, \end{gathered} \tag{7.26}$$

respectively.

This Definition holds for the invariant set $\mathcal{I}$ *of the system* ***(5.22), (5.23)*** *if and only if the set* $\mathcal{I}$ *replaces the set* $\mathcal{I}^+$ *everywhere.*

7.8 Uniform stability domain

Definition 212 ***Uniform /strict/ stability domain of the set*** $\mathcal{I}^+$ ***of the system*** *(5.22), (5.23)* ***on*** $\mathfrak{T}_i$

The ε***-uniform stability /***ε***-uniform strict stability/ domain***

$$\mathcal{D}_{US}(\varepsilon;\mathfrak{T}_i;\mathcal{I}^+)\quad /\mathcal{D}_{USS}(\varepsilon;\mathfrak{T}_i;\mathcal{I}^+)/$$

of the set $\mathcal{I}^+$ *exists on* $\mathfrak{T}_i$ *for every* $\varepsilon \in \mathfrak{R}^+$ *if and only if the conditions 1) and 2) of Definition 211 are valid and*

$$\begin{aligned} \mathcal{D}_{US}\left(\varepsilon;\mathfrak{T}_i;\mathcal{I}^+\right) &= \mathcal{D}_S\left(\varepsilon;\mathfrak{T}_i;\mathcal{I}^+\right), \ \forall\varepsilon \in \mathfrak{R}^+, \\ /\mathcal{D}_{USS}\left(\varepsilon;\mathfrak{T}_i;\mathcal{I}^+\right) &= \mathcal{D}_{SS}\left(\varepsilon;\mathfrak{T}_i;\mathcal{I}^+\right), \ \forall\varepsilon \in \mathfrak{R}^+/, \end{aligned} \tag{7.27}$$

i.e.,

$$\begin{gathered} \mathcal{D}_{US}\left(\varepsilon;\mathfrak{T}_i;\mathcal{I}^+\right) = \cap\left[\mathcal{D}_S\left(\varepsilon;t_0;\mathcal{I}^+\right) : t_0 \in \mathfrak{T}_i\right], \ \forall\varepsilon \in \mathfrak{R}^+ \Longleftrightarrow \\ Cl\mathcal{I}^+ \subset In\mathcal{D}_{US}\left(\varepsilon;\mathfrak{T}_i;\mathcal{I}^+\right), \ \forall\varepsilon \in \mathfrak{R}^+, \\ /\mathcal{D}_{USS}\left(\varepsilon;\mathfrak{T}_i;\mathcal{I}^+\right) = \cap\left[\mathcal{D}_{SS}\left(\varepsilon;t_0;\mathcal{I}^+\right) : t_0 \in \mathfrak{T}_i\right], \ \forall\varepsilon \in \mathfrak{R}^+ \Longleftrightarrow \\ Cl\mathcal{I}^+ \subset In\mathcal{D}_{USS}\left(\varepsilon;\mathfrak{T}_i;\mathcal{I}^+\right), \ \forall\varepsilon \in \mathfrak{R}^+/, \end{gathered} \tag{7.28}$$

respectively.

The uniform stability /uniform strict stability/ domain

$$\mathcal{D}_{US}\left(\mathfrak{T}_i;\mathcal{I}^+\right) \quad /\mathcal{D}_{USS}\left(\mathfrak{T}_i;\mathcal{I}^+\right)/$$

of the set $\mathcal{I}^+$ *exists on* $\mathfrak{T}_i$ *if and only if the conditions 1) through 3) of Definition 211 are satisfied and*

$$\mathcal{D}_{US}\left(\mathfrak{T}_i;\mathcal{I}^+\right) = \mathcal{D}_S\left(\mathfrak{T}_i;\mathcal{I}^+\right) \quad /\mathcal{D}_{USS}\left(\mathfrak{T}_i;\mathcal{I}^+\right) = \mathcal{D}_{SS}\left(\mathfrak{T}_i;\mathcal{I}^+\right), \tag{7.29}$$

i.e.,

$$\begin{gathered} \mathcal{D}_{US}\left(\mathfrak{T}_i;\mathcal{I}^+\right) = \cap\left[\mathcal{D}_S\left(t_0;\mathcal{I}^+\right) : t_0 \in \mathfrak{T}_i\right] \Longleftrightarrow Cl\mathcal{I}^+ \subset In\mathcal{D}_{US}\left(\mathfrak{T}_i;\mathcal{I}^+\right), \\ /\mathcal{D}_{USS}\left(\mathfrak{T}_i;\mathcal{I}^+\right) = \cap\left[\mathcal{D}_{SS}\left(t_0;\mathcal{I}^+\right) : t_0 \in \mathfrak{T}_i\right] \Longleftrightarrow Cl\mathcal{I}^+ \subset In\mathcal{D}_{USS}\left(\mathfrak{T}_i;\mathcal{I}^+\right)/, \end{gathered} \tag{7.30}$$

respectively.

The set $\mathcal{I}^+$ ***of the system*** *(5.22), (5.23)* ***is globally uniformly /strictly/ stable on*** $\mathfrak{T}_i$ *if and only if its domain of the uniform stability /strict stability/ is the whole state space,*

$$\mathcal{D}_{US}\left(\mathfrak{T}_i;\mathcal{I}^+\right) - \mathfrak{R}^n \ /\mathcal{D}_{USS}\left(\mathfrak{T}_i;\mathcal{I}^+\right) = \mathfrak{R}^n/, \tag{7.31}$$

respectively.

If and only if the above conditions hold for both $\mathfrak{T}_0$ *and* $\mathfrak{T}_i$ *replaced by* $\mathfrak{T}$*, then the term "**on*** $\mathfrak{T}_i$*" is to be omitted.*

This Definition holds for the invariant set $\mathcal{I}$ *of the system* ***(5.22), (5.23)*** *if and only if the set* $\mathcal{I}$ *replaces the set* $\mathcal{I}^+$ *everywhere.*

Comment 213 ***Equivalent form of Equations (7.30)***

Equation (7.15) /Equation (7.20)/ implies

$$\begin{aligned} \cap\left[\mathcal{D}_S\left(t;\mathfrak{T}_i;\mathcal{I}^+\right) : t \in \mathfrak{T}_i\right] &= \cap\left[\mathcal{D}_S\left(t_0;\mathcal{I}^+\right) : t_0 \in \mathfrak{T}_i\right] \\ /\cap\left[\mathcal{D}_{SS}\left(t;\mathfrak{T}_i;\mathcal{I}^+\right) : t \in \mathfrak{T}_i\right] &= \cap\left[\mathcal{D}_{SS}\left(t_0;\mathcal{I}^+\right) : t_0 \in \mathfrak{T}_i\right]/, \end{aligned}$$

respectively. Therefore, Equations (7.30) can be set in the following equivalent form:

$$\begin{gathered} \mathcal{D}_{US}\left(\mathfrak{T}_i;\mathcal{I}^+\right) = \cap\left[\mathcal{D}_S\left(t;\mathfrak{T}_i;\mathcal{I}^+\right) : t \in \mathfrak{T}_i\right] \Longleftrightarrow Cl\mathcal{I}^+ \subset In\mathcal{D}_{US}\left(\mathfrak{T}_i;\mathcal{I}^+\right), \\ /\mathcal{D}_{USS}\left(\mathfrak{T}_i;\mathcal{I}^+\right) = \cap\left[\mathcal{D}_{SS}\left(t;\mathfrak{T}_i;\mathcal{I}^+\right) : t \in \mathfrak{T}_i\right] \Longleftrightarrow Cl\mathcal{I}^+ \subset In\mathcal{D}_{USS}\left(\mathfrak{T}_i;\mathcal{I}^+\right)/, \end{gathered} \tag{7.32}$$

Comment 214 *The uniformity of the stability of* $\mathcal{I}^+$ *on* $\mathfrak{T}_i$ *means that its uniform stability domain on* $\mathfrak{T}_i$ *is independent of a particular* $t_0 \in \mathfrak{T}_i$, *i.e., that it is constant on* $\mathfrak{T}_i$.

Comment 215 *The connectedness of* $\mathcal{D}_S(t_0;\mathcal{I}^+)$ */*$\mathcal{D}_S(t_0;\mathcal{I}^+)$*/ and Condition 7.25-Condition 7.32 ensure that the uniform stability /strict stability/ domain*

$$\mathcal{D}_{US}(\mathfrak{T}_i;\mathcal{I}^+) \ \ /\mathcal{D}_{USS}(\mathfrak{T}_i;\mathcal{I}^+)/$$

is a connected neighborhood of the set $\mathcal{I}^+$ *of the system (5.22), (5.23) on* $\mathfrak{T}_i$.

7.9 Nonuniform stability

Definition 216 ***Nonuniform stability of the set*** $\mathcal{I}^+$ ***of the system*** *(5.22), (5.23)*

a) ***The set*** $\mathcal{I}^+$ ***of the system*** *(5.22), (5.23)* ***is nonuniformly /strictly/ stable on*** $\mathfrak{T}_i$ *if and only if it is /strictly/ stable at every* $t_0 \in \mathfrak{T}_i$, *the intersection of the domains* $\mathcal{D}_S(t_0;\mathcal{I}^+)$ */*$\mathcal{D}_{SS}(t_0;\mathcal{I}^+)$*/ in* $t_0 \in \mathfrak{T}_i$ *is not a neighborhood of the set* $\mathcal{I}^+$ *on* $\mathfrak{T}_i$:

$$\nexists \mathcal{N}(\mathcal{I}^+) \subseteq \mathfrak{R}^n \Longrightarrow \cap\{\mathcal{D}_S(t_0;\mathcal{I}^+) : t_0 \in \mathfrak{T}_i\} \supseteq \mathcal{N}(\mathcal{I}^+),$$
$$\nexists \mathcal{N}(\mathcal{I}^+) \subseteq \mathfrak{R}^n \Longrightarrow \cap\{\mathcal{D}_{SS}(t_0;\mathcal{I}^+) : t_0 \in \mathfrak{T}_i\} \supseteq \mathcal{N}(\mathcal{I}^+), \tag{7.33}$$

respectively, and/or the closure $Cl\mathcal{D}_S(t_0;\mathcal{I}^+)$ *of its domain* $\mathcal{D}_S(t_0;\mathcal{I}^+)$ *of stability /the closure* $Cl\mathcal{D}_{SS}(t_0;\mathcal{I}^+)$ *of its domain* $\mathcal{D}_{SS}(t_0;\mathcal{I}^+)$ *of strict stability/ at every* $t_0 \in \mathfrak{T}_i$*/ is asymptotically contractive to* $\mathcal{I}^+$ *as* $t_0 \longrightarrow sup\mathfrak{T}_i$, $t_0 \in \mathfrak{T}_i$:

$$t_0 \longrightarrow sup\mathfrak{T}_i = \infty \Longrightarrow Cl\mathcal{D}_S(t_0;\mathcal{I}^+) \longrightarrow Cl\mathcal{I}^+$$
$$/t_0 \longrightarrow sup\mathfrak{T}_i = \infty \Longrightarrow Cl\mathcal{D}_{SS}(t_0;\mathcal{I}^+) \longrightarrow Cl\mathcal{I}^+/, \tag{7.34}$$

respectively.

If and only if this holds for $\mathfrak{T}_i$ *replaced by* $\mathfrak{T}$, *then* ***the set*** $\mathcal{I}^+$ ***of the system*** *(5.22), (5.23)* ***is nonuniformly /strictly/ stable.***

b) ***The set*** $\mathcal{I}^+$ ***of the system (5.22), (5.23) is strongly nonuniformly /strictly/ stable on*** $\mathfrak{T}_i$ *if and only if it is /strictly/ stable on* $\mathfrak{T}_i$ *and the above conditions (8.17), (7.34) hold, respectively.*

This definition holds for the invariant set $\mathcal{I}$ *of the system* ***(5.22), (5.23)*** *if and only if the set* $\mathcal{I}$ *replaces* $\mathcal{I}^+$ *everywhere.*

7.10 Examples

Example 217 *Let there exist* $\varepsilon_1 \in \mathfrak{R}^+$ *and* $\varepsilon_2 \in \mathfrak{R}^+$, $\varepsilon_2 > \varepsilon_1$, *such that the boundary of* $\mathcal{N}_L(\varepsilon_1;t_0;\mathcal{I}^+)$ *is an unstable invariant set and the boundary of* $\mathcal{N}_L(\varepsilon_2;t_0;\mathcal{I}^+)$ *is a stable invariant set, where* $\mathcal{N}(\varepsilon_1;t_0;\mathcal{I}^+) \subset \mathcal{N}(\varepsilon_2;t_0;\mathcal{I}^+)$. *Then the stability domain is the whole state space:*

$$\mathcal{D}_S(t_0;\mathcal{I}^+) = \mathfrak{R}^n,$$

hence unbounded, but the strict stability domain is bounded:

$$\mathcal{D}_{SS}(t_0;\mathcal{I}^+) = \mathcal{N}_L(\varepsilon_1;t_0;\mathcal{I}^+),$$

so that

$$\mathcal{D}_{SS}(t_0;\mathcal{I}^+) \subset \mathcal{D}_S(t_0;\mathcal{I}^+).$$

Example 218 *[115, Example 2, pp. 528, 529] The first order time-varying linear system reads:*

$$\frac{dS}{dt} = -\frac{\left(1 + tsint + t^2 cost\right) exp\left(-\frac{1}{2}\pi\right)}{\frac{1}{2}\pi exp\left(-tsint\right) + texp\left(-\frac{1}{2}\pi\right)} S$$

The solution is

$$S\left(t; t_0; S_0\right) = \frac{\frac{1}{2}\pi + t_0 exp\left(-\frac{1}{2}\pi + t_0 sint_0\right)}{\frac{1}{2}\pi + texp\left(-\frac{1}{2}\pi + tsint\right)} S_0,\ t_0 \neq -\frac{\pi}{2},\ t \neq -\frac{\pi}{2},$$

so that

$$\left|S\left(t; t_0; S_0\right)\right| \longrightarrow \left\{ \begin{array}{c} \infty \ as\ t \longrightarrow \left(-\frac{\pi}{2} - 0\right),\ \forall t_0 \in]-\infty, -\frac{\pi}{2}[, \\ 0\ as\ t \longrightarrow \infty,\ \forall t_0 \in]-\frac{\pi}{2}, \infty[. \end{array} \right\}.$$

Let $\varepsilon \in \mathfrak{R}^+$ *be arbitrarily chosen. It determines* $\mathcal{N}_\varepsilon(0)$. *Let us test whether there exists* $\mathcal{N}_L(\varepsilon; t_0; 0)$ *such that it satisfies the stability definition:*

$$\forall \mathcal{N}_\varepsilon(0) \subset \mathfrak{R}^1 \Longrightarrow \exists \mathcal{N}(\varepsilon; t_0; 0)?$$

Let

$$\left|S_0\right| < \delta(\varepsilon).$$

This implies

$$\begin{aligned} \left|S\left(t; t_0; S_0\right)\right| &= \left|\frac{\frac{1}{2}\pi + t_0 exp\left(-\frac{1}{2}\pi + t_0 sint_0\right)}{\frac{1}{2}\pi + texp\left(-\frac{1}{2}\pi + tsint\right)}\right| \left|S_0\right| < \\ &< \left|\frac{\frac{1}{2}\pi + t_0 exp\left(-\frac{1}{2}\pi + t_0 sint_0\right)}{\frac{1}{2}\pi + texp\left(-\frac{1}{2}\pi + tsint\right)}\right| \delta(\varepsilon), \end{aligned}$$

which should be less than ε *for all* $t \in \mathfrak{T}_0$:

$$\left|S\left(t; t_0; S_0\right)\right| < \left|\frac{\frac{1}{2}\pi + t_0 exp\left(-\frac{1}{2}\pi + t_0 sint_0\right)}{\frac{1}{2}\pi + texp\left(-\frac{1}{2}\pi + tsint\right)}\right| \delta(\varepsilon) < \varepsilon,\ \forall t \in \mathfrak{T}_0.$$

Notice that

$$\left|\frac{\frac{1}{2}\pi + t_0 exp\left(-\frac{1}{2}\pi + t_0 sint_0\right)}{\frac{1}{2}\pi + texp\left(-\frac{1}{2}\pi + tsint\right)}\right| \leq$$
$$\leq \left\{ \begin{array}{c} \leq 1,\ \forall t_0 \in]-\frac{\pi}{2}, 0], \\ \pi^{-1}\left[\pi + 2t_0 exp\left(-\frac{1}{2}\pi + t_0 sint_0\right)\right],\ \forall t_0 \in]0, \infty[\end{array} \right\},\ \forall t \in \mathfrak{T}_0,$$

then $\mathcal{N}(\varepsilon; t_0; 0)$ *should obey*

$$\mathcal{N}(\varepsilon; t_0; 0) = \left\{S : \left[\begin{array}{c} |S| \leq \varepsilon,\ \forall t_0 \in]-\frac{\pi}{2}, 0], \\ |S| \leq \pi\left[\pi + 2t_0 exp\left(-\frac{1}{2}\pi + t_0 sint_0\right)\right]^{-1} \varepsilon,\ \forall t_0 \in]0, \infty[\end{array} \right]\right\},\ \forall \varepsilon \in \mathfrak{R}^+.$$

This means that for every neighborhood $\mathcal{N}_\varepsilon(0)$ *of the zero equilibrium state* $S_e = 0$:

a) there is not $\mathcal{N}(\varepsilon; t_0; 0)$ *that obeys the stability definition if* $t_0 \in]-\infty, -\frac{\pi}{2}]$:

$$t_0 \in]-\infty, -\frac{\pi}{2}] \Longrightarrow \mathcal{N}(\varepsilon; t_0; 0) = \phi,\ \forall \mathcal{N}_\varepsilon(0) \subset \mathfrak{R}^1.$$

b) there is $\mathcal{N}(\varepsilon;t_0;0)$ *that obeys the stability definition if* $t_0 \in]-\frac{\pi}{2},\infty[$:

$$\forall \mathcal{N}_\varepsilon(0) \subset \mathfrak{R}^1,\ \forall t_0 \in]-\frac{\pi}{2},\infty[,\ \exists \mathcal{N}(\varepsilon;t_0;0) \Longrightarrow$$

$$\mathcal{N}_L(\varepsilon;t_0;0) = \left\{ \begin{array}{c} \mathcal{N}_\varepsilon(0),\ \forall t_0 \in]-\frac{\pi}{2},0], \\ \left\langle \begin{array}{c} \mathcal{N}_{\delta_M(t_0,\varepsilon)}(0),\ \forall t_0 \in]0,\infty[, \\ \delta_M(t_0,\varepsilon) = \pi\left[\pi + 2t_0 exp\left(-\frac{1}{2}\pi + t_0 sin t_0\right)\right]^{-1}\varepsilon \end{array} \right\rangle \end{array} \right\}.$$

The zero equilibrium state $S_e = 0$ *is stable in the whole on* $\mathfrak{T}_i =]-\frac{\pi}{2},\infty[$ *and uniformly stable on every compact time set* $\mathfrak{T}_i^c \subset]-\frac{\pi}{2},\infty[$*, but it is not stable. In this example:*

$$\mathcal{D}_S(t_0;0) = \left\{ \begin{array}{c} \phi,\ t_0 \in]-\infty,-\frac{\pi}{2}], \\ \mathfrak{R}^1,\ \forall t_0 \in]-\frac{\pi}{2},\infty[, \end{array} \right\}$$

and

$$\mathcal{D}_{US}(\mathfrak{T}_i^c;0) = \left\{ \begin{array}{c} \phi,\ \mathfrak{T}_i =]-\infty,-\frac{\pi}{2}[, \\ \mathfrak{R}^1,\ \mathfrak{T}_i^c \subseteq [\beta,\gamma] \end{array} \right\},\ -\frac{\pi}{2} < \beta < \gamma < \infty.$$

Notice once more that the system is linear but time-varying.

Example 219 *Let a time-varying nonlinear system be governed by*

$$S^{(1)} + tS(1-S)(2-S) = 0. \tag{7.35}$$

Let $t_0 \in \mathfrak{T}$ *be arbitrarily chosen. The equilibrium states are*

$$S_{e1} = 0,\ S_{e2} = 1 \text{ and } S_{e3} = 2.$$

The system solution has the following form:

$$\mathcal{S}(t;t_0;S_0) = 1\left\{ \begin{array}{c} -\frac{1-S_0}{\sqrt{(1-S_0)^2+S_0(2-S_0)exp\left[-\left(t^2-t_0^2\right)\right]}},\ S_0 \le 1, \\ +\frac{S_0-1}{\sqrt{(S_0-1)^2-S_0(S_0-2)exp\left[-\left(t^2-t_0^2\right)\right]}},\ S_0 > 1 \end{array} \right\}, \tag{7.36}$$
$$\forall t \in \mathfrak{T}_0, \forall t_0 \in \mathfrak{T}.$$

Hence,

$$\mathcal{S}(t_0;t_0;S_0) = S_0,\ \forall S_0 \in \mathfrak{R}^1, \text{ and } \mathcal{S}(\infty;t_0;S_0) = \left\{ \begin{array}{c} 0,\ S_0 < 1, \\ 1,\ S_0 = 1, \\ 2,\ S_0 > 1 \end{array} \right\}. \tag{7.37}$$

Since

$$\frac{dS}{dt} = -tS(1-S)(2-S), \tag{7.38}$$

then

$$\frac{dS(t;t_0;S_0)}{dt} \left\{ \begin{array}{c} = 0,\ t = 0, \\ \left\{ \begin{array}{c} > 0,\ S < 0, \\ = 0,\ S = 0, \\ < 0,\ 0 < S < 1, \\ = 0,\ S = 1, \\ > 0,\ 1 < S < 2, \\ = 0,\ S = 2, \\ < 0,\ S > 2 \\ \forall t \in \mathfrak{T}_0, \forall t_0 \in \mathfrak{T} \cap \mathfrak{R}^+ \end{array} \right\} \end{array} \right\}. \tag{7.39}$$

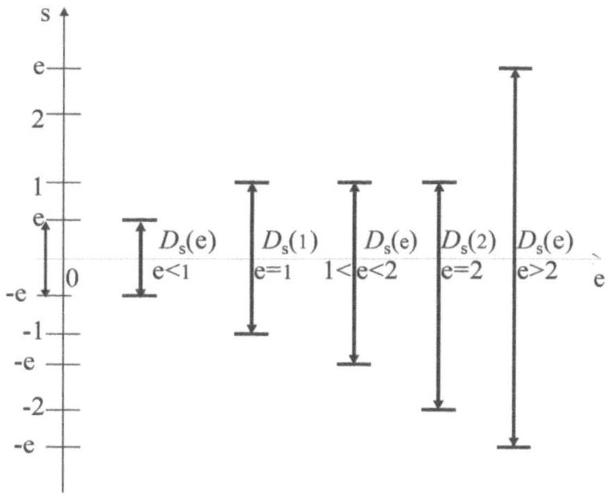

Figure 7.4: $\varepsilon-$domains $\mathcal{D}_s(\varepsilon,\mathfrak{T}_0)$ of the equilibrium vector $S_{e1}=0$; $e=\varepsilon$ in this figure. D stands for $\mathcal{D}$.

This implies

$$|\mathcal{S}(t;t_0;y_0)| \left\{ \begin{array}{c} <|S_0|,\ S_0<0,\ t>t_0, \\ =|S_0|,\ S_0=0, \\ <S_0,\ 0<S_0<1,\ S_0\neq 0,\ t>t_0, \\ =S_0,\ S_0=1, \\ \in[S_0,\ 2[,\ 1<S_0<2,\ t\geq t_0, \\ =S_0,\ S_0=2, \\ \in]2,\ S_0[,\ S_0>2,\ t>t_0 \end{array} \right\}. \\ \forall t\in\mathfrak{T}_0, \forall t_0\in\mathfrak{T}\cap\mathfrak{R}^+. \tag{7.40}$$

Let $t_0\in\mathfrak{T}\cap\mathfrak{R}_+$ and let $\mathcal{D}_S(\varepsilon,\mathfrak{T}_0)$ be the largest connected neighborhood of the equilibrium vector $S_{e1}=0$ such that

$$\forall\varepsilon>0,\ \forall(S_0\neq 0)\in\mathcal{D}_S(\varepsilon,\mathfrak{T}_0)\longrightarrow|\mathcal{S}(t;t_0;y_0)|<\varepsilon,\ \forall t\in\mathfrak{T}_0. \tag{7.41}$$

From (7.39), (7.40) and (7.41) follow:

$$\mathcal{D}_S(\varepsilon,\mathfrak{T}_0)=\left\{ \begin{array}{c} \{S:\ -\varepsilon<S<\min(1,\varepsilon)\},\ 0<\varepsilon\leq 2, \\ \mathfrak{B}_\varepsilon=\{S:|S|<\varepsilon,\},\ \varepsilon>2 \end{array} \right\}. \tag{7.42}$$

See Figure 7.4.

We define now the set $\mathcal{D}_S(\mathfrak{T}_0)$ as the union of all $\varepsilon-$domains $\mathcal{D}_S(\varepsilon,\mathfrak{T}_0)$ over ε,

$$\mathcal{D}_S(\mathfrak{T}_0)=\cup\left[\mathcal{D}_S(\varepsilon,\mathfrak{T}_0):\ \varepsilon>0\right]. \tag{7.43}$$

We call the set $\mathcal{D}_S(\mathfrak{T}_0)$ ***the domain of stability of the set*** *$\{S_{e1}\}=\{0\}$ on $\mathfrak{T}_0$. In this example,*

$$\begin{aligned} \mathcal{D}_S(\varepsilon,\mathfrak{T}_0) &= \ ]-\varepsilon,\varepsilon[=\mathfrak{B}_\varepsilon,\ \forall\varepsilon\in]0,1], \\ \mathcal{D}_S(\varepsilon,\mathfrak{T}_0) &= \ ]-\varepsilon,1],\ \forall\varepsilon\in[1,2], \\ \mathcal{D}_S(\varepsilon,\mathfrak{T}_0) &= \ ]-\varepsilon,\varepsilon[,\ \forall\varepsilon\in]2,\infty[, \end{aligned} \tag{7.44}$$

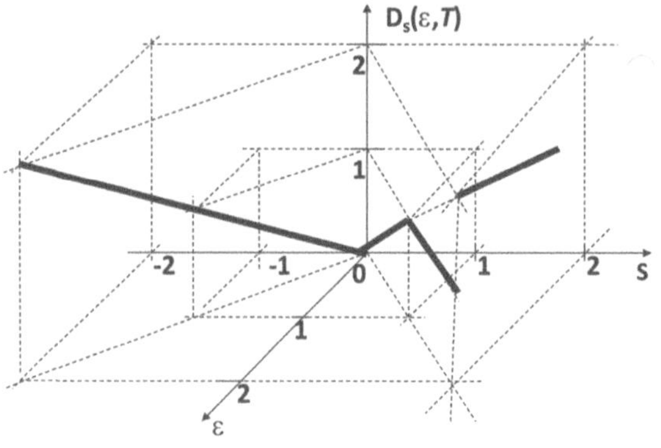

Figure 7.5: The set $\mathcal{D}_s(\varepsilon, \mathfrak{T}_0)$ is discontinuous in ε at $\varepsilon = 2$. D means $\mathcal{D}$.

so that

$$\mathcal{D}_S(\mathfrak{T}_0) = \mathfrak{R}^1. \tag{7.45}$$

Equations (7.44) show that the set $\mathcal{D}_S(\varepsilon, \mathfrak{T}_0)$ is discontinuous in ε at $\varepsilon = 2$; see Figure 7.5.

The union of the largest connected neighborhoods $\mathcal{D}_S(\varepsilon, \mathfrak{T}_0)$ that are continuous in $\varepsilon > 0$ is the domain of strict stability $\mathcal{D}_{SS}(\mathfrak{T}_0)$ of the equilibrium vector, $\mathcal{D}_{SS}(\mathfrak{T}_0) \subseteq \mathcal{D}_S(\mathfrak{T}_0)$.

Since $\mathcal{S}(t; t_0; S_0)$:

- converges to the equilibrium vector $S_{e1} = 0$ for $S_0 < 1$,

- stays at $S_{e2} = 1$ for $S_0 = 1$,

and

- diverges from both $S_{e1} = 0$ and $S_{e2} = 1$ and converges to $S_{e3} = 2$ for $S_0 \in]1, \infty[$,

then

$$\begin{aligned} \mathcal{D}_{SS}(\varepsilon, \mathfrak{T}_0) &= \ ]-\varepsilon, \varepsilon[= \mathfrak{B}_\varepsilon, \ \forall \varepsilon \in]0, 1], \\ \mathcal{D}_{SS}(\varepsilon, \mathfrak{T}_0) &= \ ]-\varepsilon, 1], \ \forall \varepsilon \in]1, 2], \\ \mathcal{D}_{SS}(\varepsilon, \mathfrak{T}_0) &= \ ]-\varepsilon, 1], \ \forall \varepsilon \in [2, \infty[. \end{aligned}$$

This implies

$$\begin{aligned} \mathcal{D}_{SS}(\mathfrak{T}_0) = \cup\left[\mathcal{D}_{SS}(\varepsilon, \mathfrak{T}_0) : \varepsilon \in \mathfrak{R}^+\right] &= \ ]-\infty, 1], \\ \mathcal{D}_{SS}(\mathfrak{T}_0) \subset \mathcal{D}_S(\mathfrak{T}_0)&. \end{aligned} \tag{7.46}$$

Exercise 220 *Verify for the system (7.35):*

$$\begin{aligned} \{S_{e1}\} &= \{0\} \text{ is stable, but not globally,} \\ \{S_{e2}\} &= \{1\} \text{ is unstable,} \\ \{S_{e3}\} &= \{2\} \text{ is stable, but not globally.} \end{aligned}$$

Determine the /strict/ stability domain of $\{S_{e3}\} = \{2\}$.

Example 221 *Let a time-varying nonlinear system be defined by*

$$(1+t^2)\frac{dS}{dt} + 2tS + (1+t^2)S\left[1 - (1+t^2)S\right]\left[2 - (1+t^2)S\right] = 0. \tag{7.47}$$

Its solution reads

$$S(t;t_0;S_0) = \frac{1}{(1+t^2)}\left\{1+ \bullet\left[\sqrt{\frac{\left[\left(1+t_0^2\right)S_0-1\right]\bullet}{\left[\left(1+t_0^2\right)S_0-1\right]^2- -\left(1+t_0^2\right)S_0\left[\left(1+t_0^2\right)S_0-2\right]e^{[-2(t-t_0)]}}}\right]^{-1}\right\},$$

$$S(t_0;t_0;S_0) \equiv S_0,$$

$$S(t;t_0;S_0)\left\{\begin{array}{c} =0 \ for\ every\ t\in\mathfrak{T},\ S_0=0, \\ \rightarrow 0\ for\ t\rightarrow\infty,\ S_0\neq 0, \end{array}\right\}$$

The set $\mathcal{I}^+$ *is the equilibrium set* $\{0\}$. *The equilibrium state* $S_e = 0$ *is unique. It is globally stable. Its domain of stability* $\mathcal{D}_S(t_0, 0)$ *is the whole space* $\mathfrak{R}^1$ *at every* $t_0 \in \mathfrak{T}$,

$$\mathcal{D}_S(0) = \mathcal{D}_S(\mathfrak{T}; 0) = \mathfrak{R}^1.$$

It is also the strict stability domain of the set $S_e = 0$ *at every initial instant* $t_0 \in \mathfrak{T}$. *In this example:*

$$\mathcal{D}_S(0) = \mathcal{D}_{SS}(0) = \mathfrak{R}^1.$$

Example 222 *A time varying nonlinear system is described by*

$$\frac{d\mathbf{S}}{dt} = \left(-\frac{\alpha+2t}{1+t^2} + \mathbf{b}^T|\mathbf{S}|\right)\mathbf{S},\ \alpha\in\mathfrak{R}^+,\ \mathbf{b}\in\mathfrak{R}^{+^n},\ \mathbf{S}\in\mathfrak{R}^n.$$

The system motions $S(t;t_0;\mathbf{S}_0)$ *obey the following for*

$$\mathfrak{V}_\alpha(t) =: \left\{\mathbf{S} : \mathbf{S}\in\mathfrak{R}^n,\ \mathbf{b}^T|\mathbf{S}| \leq \alpha\left(1+t^2\right)^{-1}\right\} \longrightarrow \mathfrak{O}\ as\ t\longrightarrow\infty :$$

$$\mathbf{S}_0 \in Cl\mathfrak{V}_\alpha(t_0) \Longrightarrow \mathbf{S}(t;t_0;\mathbf{S}_0) \in Cl\mathfrak{V}_\alpha(t),\ \forall t\in\mathfrak{T}_0,\ \forall t_0\in\mathfrak{T},$$
$$\mathbf{S}_0 \in \partial\mathfrak{V}_\alpha(t_0) \Longrightarrow \mathbf{S}(t;t_0;\mathbf{S}_0) \in \partial\mathfrak{V}_\alpha(t),\ \forall t\in\mathfrak{T}_0,\ \forall t_0\in\mathfrak{T},$$
$$\mathbf{S}_0 \in \mathfrak{R}^n\backslash Cl\mathfrak{V}_\alpha(t) \Longrightarrow \mathbf{S}(t;t_0;\mathbf{S}_0) \in \mathfrak{R}^n\backslash Cl\mathfrak{V}_\alpha(t),\ \forall t\in\mathfrak{T}_0,\ \forall t_0\in\mathfrak{T}.$$

The set $\mathcal{I}^+$ *is the singleton* $\{\mathbf{0}_2\}$. *The zero equilibrium state* $S_e = \mathbf{0}_2$ *is nonuniformly stable on* $\mathfrak{T}_0$ *with the nonuniform stability domain shown in Figure 7.6 for* $n = 2$, $\mathbf{b}^T = [b_1\ \ b_2]$ *and* $v(t, \mathbf{S}) = \left(1+t^2\right)\mathbf{b}^T|\mathbf{S}|$:

$$\begin{aligned} \mathcal{D}_{NUS}\left(t_0;\mathcal{I}^+\right) &= Cl\mathfrak{V}_\alpha(t_0) = \{\mathbf{S} : \mathbf{S}\in\mathfrak{R}^n,\ v(t_0,\mathbf{S})\leq\alpha\} = \\ &= \left\{\mathbf{S} : \mathbf{S}\in\mathfrak{R}^n,\ \mathbf{b}^T|\mathbf{S}|\leq\alpha\left(1+t_0^2\right)^{-1}\right\} \longrightarrow \mathfrak{O}\ as\ t_0\longrightarrow\infty. \end{aligned} \tag{7.48}$$

However, the zero equilibrium state $S_e = \mathbf{0}_2$ *is uniformly stable on every compact subset* $\mathfrak{T}_i^c$ *of* $\mathfrak{T}_i$ *with the uniform stability domain* $\mathcal{D}_{US}\left(\mathfrak{T}_i^c;\mathcal{I}^+\right)$,

$$\begin{aligned} \mathcal{D}_{US}\left(\mathfrak{T}_i^c;\mathcal{I}^+\right) &= \cap\left[\mathcal{D}_{NUS}\left(t_0;\mathcal{I}^+\right) : t_0\in\mathfrak{T}_i^c\right] = \\ &= \cap\left[\mathfrak{V}_\alpha(t_0) : t_0\in\mathfrak{T}_i^c\right] = \\ &= \mathfrak{V}_\alpha(t_S),\ t_S = sup\left\{t_0\in\mathfrak{T}_i^c\right\} < \infty \Longrightarrow \end{aligned}$$

$$\begin{aligned} \mathcal{D}_{US}\left(\mathfrak{T}_i^c;\mathcal{I}^+\right) &= \left\{\mathbf{S} : \mathbf{S}\in\mathfrak{R}^n,\ \left(1+t^2\right)\mathbf{b}^T|\mathbf{S}|\leq\alpha\right\} = \\ &= \left\{\mathbf{S} : \mathbf{S}\in\mathfrak{R}^n,\ \mathbf{b}^T|\mathbf{S}|\leq\alpha\left(1+t_S^2\right)^{-1}\right\}. \end{aligned}$$

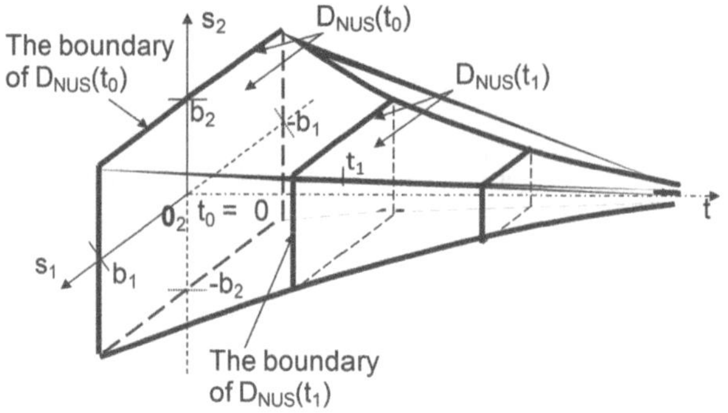

Figure 7.6: The nonuniform stability domain $\mathcal{D}_{NUS}(t_0; \mathbf{0}_n)$ (7.48). D designates $\mathcal{D}$.

7.11 Stability domain estimate

Definition 223 ***Estimate of a stability domain of the constant nonempty set $\mathcal{I}^+$***

A connected set $\mathfrak{A}(.)$ is ***an estimate*** *$\mathfrak{E}_{xx}(.\mathcal{I}^+)$* ***of a stability domain*** *$\mathcal{D}_{xx}(.;\mathcal{I}^+)$* ***of the set*** *$\mathcal{I}^+$* ***of the system*** *(5.22), (5.23) if and only if it satisfies the following conditions:*

1. *It is a connected subset of $\mathcal{D}_{xx}(.;\mathcal{I}^+)$.*
2. *It is a connected neighborhood of $\mathcal{I}^+$.*
3. *It is a positive invariant set relative to the motions of the system (5.22), (5.23).*

7.12 Conclusion

Conclusion 224 ***Set stability, motion stability and equilibrium stability***

In all set stability definitions (Definition 201 through Definition 216) the set $\mathcal{I}^+$ is a time-invariant set.

A special form of a time-invariant set $\mathcal{I}^+$ is a time-invariant singleton $\mathcal{I}^+$. Its single element is a system equilibrium state $\mathbf{S}_E$,

$$\mathcal{I}^+ = \mathcal{I}_E^+ = \{\mathbf{S}_E\}, \ \mathbf{S}_E \in \mathfrak{R}^n. \tag{7.49}$$

In this case, the set stability definition Definition 201-Definition 206, Definition 211-Definition 7.9 become the definitions of stability properties of the system equilibrium state $\mathbf{S}_E$.

The stability criteria valid for the time-invariant set $\mathcal{I}^+$ are simultaneously the stability criteria for the system equilibrium state $\mathbf{S}_E$.

The simplest form of a time-invariant set $\mathcal{I}^+$ is a time-invariant singleton $\mathcal{I}_{E0}^+$, the single element of which is the system zero equilibrium state $\mathbf{S}_E = \mathbf{0}_n$,

$$\mathcal{I}^+ = \mathcal{I}_{E0}^+ = \{\mathbf{0}_n\}, \ \mathbf{0}_n \in \mathfrak{R}^n. \tag{7.50}$$

In this case, the set stability definitions, Definition 201-Definition 206, Definition 211-Definition 7.9, become the definitions of the stability properties of the system zero equilibrium state $\mathbf{S}_E = \mathbf{0}_n$.

Chapter 8

Attraction and domains

8.1 Introduction

Attraction is another qualitative dynamical system property. It is independent of the system stability properties in general.

Attraction of the zero equilibrium state is a special case of attraction of any equilibrium state, which is further a special case of attraction of a *time*-invariant set. The study of attraction of the zero equilibrium state and of any equilibrium state will be imbedded in the study of *time*-invariant set attraction.

8.2 Attraction

Attraction of a constant set of the general nonlinear systems extends and generalizes the attraction concept.

Let us refer to Note 198, Section 7.3.

Definition 225 ***Attraction of the set $\mathcal{I}^+$ of the system (5.22), (5.23)***

The set $\mathcal{I}^+$ of the system (5.22), (5.23) *is* ***attractive at*** $t_0 \in \mathfrak{T}_i$ *if and only if there exists a connected nonempty neighborhood* $\mathcal{N}(t_0; \mathcal{I}^+)$,

$$\exists \mathcal{N}(t_0; \mathcal{I}^+) \subseteq \mathfrak{R}^n, \tag{8.1}$$

of the set $\mathcal{I}^+$ *at the initial moment* $t_0 \in \mathfrak{T}_i$ *and for every* $\varsigma \in \mathfrak{R}^+$ *there is a nonnegative real number* τ,

$$\tau = \tau(t_0, \varsigma, \mathbf{S}_0, \mathcal{I}^+) \in \mathfrak{R}_+, \tag{8.2}$$

with the property that for any $\mathbf{S}_0$ *from the neighborhood* $\mathcal{N}(t_0; \mathcal{I}^+)$ *of* $\mathcal{I}^+$ *the system real state vector* $\mathbf{S}(t; t_0; \mathbf{S}_0)$ *belongs to the* $\varsigma-$*neighborhood* $\mathcal{N}_\varsigma(\mathcal{I}^+)$ *of* $\mathcal{I}^+$ *for all time* $t \in]t_0 + \tau, \infty[$, *i.e.,*

$$\begin{gathered} \exists \mathcal{N}(t_0; \mathcal{I}^+) \subseteq \mathfrak{R}^n, \ Cl\mathcal{I}^+ \subset In\mathcal{N}(t_0; \mathcal{I}^+), \\ \forall \varsigma \in \mathfrak{R}^+, \exists \tau = \tau(t_0, \varsigma, \mathbf{S}_0, \mathcal{I}^+) \in \mathfrak{R}_+ \Longrightarrow \\ \forall \mathbf{S}_0 \in \mathcal{N}(t_0; \mathcal{I}^+) \Longrightarrow \mathbf{S}(t; t_0; \mathbf{S}_0) \in \mathcal{N}_\varsigma(\mathcal{I}^+), \ \forall t \in]t_0 + \tau, \infty[. \end{gathered} \tag{8.3}$$

If and only if this holds for every $t_0 \in \mathfrak{T}_i$, *then the set* $\mathcal{I}^+$ ***of the system (5.22), (5.23)*** *is* ***attractive on*** $\mathfrak{T}_i$.

DOI: 10.1201/9781003744528-13

If and only if additionally this holds for every $t_0 \in \mathfrak{T}$*, then* ***the set*** $\mathcal{I}^+$ ***of the system (5.22), (5.23 is attractive)****.*

This definition holds for the invariant set $\mathcal{I}$ *of the system* ***(5.22), (5.23)*** *if and only if the set* $\mathcal{I}$ *replaces* $\mathcal{I}^+$ *everywhere.*

8.3 Attraction domains

Definition 226 ***Attraction domain of the set*** $\mathcal{I}^+$ ***of the system (5.22), (5.23)***

The largest connected neighborhood $\mathcal{N}(t_0;\mathcal{I}^+)$ *(8.3) of* $\mathcal{I}^+$ *denoted by*

$$\mathcal{N}_L\left(t_0;\mathcal{I}^+\right), \tag{8.4}$$

which satisfies Definition 225, is ***the attraction domain*** $\mathcal{D}_A(t_0;\mathcal{I}^+)$ ***of the set*** $\mathcal{I}^+$ ***of the system (5.22), (5.23) at*** $t_0 \in \mathfrak{T}_i$*,*

$$\mathcal{D}_A\left(t_0;\mathcal{I}^+\right) = \mathcal{N}_L\left(t_0;\mathcal{I}^+\right). \tag{8.5}$$

$\mathcal{D}_A(t;t_0;\mathcal{I}^+)$ *is* ***the time development of the attraction domain*** $\mathcal{D}_A(t_0;\mathcal{I}^+)$ ***along the system motions*** $\mathbf{S}(;t_0;\mathbf{S}_0)$ *at the moment* $t \in \mathfrak{T}_0$*,*

$$\mathcal{D}_A\left(t;t_0;\mathcal{I}^+\right) = \left\{\mathbf{S} = \mathbf{S}(t) = \mathbf{S}(t;t_0;\mathbf{S}_0) : \ \mathbf{S}_0 \in \mathcal{D}_A\left(t_0;\mathcal{I}^+\right).\right\}, \ t \in \mathfrak{T}_0, \tag{8.6}$$

$$\mathcal{D}_A\left(t_0;t_0;\mathcal{I}^+\right) = \mathcal{D}_A\left(t_0;\mathcal{I}^+\right), \ t_0 \in \mathfrak{T}. \tag{8.7}$$

The attraction domain $\mathcal{D}_A(t;\mathfrak{T}_i;\mathcal{I}^+)$ ***of the set*** $\mathcal{I}^+$ ***on*** $\mathfrak{T}_i$ *exists if and only the following conditions hold:*

1) The set $\mathcal{I}^+$ *is attractive on* $\mathfrak{T}_i$*.*

2) $\mathcal{D}_A(t;\mathfrak{T}_i;\mathcal{I}^+)$ *is a continuous neighborhood of the set* $\mathcal{I}^+$ *for all* $t \in \mathfrak{T}_i$ *such that it is equal to the attraction domain* $\mathcal{D}_A(t_0;\mathcal{I}^+)$ *at every* $t_0 \in \mathfrak{T}_i$*,*

$$\mathcal{D}_A\left(t;\mathfrak{T}_i;\mathcal{I}^+\right) \in \mathfrak{C}\left(\mathfrak{T}_i\right), \ Cl\mathcal{I}^+ \subset In\mathcal{D}_A\left(t;\mathfrak{T}_i;\mathcal{I}^+\right), \ \forall t \in \mathfrak{T}_i,$$
$$\mathcal{D}_A\left(t_0;\mathfrak{T}_i;\mathcal{I}^+\right) = \mathcal{D}_A\left(t_0;\mathcal{I}^+\right), \ \forall t_0 \in \mathfrak{T}_i. \tag{8.8}$$

This is ***the state space attraction domain*** $\mathcal{D}_A(\mathfrak{T}_i;\mathcal{I}^+)$ ***of the set*** $\mathcal{I}^+$ ***on*** $\mathfrak{T}_i$*.*

The time evolution $Ev_{\mathfrak{T}_i}\mathcal{D}_A(t_0;\mathcal{I}^+)$ *of the attraction domain* $\mathcal{D}_A(t_0;\mathcal{I}^+)$ *in* $t_0 \in \mathfrak{T}_i$ *is* ***the integral space attraction domain*** $\mathbb{I}_A(\mathfrak{T}_i;\mathcal{I}^+)$ ***of the set*** $\mathcal{I}^+$ ***of the system (5.22), (5.23) on*** $\mathfrak{T}_i$*,*

$$\mathbb{I}_A\left(\mathfrak{T}_i;\mathcal{I}^+\right) = Ev_{\mathfrak{T}_i}\mathcal{D}_A\left(t_0;\mathcal{I}^+\right) \subseteq \mathfrak{T}_i \times \mathfrak{R}^n. \tag{8.9}$$

The attraction of the set $\mathcal{I}^+$ *at* $t_0 \in \mathfrak{T}_i$ *is* ***global (in the whole)*** *if and only if*

$$\mathcal{D}_A\left(t_0;\mathcal{I}^+\right) = \mathfrak{R}^n, \ t_0 \in \mathfrak{T}_i.$$

The attraction of the set $\mathcal{I}^+$ *is* ***global (in the whole) on*** $\mathfrak{T}_i$ *if and only if*

$$\mathcal{D}_A\left(t_0;\mathcal{I}^+\right) = \mathfrak{R}^n, \ \forall t_0 \in \mathfrak{T}_i.$$

If and only if the above conditions hold for every $t_0 \in \mathfrak{T}$*, then the term "****on*** $\mathfrak{T}_i$*" is to be omitted.*

This definition holds for the invariant set $\mathcal{I}$ *of the system* ***(5.22), (5.23)*** *if and only if the set* $\mathcal{I}$ *replaces* $\mathcal{I}^+$ *everywhere.*

8.4 Properties of attraction domains

Lemma 227 ***Positive invariance and continuity of the attraction domains of the set $\mathcal{I}^+$ of the system (5.22), (5.23)***

a) Let the set $\mathcal{I}^+$ of the system (5.22), (5.23) be attractive at $t_0 \in \mathfrak{T}_i$. Its domain of attraction $\mathcal{D}_A(t;\mathcal{I}^+)$ is a positively invariant set of the system on $\mathfrak{T}_i$.

b) Let the set $\mathcal{I}^+$ of the system (5.22), (5.23) be attractive at $t_0 \in \mathfrak{T}_i$. The time development $\mathcal{D}_A(t;t_0;\mathcal{I}^+)$ of the attraction domain $\mathcal{D}_A(t_0;\mathcal{I}^+)$ of $\mathcal{I}^+$ of the system (5.22), (5.23) is positively invariant on $\mathfrak{T}_0$ with respect to system motions, that is that

$$t_0 \in \mathfrak{T}_i,\ \mathbf{S}_0 \in \mathcal{D}_A(t_0;\mathcal{I}^+) \Longrightarrow \mathbf{S}(t;t_0;\mathbf{S}_0) \in \mathcal{D}_A(t;t_0;\mathcal{I}^+),\ \forall t \in \mathfrak{T}_0. \tag{8.10}$$

c) Let the set $\mathcal{I}^+$ of the system (5.22), (5.23) be attractive at $t_0 \in \mathfrak{T}_i$. The time development $\mathcal{D}_A(t;t_0;\mathcal{I}^+)$ of the attraction domain $\mathcal{D}_A(t_0;\mathcal{I}^+)$ is continuous in $t \in \mathfrak{T}_0$.

Proof. a) Let the set $\mathcal{I}^+$ of the system (5.22), (5.23) be attractive at $t_0 \in \mathfrak{T}_i$. Let t_0 and τ belong to $\mathfrak{T}_i$, be arbitrary, and $t_0 < \tau$. Let any $\mathbf{S}_0 \in \mathcal{D}_A(t_0;\mathcal{I}^+)$. At the moment τ the system state along the system motion is $\mathbf{S}_\tau = \mathbf{S}(\tau;\tau;\mathbf{S}_\tau)$ and $\mathbf{S}_\tau = \mathbf{S}(\tau;t_0;\mathbf{S}_0) = \mathbf{S}[\tau;\tau;\mathbf{S}(\tau;t_0;\mathbf{S}_0)]$. The choice $\mathbf{S}_0 \in \mathcal{D}_A(t_0;\mathcal{I}^+)$ guarantees that $\mathbf{S}(\sigma;t_0;\mathbf{S}_0) \longrightarrow \mathcal{I}^+$ as $\sigma \longrightarrow \infty$. This and $\mathbf{S}(\sigma;t_0;\mathbf{S}_0) = \mathbf{S}(\sigma;\tau;\mathbf{S}_\tau)$ imply $\mathbf{S}(\sigma;\tau;\mathbf{S}_\tau) \longrightarrow \mathcal{I}^+$ as $\sigma \longrightarrow \infty$, The system state $\mathbf{S}_\tau = \mathbf{S}(\tau;t_0;\mathbf{S}_0) \in \mathcal{D}_A(\tau;\mathcal{I}^+)$ due to Definition 226, which proves the positive invariance of the attraction domain $\mathcal{D}_A(t;\mathcal{I}^+)$ on $\mathfrak{T}_i$.

b) Let the set $\mathcal{I}^+$ of the system (5.22), (5.23) be attractive at $t_0 \in \mathfrak{T}_i$. The set $\mathcal{I}^+$ has the attraction domain $\mathcal{D}_A(t_0;\mathcal{I}^+)$ at $t_0 \in \mathfrak{T}_i$.Let $\mathbf{S}_0 \in \mathcal{D}_A(t_0;\mathcal{I}^+)$. This and (8.6) prove

$$\forall \mathbf{S}_0 \in \mathcal{D}_S(t_0;\mathcal{I}^+) \Longrightarrow \mathbf{S}(t;t_0;\mathbf{S}_0) \in \mathcal{D}_S(t;t_0;\mathcal{I}^+),\ \forall t \in \mathfrak{T}_0.$$

This proves the statement under b) i.e., it proves (8.10).

c) Let the set $\mathcal{I}^+$ of the system (5.22), (5.23) be attractive at $t_0 \in \mathfrak{T}_i$. Continuity of the system's motions (Property 147 and Property 148, Note 149, Section 4.11), constancy of the set $\mathcal{I}^+$ and positive invariance of $\mathcal{D}_A(t;t_0;\mathcal{I}^+)$ on $\mathfrak{T}_0$ prove the statement under c) for the following reasons. Let us assume that $\mathcal{D}_A(t;t_0;\mathcal{I}^+)$ is not continuous in $t \in \mathfrak{T}_i$. Then, in view of Definition 104 (Section 3.2.3), there are $\mathbf{S}_0^* \in \mathcal{D}_A(t_0;\mathcal{I}^+)$, $\epsilon \in \mathfrak{R}^+$, $t^* \in \mathfrak{T}_0$, and the *time* set $\mathfrak{T}_{t^*}$ composed of a *time*-sequence t_k converging to t^* as $k \longrightarrow \infty$,

$$\mathfrak{T}_{t^*} = \{t_k : t_k \in \mathfrak{T},\ k = 0, 1, 2, ...,\ k \longrightarrow \infty \Longrightarrow t_k \longrightarrow t^*\},\ t^* \in \mathfrak{T},$$

which imply the following:

$$\left[\mathcal{D}_A(t_k;t_0;\mathcal{I}^+)\, \Delta \mathcal{D}_A(t^*;t_0;\mathcal{I}^+)\right] \nsubseteq \mathcal{N}_\epsilon\left[\partial \mathcal{D}_S(t^*;t_0;\mathcal{I}^+)\right] \Longrightarrow$$
$$\|\mathbf{S}(t^*;t_0;\mathbf{S}_0^*) - \mathbf{S}(t_k;t_0;\mathbf{S}_0^*)\| > \epsilon \Longrightarrow \mathbf{S}(t;t_0;\mathbf{S}_0) \notin \mathfrak{C}(\mathfrak{T}_0),$$

which contradicts Property P4, i.e., (4.15) of Definition 117 (Section 4.2). Since the contradiction is the consequence of the assumption that the claim c) is invalid, it follows that the claim c) is valid. ∎

8.5 Uniform attraction

Uniform attraction means *the temporal uniform attraction.*

Definition 228 ***Uniform attraction of the set $\mathcal{I}^+$ of the system** (5.22), (5.23)*

*The set $\mathcal{I}^+$ of the system (5.22), (5.23) is **uniformly attractive on** $\mathfrak{T}_i$, if and only 1) and 2) are valid:*

1) The set $\mathcal{I}^+$ of the system (5.22), (5.23) is attractive at every $t_0 \in \mathfrak{T}_i$,

2) the intersection $\mathcal{D}_A(\mathfrak{T}_i;\mathcal{I}^+)$ of the attraction domain $\mathcal{D}_A(t_0;\mathcal{I}^+)$ of the set $\mathcal{I}^+$ in $t_0 \in \mathfrak{T}_i$ is a connected neighborhood of the set $\mathcal{I}^+$,

$$\mathcal{D}_A(\mathfrak{T}_i;\mathcal{I}^+) = \cap\left[\mathcal{D}_A(t_0;\mathcal{I}^+) : t_0 \in \mathfrak{T}_i\right] \Longleftrightarrow Cl\mathcal{I}^+ \subset In\mathcal{D}_A(\mathfrak{T}_i;\mathcal{I}^+), \tag{8.11}$$

such that for every $\mathbf{S}_0 \in \mathcal{D}_A(\mathfrak{T}_i;\mathcal{I}^+)$ the supremum $\tau(\mathfrak{T}_i,\varsigma,\mathbf{S}_0,\mathcal{I}^+)$ in $t_0 \in \mathfrak{T}_i$ of the minimal $\tau(t_0,\varsigma,\mathbf{S}_0,\mathcal{I}^+)$ (8.3) denoted by $\tau_m(t_0,\varsigma,\mathbf{S}_0,\mathcal{I}^+)$ is a nonnegative real number:

$$\forall(\zeta,\mathbf{S}_0) \in \mathfrak{R}^+ \times \mathcal{D}_A(\mathfrak{T}_i;\mathcal{I}^+) \Longrightarrow$$
$$\tau(\mathfrak{T}_i,\varsigma,\mathbf{S}_0,\mathcal{I}^+) = sup\left\{\tau_m(t_0,\varsigma,\mathbf{S}_0,\mathcal{I}^+) : t_0 \in \mathfrak{T}_i\right\} \in \mathfrak{R}_+. \tag{8.12}$$

The set $\mathcal{I}^+$ of the system (5.22), (5.23) is ***globally uniformly attractive in*** *$t_0 \in \mathfrak{T}_i$, for short,* ***globally uniformly attractive on*** *$\mathfrak{T}_i$, if and only if additionally the intersection $\mathcal{D}_A(\mathfrak{T}_i;\mathcal{I}^+)$ (8.11) is the whole state space),*

$$\mathcal{D}_A(\mathfrak{T}_i;\mathcal{I}^+) = \mathfrak{R}^n. \tag{8.13}$$

This definition holds for the invariant set $\mathcal{I}$ of the system ***(5.22), (5.23)*** *if and only if the set $\mathcal{I}$ replaces $\mathcal{I}^+$ everywhere.*

8.6 Uniform attraction domain

Definition 229 ***Uniform attraction domain of the set $\mathcal{I}^+$ of the system*** *(5.22), (5.23)*

The uniform attraction domain *$\mathcal{D}_{UA}(\mathfrak{T}_i;\mathcal{I}^+)$* ***of $\mathcal{I}^+$ on*** *$\mathfrak{T}_i$ exists if and only if the conditions 1) and 2) of Definition 228 are satisfied and*

$$\mathcal{D}_{UA}(\mathfrak{T}_i;\mathcal{I}^+) = \mathcal{D}_A(\mathfrak{T}_i;\mathcal{I}^+). \tag{8.14}$$

The uniform attraction of $\mathcal{I}^+$ on $\mathfrak{T}_i$ is ***global (in the whole)*** *if and only if the uniform attraction domain is the whole state space $\mathfrak{R}^n$,*

$$\mathcal{D}_{UA}(\mathfrak{T}_i;\mathcal{I}^+) = \mathfrak{R}^n$$

This definition holds for the invariant set $\mathcal{I}$ of the system ***(5.22), (5.23)*** *if and only if the set $\mathcal{I}$ replaces $\mathcal{I}^+$ everywhere.*

8.7 Nonuniform attraction

Comment 230 *The attraction characteristics $\mathcal{D}_A(t_0;\mathcal{I}^+)$ and $\tau = \tau(t_0,\varsigma,\mathbf{S}_0,\mathcal{I}^+)$ depend on the choice of the initial moment $t_0 \in \mathfrak{T}_i$. Their dependence on the initial moment $t_0 \in \mathfrak{T}_i$ can be such that*

$$\exists(\varsigma,\mathbf{S}_0) \in \mathfrak{R}^+ \times \mathcal{D}_A(t_0;\mathcal{I}^+),\ t_0 \longrightarrow \infty \Longrightarrow$$
$$\tau(t_0,\varsigma,\mathbf{S}_0,\mathcal{I}^+) \longrightarrow \infty,\ and/or\ Cl\mathcal{D}_A(t_0;\mathcal{I}^+) \longrightarrow Cl\mathcal{I}^+.$$

This means that $\mathcal{D}_A(t_0;\mathcal{I}^+)$ and/or $\tau(t_0,\varsigma,\mathbf{S}_0,\mathcal{I}^+)$ are/is nonuniform in $t_0 \in \mathfrak{T}$, respectively. In the former case, $\mathcal{D}_A(t_0;\mathcal{I}^+)$ asymptotically contracts to $\mathcal{I}^+$ as $t_0 \longrightarrow \infty$.

Definition 231 ***Nonuniform attraction of the set $\mathcal{I}^+$ of the system** (5.22), (5.23)*

*a) The set $\mathcal{I}^+$ is **nonuniformly attractive on** $\mathfrak{T}_i$ if and only if it is attractive on $\mathfrak{T}_i$ and either*

$$\exists\zeta \in \mathfrak{R}^+,\ \exists \mathbf{S}_0 \in \mathcal{D}_A\left(t_0;\mathcal{I}^+\right)\ as\ t_0 \longrightarrow sup\mathfrak{T}_i \Longrightarrow$$
$$sup\left\{\tau_m\left(t_0,\varsigma,\mathbf{S}_0,\mathcal{I}^+\right) : t_0 \in \mathfrak{T}_i\right\} = \infty, \tag{8.15}$$

or

$$\exists\zeta \in \mathfrak{R}^+,\ \exists t_0 \in \mathfrak{T}_i \Longrightarrow$$
$$sup\left\{\tau_m\left(t_0,\varsigma,\mathbf{S}_0,\mathcal{I}^+\right) : \mathbf{S}_0 \in \mathcal{D}_A\left(t_0;\mathcal{I}^+\right)\right\} = \infty, \tag{8.16}$$

or the intersection of the domain $\mathcal{D}_A\left(t_0;\mathcal{I}^+\right)$ in $t_0 \in \mathfrak{T}_i$ is not a neighborhood of the set $\mathcal{I}^+$ on $\mathfrak{T}_i$:

$$\nexists\mathcal{N}\left(\mathcal{I}^+\right) \subseteq \mathfrak{R}^n \Longrightarrow \cap\left\{\mathcal{D}_A\left(t_0;\mathcal{I}^+\right) : t_0 \in \mathfrak{T}_i\right\} \supseteq \mathcal{N}\left(\mathcal{I}^+\right), \tag{8.17}$$

or the closure of $\mathcal{D}_A\left(t_0;\mathcal{I}^+\right)$ asymptotically contracts to the closure of $\mathcal{I}^+$ as $t_0 \longrightarrow sup\mathfrak{T}_i$:

$$t_0 \in \mathfrak{T}_i,\ t_0 \longrightarrow sup\mathfrak{T}_i \Longrightarrow Cl\mathcal{D}_A\left(t_0;\mathcal{I}^+\right) \longrightarrow Cl\mathcal{I}^+. \tag{8.18}$$

*b) The set $\mathcal{I}^+$ of the system (5.22), (5.23) is **strongly nonuniformly attractive on** $\mathfrak{T}_i$ if and only if all the conditions (8.15) through (8.18) hold.*

*This definition holds for the invariant set $\mathcal{I}$ of the system **(5.22), (5.23)** if and only if the set $\mathcal{I}$ replaces $\mathcal{I}^+$ everywhere.*

8.8 Attraction domain estimate

Definition 232 ***Estimate of an attraction domain of the set $\mathcal{I}^+$***

*A set $\mathfrak{A}(.)$ is **an estimate** $\mathfrak{E}_{xx}\left(.;\mathcal{I}^+\right)$ **of an attraction domain** $\mathcal{D}_{xx}\left(.;\mathcal{I}^+\right)$ **of the constant set** $\mathcal{I}^+$ **of the system** (5.22), (5.23) if and only if it satisfies the following conditions:*

1. *It is a connected subset of $\mathcal{D}_{xx}\left(.;\mathcal{I}^+\right)$.*
2. *It is a connected neighborhood of $\mathcal{I}^+$.*
3. *It is a positive invariant set relative to the motions of the system (5.22), (5.23).*

8.9 Conclusion

Conclusion 233 ***Set attraction and equilibrium attraction***

i) A special form of a time-invariant set $\mathcal{I}^+$ is a time-invariant singleton $\mathcal{I}^+$. Its single element is a system equilibrium state $\mathbf{S}_E$,

$$\mathcal{I}^+ = \mathcal{I}_E^+ = \{\mathbf{S}_E\},\ \mathbf{S}_E \in \mathfrak{R}^n. \tag{8.19}$$

In this case, the set attraction definitions Definition 225 through Definition 231 become the definitions of attraction properties of the system equilibrium state $\mathbf{S}_E$.

The attraction criteria valid for time-invariant set $\mathcal{I}^+$ are simultaneously the attraction criteria for the system equilibrium state $\mathbf{S}_E$.

ii) The simplest form of a time-invariant set $\mathcal{I}^+$ is a time-invariant singleton $\mathcal{I}^+$, the single element of which is the system zero equilibrium state $\mathbf{S}_E = \mathbf{0}_n$,

$$\mathcal{I}^+ = \mathcal{I}_{E0}^+ = \{\mathbf{0}_n\},\ \mathbf{0}_n \in \mathfrak{R}^n. \tag{8.20}$$

In this case, the set attraction definitions Definition 225 through Definition 231 become the definitions of the attraction properties of the system zero equilibrium state $\mathbf{S}_E = \mathbf{0}_n$.

The attraction criteria valid for the time-invariant set $\mathcal{I}^+$ *are simultaneously the attraction criteria for the system zero equilibrium state* $\mathbf{S}_E = \mathbf{0}_n$.

Chapter 9

Asymptotic stability and domains

9.1 Introduction

The synthesis of the preceding chapters results in the extended and generalized asymptotic stability concept.

The study of asymptotic stability of a *time*-invariant set is simultaneously the study of asymptotic stability of any equilibrium state, hence also of asymptotic stability of the zero equilibrium state.

9.2 Asymptotic stability

Asymptotic stability joins two mutually independent qualitative dynamical system properties: stability and attraction.

Definition 234 ***Asymptotic /strict/ stability of the set $\mathcal{I}^+$ of the system (5.22), (5.23)***

1. ***The set $\mathcal{I}^+$ of the system (5.22), (5.23) is asymptotically stable at*** $t_0 \in \mathfrak{T}_i$ *if and only if it is both stable at $t_0 \in \mathfrak{T}_i$ and attractive at $t_0 \in \mathfrak{T}_i$.*

2. ***It is globally asymptotically stable at*** $t_0 \in \mathfrak{T}_i$ *if and only if it is both globally stable at $t_0 \in \mathfrak{T}_i$, respectively, and globally attractive at $t_0 \in \mathfrak{T}_i$.*

3. If and only if 1. (2.) holds for every $t_0 \in \mathfrak{T}_i$ then ***the set $\mathcal{I}^+$ of the system (5.22), (5.23) is (globally) asymptotically stable on*** $\mathfrak{T}_i$, *respectively.*

9.3 Asymptotic stability domains

AS is the abbreviation for *asymptotic stability.*

Definition 235 ***Asymptotic stability /strict asymptotic stability/ domain of the set $\mathcal{I}^+$ of the system (5.22), (5.23)***

The asymptotic stability /the strict asymptotic stability/ domain $\mathcal{D}_{AS}(t_0; \mathcal{I}^+)$ $/\mathcal{D}_{SAS}(t_0; \mathcal{I}^+)/$ ***of the set $\mathcal{I}^+$ of the system (5.22), (5.23) at the initial moment***

DOI: 10.1201/9781003744528-14

t_0 exists and it is the intersection of the stability /the strict stability/ domain $\mathcal{D}_S(t_0;\mathcal{I}^+)$ /$\mathcal{D}_{SS}(t_0;\mathcal{I}^+)$/ and the attraction domain $\mathcal{D}_A(t_0;\mathcal{I}^+)$,

$$\mathcal{D}_{AS}\left(t_0;\mathcal{I}^+\right) = \mathcal{D}_S\left(t_0;\mathcal{I}^+\right) \cap \mathcal{D}_A\left(t_0;\mathcal{I}^+\right),\ t_0 \in \mathfrak{T}_i,$$
$$/\mathcal{D}_{SAS}\left(t_0;\mathcal{I}^+\right) = \mathcal{D}_{SS}\left(t_0;\mathcal{I}^+\right) \cap \mathcal{D}_A\left(t_0;\mathcal{I}^+\right),\ t_0 \in \mathfrak{T}_i/, \tag{9.1}$$

if and only if the intersection is a neighborhood of the set $\mathcal{I}^+$,

$$\exists \mathcal{D}_{AS}\left(t_0;\mathcal{I}^+\right) \Longleftrightarrow Cl\mathcal{I}^+ \subset In\left[\mathcal{D}_S\left(t_0;\mathcal{I}^+\right) \cap \mathcal{D}_A\left(t_0;\mathcal{I}^+\right)\right],\ t_0 \in \mathfrak{T}_i,$$
$$/\exists \mathcal{D}_{SAS}\left(t_0;\mathcal{I}^+\right) \Longleftrightarrow Cl\mathcal{I}^+ \subset In\left[\mathcal{D}_{SS}\left(t_0;\mathcal{I}^+\right) \cap \mathcal{D}_A\left(t_0;\mathcal{I}^+\right)\right],\ t_0 \in \mathfrak{T}_i/ \tag{9.2}$$

respectively.

***The asymptotic stability of the set $\mathcal{I}^+$ of the system (5.22), (5.23) at the initial moment** t_0 **is global [in the whole]** if and only if the asymptotic stability domain $\mathcal{D}_{AS}(t_0;\mathcal{I}^+)$ is the whole state space $\mathfrak{R}^n$:*

$$\mathcal{D}_{AS}\left(t_0;\mathcal{I}^+\right) = \mathfrak{R}^n,\ t_0 \in \mathfrak{T}_i, \tag{9.3}$$

respectively.

*$\mathcal{D}_{AS}(t;t_0;\mathcal{I}^+)$ /$\mathcal{D}_{SAS}(t;t_0;\mathcal{I}^+)$/ is **the time development of the asymptotic stability /the strict asymptotic stability/ domain** $\mathcal{D}_{AS}(t_0;\mathcal{I}^+)$ /$\mathcal{D}_{SAS}(t_0;\mathcal{I}^+)$/ **along the system motions** $\mathcal{S}(;t_0;\mathbf{S}_0)$ at a moment t when t varies over $\mathfrak{T}_0$,*

$$\mathcal{D}_{AS}\left(t;t_0;\mathcal{I}^+\right) = \left\{\mathbf{S} : \mathbf{S} = \mathbf{S}(t) = \mathbf{S}(t;t_0;\mathbf{S}_0),\ \mathbf{S}_0 \in \mathcal{D}_{AS}\left(t_0;\mathcal{I}^+\right)\right\},\ t \in \mathfrak{T}_0$$
$$/\mathcal{D}_{SAS}\left(t;t_0;\mathcal{I}^+\right) = \left\{\mathbf{S} : \mathbf{S} = \mathbf{S}(t) = \mathbf{S}(t;t_0;\mathbf{S}_0),\ \mathbf{S}_0 \in \mathcal{D}_{SAS}\left(t_0;\mathcal{I}^+\right)\right\},\ t \in \mathfrak{T}_0/, \tag{9.4}$$
$$\mathcal{D}_{AS}\left(t_0;t_0;\mathcal{I}^+\right) = \mathcal{D}_{AS}\left(t_0;\mathcal{I}^+\right),\ /\mathcal{D}_{SAS}\left(t_0;t_0;\mathcal{I}^+\right) = \mathcal{D}_{SAS}\left(t_0;\mathcal{I}^+\right)/,\ t_0 \in \mathfrak{T}, \tag{9.5}$$

respectively.

***The asymptotic stability /the strict asymptotic stability/ domain** $\mathcal{D}_{AS}(t;\mathfrak{T}_i;\mathcal{I}^+)$ /$\mathcal{D}_{SAS}(t;\mathfrak{T}_i;\mathcal{I}^+)$/ **of the set** $\mathcal{I}^+$ **on** $\mathfrak{T}_i$ exists and it is the intersection of the /strict/ stability domain $\mathcal{D}_S(t;\mathfrak{T}_i;\mathcal{I}^+)$ /$\mathcal{D}_{SS}(t;\mathfrak{T}_i;\mathcal{I}^+)$/ and the attraction domain $\mathcal{D}_A(t;\mathfrak{T}_i;\mathcal{I}^+)$ on $\mathfrak{T}_i$ if and only if the intersection is a neighborhood of the set $\mathcal{I}^+$,*

$$\exists \mathcal{D}_{AS}\left(t;\mathfrak{T}_i;\mathcal{I}^+\right) \Longleftrightarrow Cl\mathcal{I}^+ \subset In\left[\mathcal{D}_S\left(t;\mathfrak{T}_i;\mathcal{I}^+\right) \cap \mathcal{D}_A\left(\mathfrak{T}_i;\mathcal{I}^+\right)\right],\ \forall t \in \mathfrak{T}_i \Longrightarrow$$
$$\mathcal{D}_{AS}\left(t;\mathfrak{T}_i;\mathcal{I}^+\right) = \mathcal{D}_S\left(t;\mathfrak{T}_i;\mathcal{I}^+\right) \cap \mathcal{D}_A\left(t;\mathfrak{T}_i;\mathcal{I}^+\right),\ \forall t \in \mathfrak{T}_i$$
$$/\exists \mathcal{D}_{SAS}\left(t;\mathfrak{T}_i;\mathcal{I}^+\right) \Longleftrightarrow Cl\mathcal{I}^+ \subset In\left[\mathcal{D}_{SS}\left(t;\mathfrak{T}_i;\mathcal{I}^+\right) \cap \mathcal{D}_A\left(t;\mathfrak{T}_i;\mathcal{I}^+\right)\right],\ \forall t \in \mathfrak{T}_i \Longrightarrow$$
$$\mathcal{D}_{SAS}\left(t;\mathfrak{T}_i;\mathcal{I}^+\right) = \mathcal{D}_{SS}\left(t;\mathfrak{T}_i;\mathcal{I}^+\right) \cap \mathcal{D}_A\left(t;\mathfrak{T}_i;\mathcal{I}^+\right),\ \forall t \in \mathfrak{T}_i/, \tag{9.6}$$

respectively.

The integral space asymptotic /strict/ stability domain

$$\mathbb{I}_{AS}\left(\mathfrak{T}_i;\mathcal{I}^+\right) \quad /\mathbb{I}_{SAS}\left(\mathfrak{T}_i;\mathcal{I}^+\right)/$$

***of** $\mathcal{I}^+$ **on** $\mathfrak{T}_i$ is the time evolution $Ev_{\mathfrak{T}_i}\mathcal{D}_{AS}(t_0;\mathcal{I}^+)$ /$Ev_{\mathfrak{T}_i}\mathcal{D}_{SAS}(t_0;\mathcal{I}^+)$/ of the asymptotic stability /strict asymptotic stability/ domain $\mathcal{D}_S(t_0;\mathcal{I}^+)$ /$\mathcal{D}_{SS}(t_0;\mathcal{I}^+)$/ in t_0 over $\mathfrak{T}_i$,*

$$\mathbb{I}_{AS}\left(\mathfrak{T}_i;\mathcal{I}^+\right) = Ev_{\mathfrak{T}_i}\mathcal{D}_{AS}\left(t_0;\mathcal{I}^+\right) \subseteq \mathfrak{T}_i \times \mathfrak{R}^n,$$
$$/\mathbb{I}_{SAS}\left(\mathfrak{T}_i;\mathcal{I}^+\right) = Ev_{\mathfrak{T}_i}\mathcal{D}_{SAS}\left(t_0;\mathcal{I}^+\right) \subseteq \mathfrak{T}_i \times \mathfrak{R}^n/, \tag{9.7}$$

respectively.

The asymptotic stability of the set $\mathcal{I}^+$ of the system (5.22), (5.23) is global [in the whole] at $t_0 \in \mathfrak{T}_i$ *if and only if $\mathcal{I}^+$ is asymptotically stable and the asymptotic stability domain $\mathcal{D}_{AS}(t_0;\mathcal{I}^+)$ of $\mathcal{I}^+$ at $t_0 \in \mathfrak{T}_i$ is the whole state space $\mathfrak{R}^n$:*

$$\mathcal{D}_{AS}\left(t_0;\mathcal{I}^+\right) = \mathfrak{R}^n \ , \ t_0 \in \mathfrak{T}_i.$$

respectively.

The asymptotic stability of the set $\mathcal{I}^+$ of the system (5.22), (5.23) on $\mathfrak{T}_i$ is global /in the whole/ *if and only if the asymptotic stability domain $\mathcal{D}_{AS}(t_0;\mathcal{I}^+)$ is the whole state space $\mathfrak{R}^n$ for every $t_0 \in \mathfrak{T}_i$:*

$$\mathcal{D}_{AS}\left(t_0;\mathcal{I}^+\right) = \mathfrak{R}^n, \ \forall t_0 \in \mathfrak{T}_i, \tag{9.8}$$

respectively.

*If and only if the preceding conditions hold for $\mathfrak{T}_i = \mathfrak{T}$ then the expression "**on** $\mathfrak{T}_i$" is to be omitted.*

*This definition holds for the invariant set $\mathcal{I}$ of the system **(5.22), (5.23)** if and only if the set $\mathcal{I}$ replaces $\mathcal{I}^+$ everywhere.*

9.4 Properties of *AS* domains

Lemma 236 ***Positive invariance and continuity of the /strict/ asymptotic stability domain of the set $\mathcal{I}^+$ of the system (5.22), (5.23)***

Let the set $\mathcal{I}^+$ of the system (5.22), (5.23) be asymptotically stable on $\mathfrak{T}_i$.

a) The domain $\mathcal{D}_{AS}(t;\mathcal{I}^+)$ of asymptotic stability of the set $\mathcal{I}^+$ is a positively invariant set of the system on $\mathfrak{T}_i$.

*b) The time development $\mathcal{D}_{AS}(t;t_0;\mathcal{I}^+)$ /$\mathcal{D}_{ASS}(t;t_0;\mathcal{I}^+)$/ of the asymptotic /strict asymptotic/ stability domain $\mathcal{D}_{AS}(t_0;\mathcal{I}^+)$ /$\mathcal{D}_{ASS}(t_{0i};\mathcal{I}^+)$/ of $\mathcal{I}^+$ of the system **(5.22), (5.23)** is positively invariant on $\mathfrak{T}_0$ with respect to system motions, that is,*

$$t_0 \in \mathfrak{T}_i,\ \mathbf{S}_0 \in \mathcal{D}_{AS}\left(t_0;\mathcal{I}^+\right) \quad /t_0 \in \mathfrak{T}_i,\ \mathbf{S}_0 \in \mathcal{D}_{SAS}\left(t_0;\mathcal{I}^+\right)/ \Longrightarrow$$
$$\mathbf{S}(t;t_0;\mathbf{S}_0) \in \mathcal{D}_{AS}\left(t;t_0;\mathcal{I}^+\right) \quad /\mathbf{S}(t;t_0;\mathbf{S}_0) \in \mathcal{D}_{SAS}\left(t;t_0;\mathcal{I}^+\right)/,\ \forall t \in \mathfrak{T}_0, \tag{9.9}$$

respectively.

c) The time development $\mathcal{D}_{AS}(t;t_0;\mathcal{I}^+)$ /$\mathcal{D}_{ASS}(t;t_0;\mathcal{I}^+)$/ of the asymptotic /the strict asymptotic/ stability domain $\mathcal{D}_{AS}(t_0;\mathcal{I}^+)$ /$\mathcal{D}_{SAS}(t_0;\mathcal{I}^+)$/ is continuous in $t \in \mathfrak{T}_0$, respectively.

Proof. Definition 204, Section 7.4, Definition 206, Section 7.5, Lemma 208, Section 7.6, and Lemma 227, Section 8.2 imply Lemma 236. ■

9.5 Uniform asymptotic stability

Definition 237 ***Uniform asymptotic stability of the set $\mathcal{I}^+$ of the system (5.22), (5.23)***

The set $\mathcal{I}^+$ of the system (5.22), (5.23) is uniformly asymptotically stable on $\mathfrak{T}_i$ *if and only if it is both uniformly stable on $\mathfrak{T}_i$ and uniformly attractive on $\mathfrak{T}_i$.*

If and only if the preceding conditions hold for $\mathfrak{T}_i = \mathfrak{T}$ then ***the expression "on $\mathfrak{T}_i$" is to be omitted.***

*This definition holds for the invariant set $\mathcal{I}$ of the system **(5.22), (5.23)** if and only if the set $\mathcal{I}$ replaces the set $\mathcal{I}^+$ everywhere.*

9.6 Uniform AS domain

Definition 238 ***Uniform asymptotic /strict asymptotic/ stability domain of the set $\mathcal{I}^+$ of the system (5.22), (5.23)***

The uniform asymptotic /strict asymptotic/ stability domain

$$\mathcal{D}_{UAS}\left(\mathfrak{T}_i;\mathcal{I}^+\right) \quad /\mathcal{D}_{USAS}\left(\mathfrak{T}_i;\mathcal{I}^+\right)/$$

of the set $\mathcal{I}^+$ of the system (5.22), (5.23) on $\mathfrak{T}_i$ *exists and it is the intersection of the system uniform /strict/ stability domain $\mathcal{D}_{US}\left(\mathfrak{T}_i;\mathcal{I}^+\right)$ $/\mathcal{D}_{USS}\left(\mathfrak{T}_i;\mathcal{I}^+\right)/$ and the uniform attraction domain $\mathcal{D}_{UA}\left(\mathfrak{T}_i;\mathcal{I}^+\right)$,*

$$\mathcal{D}_{UAS}\left(\mathfrak{T}_i;\mathcal{I}^+\right)=\mathcal{D}_{US}\left(\mathfrak{T}_i;\mathcal{I}^+\right)\cap\mathcal{D}_{UA}\left(\mathfrak{T}_i;\mathcal{I}^+\right)$$
$$/\mathcal{D}_{USAS}\left(\mathfrak{T}_i;\mathcal{I}^+\right)=\mathcal{D}_{USS}\left(\mathfrak{T}_i;\mathcal{I}^+\right)\cap\mathcal{D}_{UA}\left(\mathfrak{T}_i;\mathcal{I}^+\right)/, \tag{9.10}$$

if and only if the intersection is connected neighborhood of the set $\mathcal{I}^+$,

$$\exists\mathcal{D}_{UAS}\left(\mathfrak{T}_i;\mathcal{I}^+\right)\Longleftrightarrow Cl\mathcal{I}^+\subset In\left[\mathcal{D}_{US}\left(\mathfrak{T}_i;\mathcal{I}^+\right)\cap\mathcal{D}_{UA}\left(\mathfrak{T}_i;\mathcal{I}^+\right)\right]$$
$$/\exists\mathcal{D}_{USAS}\left(\mathfrak{T}_i;\mathcal{I}^+\right)\Longleftrightarrow Cl\mathcal{I}^+\subset In\left[\mathcal{D}_{USS}\left(\mathfrak{T}_i;\mathcal{I}^+\right)\cap\mathcal{D}_{UA}\left(\mathfrak{T}_i;\mathcal{I}^+\right)\right]/, \tag{9.11}$$

respectively.

The uniform asymptotic /strict asymptotic/ stability of the set $\mathcal{I}^+$ of the system (5.22), (5.23) on $\mathfrak{T}_i$ is global [in the whole] *if and only if the uniform asymptotic /strict asymptotic/ stability domain $\mathcal{D}_{UAS}\left(\mathfrak{T}_i;\mathcal{I}^+\right)$ $/\mathcal{D}_{USAS}\left(\mathfrak{T}_i;\mathcal{I}^+\right)/$ is the whole state space $\mathfrak{R}^n$:*

$$\mathcal{D}_{UAS}\left(\mathfrak{T}_i;\mathcal{I}^+\right)=\mathfrak{R}^n \quad /\mathcal{D}_{USAS}\left(\mathfrak{T}_i;\mathcal{I}^+\right)=\mathfrak{R}^n/,$$

respectively.

If and only if the preceding conditions hold for $\mathfrak{T}_i=\mathfrak{T}$, then ***the expression "on $\mathfrak{T}_i$" is to be omitted.***

9.7 Nonuniform asymptotic stability

Definition 239 ***Nonuniform asymptotic stability of the set $\mathcal{I}^+$ of the system (5.22), (5.23)***

The set $\mathcal{I}^+$ of the system (5.22), (5.23) is nonuniformly asymptotically stable on $\mathfrak{T}_i$ *if and only if it is asymptotically stable on $\mathfrak{T}$ and any of the following conditions holds:*

1) The settling time $\tau_m\left[t_0,\varsigma,\mathbf{S}_0,\mathcal{I}^+\right]$ escapes to infinity on $\mathfrak{T}_i$ for some pair

$$(\zeta,\mathbf{S}_0)\in\mathfrak{R}^+\times\mathcal{D}_{AS}\left(t_0;\mathcal{I}^+\right) ,$$

i.e.,

$$\exists(\zeta,\mathbf{S}_0)\in\mathfrak{R}^+\times\mathcal{D}_{AS}\left(t_0;\mathcal{I}^+\right):$$
$$t_0\longrightarrow sup\mathfrak{T}_i\Longrightarrow sup\left\{\tau_m\left[t_0,\varsigma,\mathbf{S}_0,\mathcal{I}^+\right]:t_0\in\mathfrak{T}_i\right\}=\infty, \tag{9.12}$$

respectively.

2) The settling time $\tau_m\left[t_0,\varsigma,\mathbf{S}_0,\mathcal{I}^+\right]$ escapes to infinity on the domain of asymptotic stability $\mathcal{D}_{AS}\left(t_0;\mathcal{I}^+\right)$ for some $(\zeta,t_0)\in\mathfrak{R}^+\times\mathfrak{T}_i$.

$$\exists(\zeta,t_0)\in\mathfrak{R}^+\times\mathfrak{T}_i\Longrightarrow$$
$$sup\left\{\tau_m\left(t_0,\varsigma,\mathbf{S}_0,\mathcal{I}^+\right):\mathbf{S}_0\in\mathcal{D}_{AS}\left(t_0;\mathcal{I}^+\right)\right\}=\infty. \tag{9.13}$$

respectively.

3) The intersection of the domains $\mathcal{D}_{AS}\left(t_0;\mathcal{I}^+\right)$ */*$\mathcal{D}_{SAS}\left(t_0;\mathcal{I}^+\right)$*/ in* $t_0 \in \mathfrak{T}_i$ *is not a neighborhood of the set* $\mathcal{I}^+$ *:*

$$\nexists\mathcal{N}\left(\mathcal{I}^+\right) \subseteq \mathfrak{R}^n \Longrightarrow \cap\left\{\mathcal{D}_{AS}\left(t_0;\mathcal{I}^+\right) : t_0 \in \mathfrak{T}_i\right\} \supseteq \mathcal{N}\left(\mathcal{I}^+\right),$$
$$\nexists\mathcal{N}\left(\mathcal{I}^+\right) \subseteq \mathfrak{R}^n \Longrightarrow \cap\left\{\mathcal{D}_{SAS}\left(t_0;\mathcal{I}^+\right) : t_0 \in \mathfrak{T}_i\right\} \supseteq \mathcal{N}\left(\mathcal{I}^+\right), \tag{9.14}$$

respectively, and/or the closure of $\mathcal{D}_{AS}\left(t_0;\mathcal{I}^+\right)$ */*$\mathcal{D}_{SAS}t_0;\mathcal{I}^+$*/ asymptotically contracts to the closure of* $\mathcal{I}^+$ *as* $t_0 \longrightarrow \infty$ *:*

$$t_0 \in \mathfrak{T}_i,\ t_0 \longrightarrow sup\mathfrak{T}_i = \infty \Longrightarrow Cl\mathcal{D}_{AS}\left(t_0;\mathcal{I}^+\right)\ /Cl\mathcal{D}_{SAS}\left(t_0;\mathcal{I}^+\right)/ \longrightarrow Cl\mathcal{I}^+. \tag{9.15}$$

The set $\mathcal{I}^+$ ***of the system (5.22), (5.23) is strongly nonuniformly asymptotically stable on*** $\mathfrak{T}_i$ *if and only if it is asymptotically stable on* $\mathfrak{T}_i$ *and all the above conditions (9.12)-(9.15) hold, respectively.*

If and only if the preceding conditions hold for $\mathfrak{T}_i = \mathfrak{T}$*, then the expression "**on** $\mathfrak{T}_i$" is to be omitted.*

This definition holds for the invariant set $\mathcal{I}$ *of the system **(5.22), (5.23)** if and only if the set* $\mathcal{I}$ *replaces* $\mathcal{I}^+$ *everywhere.*

9.8 Estimate of *AS* domain

Definition 240 ***Estimate of an asymptotic stability domain of the set*** $\mathcal{I}^+$

A set $\mathfrak{A}(.)$ *is **an estimate*** $\mathfrak{E}_{xx}\left(.;\mathcal{I}^+\right)$ ***of an asymptotic stability domain*** $\mathcal{D}_{xx}\left(.;\mathcal{I}^+\right)$ ***of the set*** $\mathcal{I}^+$ ***of the system (5.22), (5.23)*** *if and only if it satisfies the following conditions:*

1. It is a connected subset of $\mathcal{D}_{xx}\left(.;\mathcal{I}^+\right)$*.*

2. It is a connected neighborhood of $\mathcal{I}^+$*.*

*3. It is a positive invariant set relative to the motions of the system **(5.22), (5.23)**.*

9.9 Conclusion

Conclusion 241 ***Set asymptotic stability and equilibrium state asymptotic stability***

i) A special form of the set $\mathcal{I}^+$ *is a time-invariant singleton. Its single element is a system equilibrium state* $\mathbf{S}_E$*,*

$$\mathcal{I}^+ = \mathcal{I}_E^+ = \left\{\mathbf{S}_E\right\},\ \mathbf{S}_E \in \mathfrak{R}^n. \tag{9.16}$$

In this case, the set asymptotic stability definitions: Definition 234 through Definition 239 become the definitions of asymptotic stability properties of a system equilibrium state $\mathbf{S}_E$*.*

ii) The simplest form of the set $\mathcal{I}^+$ *is a time-invariant singleton, the single element of which is the system zero equilibrium state* $\mathbf{S}_E = \mathbf{0}_n$*,*

$$\mathcal{I}^+ = \mathcal{I}_{E0}^+ = \left\{\mathbf{0}_n\right\},\ \mathbf{0}_n \in \mathfrak{R}^n. \tag{9.17}$$

In this case, the set asymptotic stability definitions Definition 234 through Definition 239 become the definitions of the asymptotic stability properties of the system zero equilibrium state $\mathbf{S}_E = \mathbf{0}_n$*.*

Part VI

CONSISTENT STABILITY METHODOLOGY

Chapter 10

General consideration

10.1 Physical origin of the Lyapunov method

Energy and/or matter transmissions and/or external disturbances are the inherent causes of *time* variations of physical variable values.

When the input vector and the initial conditions are nominal, then for the system to stay in the nominal regime it is mainly necessary and sufficient to guarantee that energy and matter flows are nominal. If the input vector is nominal all the *time*, but some or all initial conditions are not nominal, then energy and matter flows are not nominal. The goal is then to ensure that their real flows stay close to the nominal ones, and additionally, for better system behavior, to achieve the convergence of the former to the latter.

Let the values of physical variables be divided by their units. The so-obtained *normalized* values of the variables *preserve their numerical values* but they are *dimensionless*. Physical variables with dimensionless values are called *dimensionless variables*. We use only them throughout the book.

Let $E(t)$ denote the (dimensionless) instantaneous accumulated energy and $M(t)$ be the (dimensionless) instantaneous accumulated matter contained in the system at a moment $t \in \mathfrak{T}$. Let $\sigma \in \mathfrak{T}$, $\sigma < t$, $t = \sigma + \varepsilon$, $\varepsilon \longrightarrow 0^+$.

Let it be assumed that the initial (at t_0) accumulated matter $M(t_0)$ is nominal and rests the nominal $M_N(t)$ all the time: $M(t) = M_N(t), \forall t \in \mathfrak{T}$.

If the real energy $E(\tau)$ is bigger than the nominal $E_N(\tau)$ at any $\tau \in \mathfrak{T}_0$, then $E(t)$ should not increase faster than $E_N(t)$, or better, $E(t)$ should increase slower than $E_N(t)$; i.e., its *time* derivative $E^{(1)}(t)$ should not be bigger than *time* derivative $E_N^{(1)}(t)$ of $E_N(t)$, or better, the former should be less than the latter,

$$E(\tau) > E_N(\tau) \text{ } requests \text{ } E^{(1)}(t) \leq E_N^{(1)}(t), \text{ } or \text{ } better$$
$$E^{(1)}(t) < E_N^{(1)}(t), \text{ } t = \tau + \varepsilon; \text{ } \varepsilon \longrightarrow 0^+ \Longrightarrow t \longrightarrow \tau^+. \tag{10.1}$$

If the real $E(\tau)$ is smaller than the nominal $E_N(\tau)$, then $E(t)$ should not decrease faster than $E_N(t)$, or better, $E(t)$ should decrease slower than $E_N(t)$; i.e., its *time* derivative $E^{(1)}(t)$ should not be smaller than $E_N^{(1)}(t)$, or better, $E^{(1)}(t)$ should be bigger than $E_N^{(1)}(t)$,

$$E(\tau) < E_N(\tau) \Longrightarrow E^{(1)}(t) \geq E_N^{(1)}(t), \text{ } or \text{ } better$$
$$E^{(1)}(t) > E_N^{(1)}(t), \text{ } t = \tau + \varepsilon; \text{ } \varepsilon \longrightarrow 0^+ \Longrightarrow t \longrightarrow \tau^+. \tag{10.2}$$

DOI: 10.1201/9781003744528-16

If the real $E(\tau)$ is equal to the nominal $E_N(\tau)$, then $E(t)$ should rest equal to $E_N(t)$; i.e., its *time* derivative $E^{(1)}(t)$ should be equal to $E_N^{(1)}(t)$,

$$E(\tau) = E_N(\tau) \Longrightarrow E^{(1)}(t) = E_N^{(1)}(t), \quad t = \tau + \varepsilon;\ \varepsilon \longrightarrow 0^+ \Longrightarrow t \longrightarrow \tau^+. \tag{10.3}$$

We can express this analysis in the following compact mathematical form:

$$[E(\tau) \ - E_N(\tau)]\left[E^{(1)}(\tau) \ - \ E_N^{(1)}(\tau)\right] \le 0, \tag{10.4}$$

or better,

$$[E(\tau) - E_N(\tau)]\left[E^{(1)}(\tau) - E_N^{(1)}(v)\right]\left\{ \begin{array}{c} <0, \ if \ [E(\tau) - E_N(\tau)] \neq 0, \\ =0, \ if \ [E(\tau) - E_N(\tau)] = 0 \end{array} \right\}. \tag{10.5}$$

When we consider the *limit* as $\varepsilon \longrightarrow 0^+$, then $t \longrightarrow \tau^+$ and the preceding conditions (10.1) through (10.5) yield in the limit, when $t = \tau$:

$$[E(t) - E_N(t)]\left[E^{(1)}(t) - E_N^{(1)}(t)\right] \le 0, \tag{10.6}$$

or better,

$$[E(t) - E_N(t)]\left[E^{(1)}(t) - E_N^{(1)}(t)\right]\left\{ \begin{array}{c} <0, \ if \ [E(t) - E_N(t)] \neq 0, \\ =0, \ if \ [E(t) - E_N(t)] = 0 \end{array} \right\}. \tag{10.7}$$

The left-hand sides of these expressions represent the *time* derivative of

$$[E(t) - E_N(t)]^2/2$$

so that we can them set into the following forms:

$$\frac{d}{dt}\left\{\frac{1}{2}[E(t) - E_N(t)]^2\right\} \le 0, \tag{10.8}$$

or better, (after multiplying (10.8) with 2,

$$\frac{d}{dt}\left\{[E(t) - E_N(t)]^2\right\}\left\{ \begin{array}{c} <0, \ if \ [E(t) - E_N(t)] \neq 0, \\ =0, \ if \ [E(t) - E_N(t)] = 0 \end{array} \right\}. \tag{10.9}$$

Lyapunov generalized the function $E(.)$ in [335]. He introduced the concepts of *sign semidefinite* and *sign definite functions* denoted by $V(.)$, or by $v(.)$, which play the role of the squared difference function $[E(t) - E_N(t)]^2$ in the general mathematical setting. Roughly speaking, just to explain the essence of the idea, we can define $V(.)$ by

$$V(t) = [E(t) - E_N(t)]^2. \tag{10.10}$$

To summarize, the appropriate function $V(.)$ and its total *time* derivative along system motions $V^{(1)}(.)$ should obey

$$V(t, ...) \ge 0, \frac{dV(t, ...)}{dt} \le 0, \tag{10.11}$$

or better,

$$\frac{dV(t, ...)}{dt}\left\{ \begin{array}{c} <0, \ if \ V(t, ...) > 0, \\ =0, \ if \ V(t, ...) = 0 \end{array} \right\}. \tag{10.12}$$

Such a function $V(.)$ is a *Lyapunov function* of the system.

If the accumulated mass $M(t)$ of the system is also *time*-varying, then the above consideration is directly applicable to the mass. Then the physical nature of the system induces the vector $\mathbf{V}(t)$,

$$\mathbf{V}(t) = \begin{bmatrix} [E(t) - E_N(t)]^2 \\ [M(t) - M_N(t)]^2 \end{bmatrix}. \tag{10.13}$$

This is a physical justification for *the vector Lyapunov Function* concept established coincidentally by Bellman [27] in the framework of linear systems and by Matrosov [363] established in the general setting of *time*-varying nonlinear systems. We can also use the scalar Lyapunov function in the form $V(t) = [E(t) - E_N(t)]^2 + [M(t) - M_N(t)]^2$ instead of $\mathbf{V}(t)$, (10.13).

10.2 Lyapunov method

The famous Lyapunov method and methodologies, [335], have been developed for stability studies and stabilizing control synthesis of the ISO nonlinear systems. The literature on it is particularly huge, e.g., [1]-[3], [5], [6], [11], [17]-[21], [26], [27], [35], [60], [61], [63], [68], [70], [71], [73], [74], [76], [83], [103], [104], [109], [128]-[140], [127]-[117], [118]-[138], [115]-[134], [153]-[160], [175]-[177], [207], [219]-[224], [247], [253], [257], [262], [265], [270], [271], [272], [278], [296], [301], [307], [310], [311], [312], [322], [323], [325], [326], [333], [338], [339], [340], [341], [342]-[352], [355], [356], [357]-[365], [366]-[368], [379]-[383], [391], [396], [397], [411], [438], [440], [445], [446], [449], [456], [457], [460]-[474], [481], [495], [498]-[502], [509]-[515], [521]-[524], [534], [541].

Lyapunov distinguished two general methods related to dynamical systems. One of them, known as *the first Lyapunov method,* demands solution of the system mathematical model and study of the system properties by analyzing its solutions *for every initial condition.* This method can be effective for *time*-invariant linear systems, but it is generally inapplicable to the nonlinear systems. Another Lyapunov method, known as *the second* (or *direct*) *Lyapunov method*, for short, the *Lyapunov method*, establishes two methodologies and conditions how to study properties of linear and nonlinear dynamical systems without a need to solve their mathematical models, hence, without knowing their solutions.

10.3 Main original Lyapunov results

10.3.1 Main results for linear systems

Let us present Lyapunov's original results on stability of *time*-invariant linear systems by using the notation accepted in this book (for their presentation by using Lyapunov's original notation see [230, Section 13.3, pp. 275-286]). They concern the ISO linear systems in the free regime and in terms of the state vector deviation $\mathbf{x} = \mathbf{X} - \mathbf{X}_d$ described by:

$$\frac{d\mathbf{x}}{dt} = \mathbf{x}^{(1)} = A\mathbf{x},\ \mathbf{x} = \begin{bmatrix} x_1 & x_2 & ... & x_n \end{bmatrix}^T \in \mathfrak{R}^n. \tag{10.14}$$

Theorem 242 ***Original Lyapunov's Theorem I on the ISO linear systems in the free regime [335, pp. 75, 76 in the Russian edition]***

When the eigenvalues s_i of the matrix A of the time-invariant linear system (10.14) are such that, for every given natural number m, the relationships of the following form

$$m_1 s_1 + m_2 s_2 + ... + m_n s_n = 0,$$

are impossible, in which all m_i are nonnegative integers, the sum of which is m, then it is always possible to find exactly one full homogeneous function V of the power m of variables x_i, which fulfills the equation

$$grad^T V(\mathbf{x}) A\mathbf{x} = U(\mathbf{x}) \tag{10.15}$$

for an arbitrarily given fully homogenous function U of the variables x_i also of the power m.

This theorem is the basis for Lyapunov's following fundamental theorem on stability of the *time*-invariant linear system (10.14):

Theorem 243 ***Original Lyapunov's Theorem II on the ISO linear systems in the free regime [335, pp. 76, 77 in the Russian edition]***

When the real parts of all eigenvalues of the matrix A of the time-invariant linear system (10.14) are negative and when in Equation (10.15) the function U is sign definite of any even power m, then the form V of the power m, which obeys the same equation, is also sign definite and with the sign opposite to the sign of U.

This theorem presents the necessary conditions for stability of the system (10.14). The sufficiency of the conditions of the theorem follows from Lyapunov's general conditions presented in the following Theorem and Note 245 244.

10.3.2 Main results for nonlinear systems

Theorem 244 ***Original Lyapunov's Theorem I on the ISO nonlinear systems in the free regime [335, pp. 59, 60 in the Russian edition]***

If the differential equations of the perturbed motions are such, that it is possible to find a sign definite function $V(.)$, the derivative $V^{(1)}(.)$ of which due to those equations is either a sign semidefinite function with the sign inverse to the sign of $V(.)$, or identically equal to zero, the unperturbed motion is stable.

Note 245 ***Original Lyapunov's Note II on the ISO nonlinear systems in the free regime [335, p. 61 in the Russian edition]***

If the function $V(.)$, which satisfies the conditions of Theorem 244, simultaneously permits the infinitesimally small upper bound, and its derivative is a sign definite function, then it is possible to prove that every perturbed motion, sufficiently close to the unperturbed one, will converge to it asymptotically.

Comment 246 *Every time-varying decrescent function has the infinitesimally small upper bound and vice versa. The difference is only terminological.*

10.4 Lyapunov methodologies

10.4.1 Consistent Lyapunov methodology: linear systems

The Lyapunov stability methodology established for *time*-invariant linear systems begins with an arbitrary selection of a negative definite function and continues with the test of the properties of the solution of the corresponding differential equation. Its left-hand side contains the terms of the system mathematical model, and its right-hand side contains any accepted negative definite function. The solution function is called *the Lyapunov function of the system.* The test completely resolves the problem of asymptotic stability and the exact determination of the Lyapunov function in the framework of the *time*-invariant linear systems. This is **the direct Lyapunov methodology**. Lyapunov established it only for *time*-invariant linear systems.

Original Lyapunov's Theorem 242-Theorem 244, Note 245 (Section 10.4.2) have enabled the following well-known basic results.

Let us remember that the *time*-invariant linear system (10.14) is **stable** if and only if its zero equilibrium state $\mathbf{x} = \mathbf{0}_n$ is globally (in the whole) asymptotically stable.

Theorem 247 ***The full Lyapunov solution of the linear system stability problem***
In order for the time-invariant linear system (10.14) to be stable, it is necessary and sufficient that for every positive definite quadratic form $G(\mathbf{x}) = \mathbf{x}^T G\mathbf{x}$, $G = G^T \in \mathfrak{R}^{n\times n}$, *the solution function* $V(.) : \mathfrak{R}^n \longrightarrow \mathfrak{R}^{n\times n}$ *of the following differential equation taken along system motions is unique and also of positive definite quadratic form* $H(\mathbf{x}) = \mathbf{x}^T H\mathbf{x}$, $H = H^T \in \mathfrak{R}^{n\times n}$:

$$G(\mathbf{x}) = \mathbf{x}^T G\mathbf{x} > 0,\ \forall (\mathbf{x} \neq \mathbf{0}) \in \mathfrak{R}^n \ \wedge G = G^T \in \mathfrak{R}^{n\times n} \Longrightarrow$$
$$V(\mathbf{x}) = H(\mathbf{x}) = \mathbf{x}^T H\mathbf{x} > 0,\ \forall (\mathbf{x} \neq \mathbf{0}) \in \mathfrak{R}^n \ \wedge H = H^T \in \mathfrak{R}^{n\times n}$$
$$\Longleftrightarrow \frac{dV(\mathbf{x})}{dt} = -G(\mathbf{x}),\ \forall \mathbf{x} \in \mathfrak{R}^n.$$

This theorem has the following matrix form called **the fundamental Lyapunov matrix theorem.**

Theorem 248 ***The fundamental Lyapunov matrix theorem***
In order for the time-invariant linear system (10.14) to be stable, it is necessary and sufficient that for every positive definite $n \times n$ *symmetric matrix* G, *the matrix equation (10.16)*

$$A^T H + HA = -G, \tag{10.16}$$

has the unique solution $n \times n$ *matrix* H, *which is also symmetric and positive definite.*

Theorem 247 and Theorem 248 determine **the Lyapunov direct methodology for *time*-invariant linear systems**, i.e., they specify:

1. the direct methodology to get the complete solution to the linear system stability problem, which starts with an arbitrary choice of a positive definite quadratic form $G(\mathbf{x}) = \mathbf{x}^T G\mathbf{x}$, i.e., with an arbitrary selection of a positive definite symmetric matrix G and continuous with solving the differential equation $V^{(1)}(\mathbf{x}) = -G(\mathbf{x})$, equivalently with solving the linear algebraic Lyapunov matrix equation (10.16),
2. the necessary and sufficient conditions for the linear system stability,
3. the complete procedure for finding the system Lyapunov function in one step.

10.4.2 Inconsistent Lyapunov methodology: nonlinear systems

Lyapunov himself established in 1892 [335] another stability methodology in order to apply his method to the stability study of nonlinear systems. The methodology is **indirect**, i.e., **inverse to**, and **inconsistent with**, the direct methodology valid for the linear *time*-invariant systems. The conditions of the *Inconsistent Lyapunov methodology* are expressed in terms of *the existence* of a positive definite function with a negative (semi)definite total *time* derivative along the system motions. It starts with a choice of a positive definite function, without giving any hint or guideline for selecting it for a given nonlinear system or for a class of nonlinear systems. If, fortunately, it happens (by chance) that the total *time* derivative of the selected function along system motions is negative (semidefinite) definite, then the equilibrium state is (stable) asymptotically stable, respectively. The problem of finding a suitable function to start with has been the crucial drawback of this *inverse methodology*, that is, ***Inconsistent Lyapunov (Stability) Methodology***, i.e., ***Inconsistent Stability Methodology*** relative to the direct Lyapunov methodology.

In the classical literature on the Lyapunov stability theory and the Lyapunov method, a sample of which is cited above, the Lyapunov methodology for *time*-invariant linear systems was not broadened to the nonlinear systems possibly until the publication in 1990 of the paper [157].

10.4.3 Consistent methodology: nonlinear systems

In order to resolve the three fundamental problems of the Lyapunov method, another methodology for the stability study of nonlinear systems has been established and developed since 1990 for ISO nonlinear systems. It extends and generalizes the Lyapunov methodology for *time*-invariant linear systems to nonlinear systems. It is the ***Consistent Lyapunov Methodology*** (CLM) i.e., ***Consistent Stability Methodology*** (CSM). It is consistent with Lyapunov's primary, i.e., direct, stability methodology for *time*-invariant linear systems.

The paper [157], 1990, established the *Consistent Lyapunov Methodology*, i.e., *Consistent Stability Methodology* for stability studies of ISO nonlinear systems. CLM (CSM) was further developed in [128], [140], [127], [125], [129], [139], [142], [146], [136], [154], [155], [156], [158], [159], [160], [176], [189], [175], [220], [223], [224]. Its link with bond-graphs is in [78]. CLM (CSM) enables us to discover and prove the complete conditions in the framework of both *time*-invariant and *time*-varying nonlinear systems for:

- the necessary and sufficient conditions for asymptotic and exponential stability of an equilibrium state or of a set of the system states, which are not expressed in terms of the existence of a Lyapunov function,
- the direct determination of the Lyapunov function of the system,
- the direct determination of the exact domain of the asymptotic stability,
- the need only for the single application of the methodology to get the final result,
- the conditions should not be expressed in terms of the existence of a Lyapunov function (which characterizes the original Lyapunov inconsistent methodology).

Comment 249 ***Crucial developments of the Lyapunov method and methodologies***
The book presents, in the subsequent chapters, the generalization and the extension of the Lyapunov method, the Lyapunov methodology, and the Consistent Stability Methodology to stability properties of all five, large classes of time-varying nonlinear dynamical systems.

Chapter 11

Time-invariant sets and functions

11.1 Introduction

This chapter refines and broadens **[232, Chapter 20, pp. 149-162]**

11.2 Comparison functions

Hahn [257] introduced several classes of *comparison functions* in order to simplify the mathematical treatment of stability problems. A brief account of them follows by referring to [257].

Let $\varphi(.) : \mathfrak{R}_+ \longrightarrow \mathfrak{R}_+$ be a notation for a comparison function. Notice that the comparison functions are *time*-invariant (*time*-independent) scalar functions. This holds also for the norm functions $\|.\| : \mathfrak{R}^n \longrightarrow \mathfrak{R}_+$ and for the distance function $d(.) : \mathfrak{R}^n \times 2^{\mathfrak{R}^n} \longrightarrow \mathfrak{R}_+$.

The sign function $sign(.) : \mathfrak{R} \to \{-1, 0, 1\}$ is *a scalar signum function; see Figure 11.1,*

$$sign\ x = |x|^{-1} x \ if \ x \neq 0, \ and \ sign\ 0 = 0 \tag{11.1}$$

Definition 250 ***Classes of the comparison functions***
A function $\varphi(.) : \mathfrak{R}_+ \longrightarrow \mathfrak{R}_+$ *belongs to the following:*

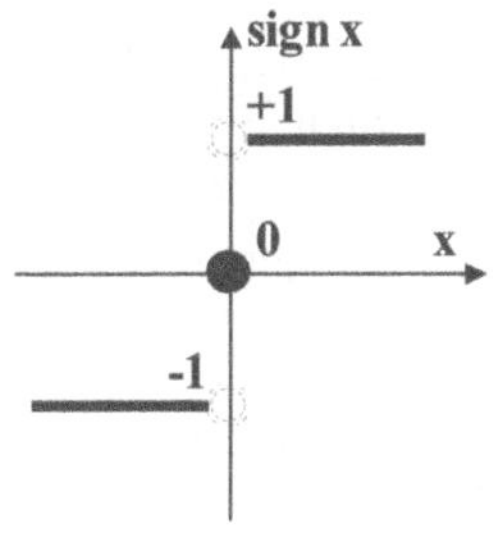

Figure 11.1: The signum function.

DOI: 10.1201/9781003744528-17

(i) The class $\mathcal{K}_{[0,\alpha[}$, $0<\alpha\leq\infty$, *if and only if it is defined, continuous, and strictly increasing on* $[0,\alpha[$, *and* $\varphi(0)=0$.

(ii) The class $\mathcal{K}$ *if and only if (i) holds for* $\alpha=\infty: \mathcal{K}=\mathcal{K}_{[0,\infty[}$.

(iii) The class $\mathcal{KR}$ *if and only if (ii) holds and* $\zeta\longrightarrow\infty$ *implies* $\varphi(\zeta)\longrightarrow\infty$.

(iv) The class $\mathcal{L}_{[0,\alpha[}$, $0<\alpha\leq\infty$, *if and only if it is defined, continuous and strictly decreasing on* $[0,\alpha[$, *and* $\zeta\longrightarrow\infty$ *implies lim* $[\varphi(\zeta)]=0$.

(v) The class $\mathcal{L}$ *if and only if (iv) holds for* $\alpha=\infty$: $\mathcal{L}=\mathcal{L}_{[0,\infty[}$.

Example 251 *The following function* $f(.):\mathfrak{R}_+\longrightarrow\mathfrak{R}_+$:

1. $f(\zeta)=sign\ \zeta$ *is not a comparison function. It is not continuous at* $\zeta=0$ *and it is not strictly increasing or decreasing.*

2. $f(\zeta)=(sign\ \zeta)\,sin\ \zeta=|sin\ \zeta|$ *is a comparison function* $\varphi(.):\mathfrak{R}_+\longrightarrow\mathfrak{R}_+$ *of the class* $\mathcal{K}_{[0,\frac{\pi}{2}[}$, $\alpha=\frac{\pi}{2}$, *but not of the class* $\mathcal{K}$ *because it is not strictly increasing at* $\zeta=\frac{\pi}{2}$ *and it is decreasing on* $]\frac{\pi}{2},\pi[$.

3. $f(\zeta)=1-e^{-\zeta}$ *is a comparison function* $\varphi(.):\mathfrak{R}_+\longrightarrow\mathfrak{R}_+$ *of the class* $\mathcal{K}$, *but not of the class* $\mathcal{KR}$.

4. $f(\zeta)=\zeta\sqrt{\zeta}$ *is a comparison function* $\varphi(.):\mathfrak{R}_+\longrightarrow\mathfrak{R}_+$ *of the class* $\mathcal{KR}$.

5. $f(\zeta)=1-\zeta$ *is a comparison function* $\varphi(.):\mathfrak{R}_+\longrightarrow\mathfrak{R}_+$ *of the class* $\mathcal{L}_{[0,1[}$, $\alpha=1$.

6. $f(\zeta)=e^{-\zeta}$ *is a comparison function* $\varphi(.):\mathfrak{R}_+\longrightarrow\mathfrak{R}_+$ *of the class* $\mathcal{L}$.

7. $f(\zeta)=cos\ \zeta$ *is not a comparison function. It is not strictly increasing or decreasing at* $\zeta=0$.

The inverse function of a comparison function $\varphi(.):\mathfrak{R}_+\longrightarrow\mathfrak{R}_+$ is $\varphi^I(.):\mathfrak{R}_+\longrightarrow\mathfrak{R}_+$, $\varphi^I[\varphi(\zeta)]\equiv\zeta$ and $\varphi[\varphi^I(\xi)]\equiv\xi$, where $\xi=\varphi(\zeta)$.

Example 252 *The following function* $f(.):\mathfrak{R}_+\longrightarrow\mathfrak{R}_+$:

1. $f(\xi)=arcsin\ \xi=|arcsin\ \xi|$, $\xi\in[0,1[$, *is the inverse to the comparison function* $\varphi(\zeta)=(sign\ \zeta)\,sin\ \zeta=|sin\ \zeta|$ *for* $\zeta\in[0,\frac{\pi}{2}[$ *because* $f[\varphi(\zeta)]=|arcsin\ \varphi(\zeta)|=|arcsin\ (|sin\ \zeta|)|=|arcsin\ (sin\ \zeta)|=|\zeta|=\zeta$.

2. $f(\xi)=-\ln(1-\xi)$ *is the inverse to the comparison function* $\varphi(\zeta)=1-e^{-\zeta}$ *because* $f[\varphi(\zeta)]=-\ln/1-\varphi(\zeta)/=-\ln/1-(1-e^{-\zeta})/=-\ln e^{-\zeta}=\zeta$.

3. $f(\xi)=\xi^{2/3}$ *is the inverse to the comparison function* $\varphi(\zeta)=\zeta\sqrt{\zeta}=\zeta^{3/2}$ *because* $f[\varphi(\zeta)]=\varphi^{2/3}(\zeta)=\left[\zeta^{3/2}\right]^{2/3}=\zeta$.

4. $f(\xi)=1-\xi$ *is the inverse to the comparison function* $\varphi(\zeta)=1-\zeta$ *because* $f[\varphi(\zeta)]=1-\varphi(\zeta)=1-(1-\zeta)=\zeta$.

5. $f(\xi)=|\ln\xi|$ *is the inverse to the comparison function* $\varphi(\zeta)=e^{-\zeta}$ *because* $f[\varphi(\zeta)]=|\ln\varphi(\zeta)|=\left|\ln e^{-\zeta}\right|=|-\zeta|=|\zeta|=\zeta$, $\zeta\geq0$.

Proposition 253 ***Properties of the comparison functions***

(i) If $\varphi(.)\in\mathcal{K}$ *and* $\psi(.)\in\mathcal{K}$, *then* $\varphi[\psi(.)]\in\mathcal{K}$.

(ii) If $\varphi(.)\in\mathcal{K}$ *and* $\psi(.)\in\mathcal{L}$, *then* $\varphi[\psi(.)]\in\mathcal{L}$.

(iii) If $\varphi(.)\in\mathcal{K}_{[0,\alpha[}$ *and* $\varphi(\alpha)=\xi$, *then* $\varphi^I(.)\in\mathcal{K}_{[0,\xi[}$.

(iv) If $\varphi(.)\in\mathcal{K}_{[0,\alpha[}$ *and lim* $[\varphi(\zeta):\zeta\longrightarrow\infty]=\mu$, *then* $\varphi^I(.)$ *is not defined on* $]\mu,\infty]$.

(v) If $\varphi(.)\in\mathcal{K}_{[0,\alpha[}$, $\psi(.)\in\mathcal{K}_{[0,\alpha[}$, *and* $\varphi(\zeta)>\psi(\zeta)$ *on* $[0,\alpha[$, *then* $\varphi^I(\zeta)<\psi^I(\zeta)$ *on* $[0,\beta[$, *where* $\beta=\psi(\alpha)$.

Definition 254 ***Classes of the compound comparison functions***

A function $\varphi(.):\mathfrak{R}_+\times\mathfrak{R}_+\longrightarrow\mathfrak{R}_+$ *belongs to:*

(i) The class $\mathcal{KK}_{[0;\alpha,\beta[}$, *if and only if both* $\varphi(*,\zeta)\in\mathcal{K}_{[0,\alpha[}$ *for every* $\zeta\in[0,\beta[$ *and* $\varphi(\zeta,*)\in\mathcal{K}_{[0,\beta[}$ *for every* $\zeta\in[0,\alpha[$.

(ii) The class $\mathcal{KK}$ if and only if (i) holds for $\alpha = \beta = \infty$.

(iii) The class $\mathcal{KL}_{[0;\alpha,\beta[}$ if and only if both $\varphi(*,\zeta) \in \mathcal{K}_{[0,\alpha[}$ for every $\zeta \in [0,\beta[$ and $\varphi(\zeta,*) \in \mathcal{L}_{[0,\beta[}$ for every $\zeta \in [0,\alpha[$.

(iv) The class $\mathcal{KL}$ if and only if (iii) holds for $\alpha = \beta = \infty$.

11.3 Semidefinite functions and constant sets

Lyapunov theorems determine the conditions on the function $V(.)$ called in general ***the Lyapunov function*** and on its total *time* derivative $V^{(1)}(.)$ along the system motions in order for the system motions to have the requested qualitative properties. The statements of the conditions are in terms of *sign semidefinite functions, sign-definite functions (for short definite functions), radially unbounded functions,* and *functions with the infinitesimally small upper bound (Russian terminology)* also called *decrescent functions (English terminology).*

Condition 255 *In what follows, the set $\mathfrak{S}$ is a fixed time-invariant connected nonempty subset of $\mathfrak{R}^n$, $\mathfrak{S} \subset \mathfrak{R}^n$, the boundary $\partial\mathfrak{S}$ of which is nonempty, $\partial\mathfrak{S} \neq \phi$.*

By referring to Lyapunov [335, p. 57], as well as to [195, Definition 10, pp. 22, 23] and [247, Definition 4.1, pp. 94, 95] we accept:

Definition 256 ***Time-invariant semidefinite function relative to a constant set $\mathfrak{S}$.***

Let a set $\mathfrak{S}$ obey Condition 255. A time-invariant function $W(.) : \mathfrak{R}^n \longrightarrow \mathfrak{R}$ is:

a) ***Positive semidefinite relative to the set $\mathfrak{S}$ on $\mathfrak{T}_0$*** *if and only if there is a time-invariant connected neighborhood $\mathcal{N}(\mathfrak{S})$ of $\mathfrak{S}$ such that*

1. $W(.)$ is defined and continuous on $\mathcal{N}(\mathfrak{S})$,

$$W(\mathbf{S}) \in \mathfrak{C}[\mathcal{N}(\mathfrak{S})], \tag{11.2}$$

2. $W(.)$ is nonnegative on $\mathcal{N}(\mathfrak{S})\setminus\mathfrak{S}$,

$$W(\mathbf{S}) \geq 0, \ \forall \mathbf{S} \in \mathcal{N}(\mathfrak{S})\setminus\mathfrak{S}, \tag{11.3}$$

3. $W(.)$ vanishes on the boundary $\partial\mathfrak{S}$ of $\mathfrak{S}$,

$$W(\mathbf{S}) = 0, \ \ \forall \mathbf{S} \in \ \partial\mathfrak{S}, \tag{11.4}$$

4. $W(.)$ is nonpositive on the interior $In\mathfrak{S}$ of $\mathfrak{S}$,

$$W(\mathbf{S}) \leq 0, \ \forall \mathbf{S} \in In\mathfrak{S}, \tag{11.5}$$

5. There is $S^* \in \mathcal{N}(\mathfrak{S})\setminus\mathfrak{S}$ such that $W(\mathbf{S}^*)$ is positive,

$$\exists \mathbf{S}^* \in \mathcal{N}(\mathfrak{S})\setminus\mathfrak{S} \Longrightarrow W(\mathbf{S}^*) > 0. \tag{11.6}$$

b) ***Globally positive semidefinite relative to $\mathfrak{S}$*** *if and only if a) holds for $\mathcal{N}(\mathfrak{S}) = \mathfrak{R}^n$,*

c) ***(Globally) Negative semidefinite relative to $\mathfrak{S}$ on*** *if and only if $[-W(.)]$ is (globally) positive semidefinite relative to $\mathfrak{S}$, respectively,*

d) ***Positive semidefinite relative to $\mathfrak{S}$ on a set $\mathfrak{A}$*** *if and only if $\mathfrak{A}$ is a neighborhood of $\mathfrak{S}$ such that the conditions under a) hold for $\mathcal{N}(\mathfrak{S}) = \mathfrak{A}$,*

e) ***Negative semidefinite relative to $\mathfrak{S}$ on a set $\mathfrak{A}$*** *if and only if $[-W(.)]$ is positive semidefinite relative to $\mathfrak{S}$ on the set $\mathfrak{A}$.*

*The expression "**relative to $\mathfrak{S}$**" is to be omitted if and only if $\mathfrak{S} = \mathfrak{O} = \{\mathbf{0}_n\}$.*

Note 257 *Continuity of a function is its crucial necessary property to be semidefinite. However, it need not be differentiable in order to be semidefinite.*

Definition 258 ***Time-varying semidefinite function relative to a constant set $\mathfrak{S}$ on $\mathfrak{T}_0$***

Let a set $\mathfrak{S}$ obey Condition 255. A time-varying function $W(.) : \mathfrak{T}_0 \times \mathfrak{R}^n \longrightarrow \mathfrak{R}$ is:

a) ***Positive semidefinite relative to the set $\mathfrak{S}$ on $\mathfrak{T}_0$*** *if and only if there is a time-invariant connected neighborhood $\mathcal{N}(\mathfrak{S})$ of $\mathfrak{S}$ such that*

1. $W(.)$ is defined and continuous on $\mathfrak{T}_0 \times \mathcal{N}(\mathfrak{S})$,

$$W(t, \mathbf{S}) \in \mathfrak{C}[\mathfrak{T}_0 \times \mathcal{N}(\mathfrak{S})], \tag{11.7}$$

2. $W(.)$ is nonnegative on $\mathfrak{T}_0 \times \mathcal{N}(\mathfrak{S}) \backslash \mathfrak{S}$,

$$W(t, \mathbf{S}) \geq 0, \ \forall (t, \mathbf{S}) \in \mathfrak{T}_0 \times \mathcal{N}(\mathfrak{S}) \backslash \mathfrak{S}, \tag{11.8}$$

3. $W(.)$ vanishes on the boundary $\partial\mathfrak{S}$ of $\mathfrak{S}$,

$$W(t, \mathbf{S}) = 0, \ \ \forall (t, \mathbf{S}) \in \mathfrak{T}_0 \times \partial\mathfrak{S}, \tag{11.9}$$

4. $W(.)$ is nonpositive on the interior $In\mathfrak{S}$ of $\mathfrak{S}$,

$$W(t, \mathbf{S}) \leq 0, \ \forall (t, \mathbf{S}) \in \mathfrak{T}_0 \times In\mathfrak{S}, \tag{11.10}$$

5. There is $S^* \in \mathcal{N}(\mathfrak{S}) \backslash \mathfrak{S}$ such that $W(t, \mathbf{S}^*)$ is positive on $\mathfrak{T}_0$,

$$\exists \mathbf{S}^* \in \mathcal{N}(\mathfrak{S}) \backslash \mathfrak{S} \Longrightarrow W(t, \mathbf{S}^*; \mathfrak{S}) > 0, \ \ \forall t \in \mathfrak{T}_0. \tag{11.11}$$

b) ***Globally positive semidefinite relative to $\mathfrak{S}$ on $\mathfrak{T}_0$*** *if and only if a) holds for $\mathcal{N}(\mathfrak{S}) = \mathfrak{R}^n$,*

c) ***(Globally) Negative semidefinite relative to $\mathfrak{S}$ on $\mathfrak{T}_0$*** *if and only if $-W(.)$ is (globally) positive semidefinite relative to $\mathfrak{S}$ on $\mathfrak{T}_0$, respectively,*

d) ***Positive semidefinite relative to $\mathfrak{S}$ on $\mathfrak{T}_0$ and on a set $\mathfrak{A}$*** *if and only if $\mathfrak{A}$ is a neighborhood of $\mathfrak{S}$ such that the conditions under a) hold for $\mathcal{N}(\mathfrak{S}) = \mathfrak{A}$,*

e) ***Negative semidefinite relative to $\mathfrak{S}$ on $\mathfrak{T}_0$ and on a set $\mathfrak{A}$*** *if and only if $-W(.)$ is positive semidefinite relative to $\mathfrak{S}$ on the set $\mathfrak{A}$.*

*The expression "**on $\mathfrak{T}_0$**" is to be omitted if and only if $\mathfrak{T}_0 = \mathfrak{T}$.*

*The expression "**relative to $\mathfrak{S}$**" is to be omitted if and only if $\mathfrak{S} = \mathfrak{O} = \{\mathbf{0}_n\}$.*

Note 259 *Continuity of a function is its crucial necessary property to be semidefinite. However, it need not be differentiable in order to be semidefinite.*

Note 260 ***Extension of Definition 258 to time-varying neighborhoods of constant sets***

For the extension of the above definition to the class of time-varying semidefinite functions on time-varying neighborhoods $\mathcal{N}(t; \mathfrak{S})$ of $\mathfrak{S}$ we apply the above definition to the set $\mathcal{N}_M(\mathfrak{T}_0, \mathfrak{S})$, which is then the union of the sets $\mathcal{N}(t; \mathfrak{S})$ for all $t \in \mathfrak{T}_0$,

$$\mathcal{N}(\mathfrak{T}_0, \mathfrak{S}) = \mathcal{N}_M(\mathfrak{T}_0, \mathfrak{S}) = \cup [\mathcal{N}(t; \mathfrak{S}) : t \in \mathfrak{T}_0] \subset \mathfrak{R}^n. \tag{11.12}$$

11.4 Definite functions and constant sets

In view of Lyapunov [335, p. 57], as well as [195, Definition 11, p. 23] and [247, Definition 4.4, pp. 96, 97], we introduce:

Definition 261 ***Time-invariant definite function relative to a constant set*** $\mathfrak{S}$

Let a set $\mathfrak{S}$ *obey Condition 255. A function* $W(.) : \mathfrak{R}^n \longrightarrow \mathfrak{R}$ *is:*

a) ***Positive definite relative to the set*** $\mathfrak{S}$ *if and only if there is a time-invariant connected neighborhood* $\mathcal{N} = \mathcal{N}(\mathfrak{S})$ *of* $\mathfrak{S}$ *such that*

1. $W(.)$ *is defined and continuous on* $\mathcal{N}$,

$$W(\mathbf{S};\mathfrak{S}) \in \mathfrak{C}[\mathcal{N}(\mathfrak{S})]. \tag{11.13}$$

2. $W(.)$ *is positive on* $\mathcal{N}(\mathfrak{S})$ *out of the set* $\mathfrak{S}$,

$$W(\mathbf{S};\mathfrak{S}) > 0, \forall \mathbf{S} \in \mathcal{N}(\mathfrak{S}) \setminus \mathfrak{S}. \tag{11.14}$$

3. $W(.)$ *vanishes on the boundary* $\partial\mathfrak{S}$ *of the set* $\mathfrak{S}$,

$$W(\mathbf{S};\mathfrak{S}) = 0, \ \forall \mathbf{S} \in \partial\mathfrak{S}. \tag{11.15}$$

4. $W(.)$ *is nonpositive on the interior* $In\mathfrak{S}$ *of the set* $\mathfrak{S}$,

$$W(\mathbf{S};\mathfrak{S}) \leq 0, \ \forall \mathbf{S} \in In\mathfrak{S}. \tag{11.16}$$

5. There is $S^* \in \mathfrak{S}$ *such that the minimal value of the function* $W(.)$ *on* $\mathfrak{R}^n$ *is at* S^*,

$$min \ [W(\mathbf{S};\mathfrak{S}) : \ \mathbf{S} \in \mathfrak{R}^n] = W(\mathbf{S}^*;\mathfrak{S}). \tag{11.17}$$

b) ***Globally positive definite relative to*** $\mathfrak{S}$ *if and only if a) holds for* $\mathcal{N}(\mathfrak{S}) = \mathfrak{R}^n$.

c) ***(Globally) Negative definite relative to*** $\mathfrak{S}$ *if and only if* $-W(.)$ *is (globally) positive definite, respectively.*

d) ***Positive definite on a set*** $\mathfrak{A}$ ***relative to*** $\mathfrak{S}$ *if and only if* $\mathfrak{A}$ *is a neighborhood of* $\mathfrak{S}$ *such that the conditions under a) hold for* $\mathcal{N}(\mathfrak{S}) = \mathfrak{A}$,

f) ***Negative definite on a set*** $\mathfrak{A}$ ***relative to*** $\mathfrak{S}$ *if and only if* $-W(.)$ *is positive definite on the set* $\mathfrak{A}$ *relative to* $\mathfrak{S}$.

*The expression "**relative to*** $\mathfrak{S}$*" is to be omitted if and only if* $\mathfrak{S} = \mathfrak{O} = \{\mathbf{0}_n\}$.

By following Hahn [257] we state the following.

Lemma 262 ***Comparison functions and time-invariant positive definite functions relative to*** $\mathfrak{S}$

Necessary and sufficient for a time-invariant function $W(.) : \mathfrak{R}^n \longrightarrow \mathfrak{R}$ *to be positive definite relative to* $\mathfrak{S}$ *on a connected neighborhood* $\mathcal{N}(\mathfrak{S})$ *of* $\mathfrak{S}$ *is the existence of comparison functions* $\varphi_i(.) \in \mathcal{K}_{[0,\alpha[}, i = 1, 2$, $\alpha = sup\{d(\mathbf{S},\mathfrak{S}) : \mathbf{S} \in \mathcal{N}(\mathfrak{S})\}$, *such that both* $W(\mathbf{S};\mathfrak{S}) \in \mathfrak{C}[\mathcal{N}(\mathfrak{S})]$ *and*

$$\varphi_2[d(\mathbf{S},\mathfrak{S});\mathfrak{S}] \geq W(\mathbf{S};\mathfrak{S}) \geq \varphi_1[d(\mathbf{S},\mathfrak{S});\mathfrak{S}], \ \forall \mathbf{S} \in \mathcal{N}(\mathfrak{S}) \setminus In\mathfrak{S}. \tag{11.18}$$

Definition 261 enables us to define *time*-varying sign definite functions.

Definition 263 ***Time-varying definite function relative to a constant set $\mathfrak{S}$ on $\mathfrak{T}_0 \subseteq \mathfrak{T}$ for any fixed $t_0 \in \mathfrak{T}_i$***

Let a set $\mathfrak{S}$ obey Condition 255. A function $V(.) : \mathfrak{T}_0 \times \mathfrak{R}^n \longrightarrow \mathfrak{R}$ is:

a) ***Positive definite relative to the set $\mathfrak{S}$ on $\mathfrak{T}_0$*** *for any fixed $t_0 \in \mathfrak{T}_i$ if and only if there are a time-invariant connected neighborhood $\mathcal{N}(\mathfrak{S})$ of $\mathfrak{S}$ and a time-invariant positive definite function $W(.)$ on $\mathcal{N}(\mathfrak{S})$ such that they obey 1) through 4):*

1) $V(.)$ is defined and continuous on $\mathfrak{T}_0 \times \mathcal{N}(\mathfrak{S})$,

$$V(t, \mathbf{S}) \in \mathfrak{C}[\mathfrak{T}_0 \times \mathcal{N}(\mathfrak{S})] \tag{11.19}$$

2) $V(.)$ satisfies the following:

$$V(t, \mathbf{S}) \geq W(\mathbf{S}), \forall (t, \mathbf{S}) \in \mathfrak{T}_0 \times \mathcal{N}(\mathfrak{S}) \tag{11.20}$$

3) $V(.)$ vanishes on the boundary $\partial\mathfrak{S}$ of $\mathfrak{S}$ at every $t \in \mathfrak{T}_0$,

$$V(t, \mathbf{S}) = 0, \ \forall \mathbf{S} \in \partial\mathfrak{S}, \ \forall t \in \mathfrak{T}_0. \tag{11.21}$$

4) $V(.)$ is nonpositive on the interior $In\mathfrak{S}$ of the set $\mathfrak{S}$ at every $t \in \mathfrak{T}_0$,

$$V(t, \mathbf{S}) \leq 0, \ \forall (t, \mathbf{S}) \in \mathfrak{T}_0 \times In\mathfrak{S}. \tag{11.22}$$

5) There is a pair $(t^*, \mathbf{S}^*) \in \mathfrak{T}_0 \times \mathfrak{S}$ such that the minimal value of the function $V(.)$ on $\mathfrak{T}_0 \times \mathfrak{R}^n$ is at $(t^*, \mathbf{S}^*) \in \mathfrak{T}_0 \times \mathfrak{S}$,

$$\begin{gathered} \exists (t^*, \mathbf{S}^*) \in \mathfrak{T}_0 \times \mathfrak{S} \Longrightarrow \\ min\ [V(t, \mathbf{S}) : \ (t, \mathbf{S}) \in \mathfrak{T}_0 \times \mathfrak{R}^n] = V(t^*, \mathbf{S}^*; \mathfrak{S}). \end{gathered} \tag{11.23}$$

b) A function $V(.) : \mathfrak{T}_0 \times \mathfrak{R}^n \longrightarrow \mathfrak{R}$ is ***positive definite relative to the set $\mathfrak{S}$ on $\mathfrak{T}_0 \times \mathcal{N}(\mathfrak{S})$*** *if and only if all conditions 1)-5) under a) hold.*

c) ***Globally positive definite relative to $\mathfrak{S}$ on $\mathfrak{T}_0$*** *if and only if the function $W(.)$ is globally positive definite and the conditions under a) hold for $\mathcal{N}(\mathfrak{S}) = \mathfrak{R}^n$.*

d) ***(Globally) Negative definite relative to $\mathfrak{S}$ on $\mathfrak{T}_0$*** *if and only if $-V(.)$ is (globally) positive definite, respectively.*

e) ***Positive definite relative to $\mathfrak{S}$ on $\mathfrak{T}_0$ and on a set $\mathfrak{A}$*** *if and only if $\mathfrak{S}$ is a neighborhood of $S = \mathbf{0}_n$ such that the conditions under a) hold for $\mathcal{N}(\mathfrak{S}) = \mathfrak{A}$,*

f) ***Negative definite relative to $\mathfrak{S}$ on $\mathfrak{T}_0$ and on a set $\mathfrak{A}$*** *if and only if $-V(.)$ is positive definite on $\mathfrak{T}_0$ and on the set $\mathfrak{A}$.*

The expression " ***on $\mathfrak{T}_0$*** *" is to be omitted if and only if $\mathfrak{T}_0 = \mathfrak{T}$.*

*The expression "****relative to $\mathfrak{S}$****" is to be omitted if and only if $\mathfrak{S} = \mathfrak{O} = \{\mathbf{0}_n\}$.*

Note 264 *Continuity of a function is its crucial obligatory property to be definite. However, it need not be differentiable in order to be definite.*

Lemma 262 and Definition 263 imply:

Lemma 265 ***Comparison functions and time-varying positive definite functions relative to a constant set $\mathfrak{S}$ on $\mathfrak{T}_0 \times \mathcal{N}(\mathfrak{S})$***

Necessary and sufficient for a time-varying function $V(.) : \mathfrak{T}_0 \times \mathfrak{R}^n \longrightarrow \mathfrak{R}$ to be positive definite relative to $\mathfrak{S}$ on $\mathfrak{T}_0 \times \mathcal{N}(\mathfrak{S})$, where $\mathcal{N}(\mathfrak{S})$ is a time-invariant connected neighborhood of $\mathfrak{S}$, is that (i) to (v) hold:

(i) $V(.)$ *is defined and continuous on* $\mathfrak{T}_0 \times \mathcal{N}(\mathfrak{S})$,

$$V(t,\mathbf{S}) \in \mathfrak{C}[\mathfrak{T}_0 \times \mathcal{N}(\mathfrak{S})], \tag{11.24}$$

(ii) $V(.)$ *vanishes on the boundary* $\partial\mathfrak{S}$ *of* $\mathfrak{S}$ *at every* $t \in \mathfrak{T}_0$,

$$V(t,\mathbf{S}) = 0, \ \forall \mathbf{S} \in \partial\mathfrak{S}, \ \forall t \in \mathfrak{T}_0. \tag{11.25}$$

(iii) $V(.)$ *is nonpositive on the interior* $In\mathfrak{S}$ *of the set* $\mathfrak{S}$ *at every* $t \in \mathfrak{T}_0$,

$$V(t,\mathbf{S}) \leq 0, \ \forall (t,\mathbf{S}) \in \mathfrak{T}_0 \times In\mathfrak{S}. \tag{11.26}$$

(iv) There is a pair $(t^*,\mathbf{S}^*) \in \mathfrak{T}_0 \times \mathfrak{S}$ *such that the minimal value of the function* $V(.)$ *on* $\mathfrak{T}_0 \times \mathfrak{R}^n$ *is at* $(t^*,\mathbf{S}^*) \in \mathfrak{T}_0 \times \mathfrak{S}$,

$$\begin{gathered} \exists (t^*,\mathbf{S}^*) \in \mathfrak{T}_0 \times \mathfrak{S} \Longrightarrow \\ min \ [V(t,\mathbf{S}) : \ (t,\mathbf{S}) \in \mathfrak{T}_0 \times \mathfrak{R}^n] = V(t^*,\mathbf{S}^*). \end{gathered} \tag{11.27}$$

(v) There exists a comparison function $\varphi(.;\mathfrak{S}) \in \mathcal{K}_{[0,\alpha[}$,

$$\alpha = sup \ \{d(\mathbf{S},\mathfrak{S}) : \ \mathbf{S} \in \mathcal{N}(\mathfrak{S})\},$$

such that

$$V(t,\mathbf{S}) \geq \varphi[d(\mathbf{S},\mathfrak{S});\mathfrak{S}], \ \forall (t,\mathbf{S}) \in \mathfrak{T}_0 \times \mathcal{N}(\mathfrak{S}) \setminus In\mathfrak{S}. \tag{11.28}$$

11.5 Positive definite function induces sets

Every positive definite function relative to a *time*-invariant, i.e., constant, connected subset $\mathfrak{S}$ of $\mathfrak{R}^n$ with a nonempty boundary $\partial\mathfrak{S}$ induces sets denoted by $\mathfrak{V}_\xi(t;t_0)$, $\mathfrak{V}_\xi(t;t_0) \subseteq \mathfrak{R}^n$, which are determined by:

Definition 266 ***The set*** $\mathfrak{V}_\xi(t;t_0)$ ***induced by both the positive definite scalar function*** $V(.;\mathfrak{S})$ ***and its scalar value*** ξ

Let $\xi \in \mathfrak{R}^+$ *or* $\xi = \infty$. *Let the function* $V(.) : \mathfrak{T}_0 \times \mathfrak{R}^n \longrightarrow \mathfrak{R}$ *be positive definite relative to the set* $\mathfrak{S}$ *on* $\mathfrak{T}_0$.

The set $\mathfrak{V}_\xi(t;t_0;\mathfrak{S})$, *for short* $\mathfrak{V}_\xi(t;t_0)$,

$$\mathfrak{V}_\xi(t;t_0;\mathfrak{S}) \equiv \mathfrak{V}_\xi(t;t_0) \subseteq \mathfrak{R}^n, \tag{11.29}$$

is ***induced by the function*** $V(.)$ ***relative to the set*** $\mathfrak{S}$ ***on*** $\mathfrak{T}_0$ *if and only if it is the largest open connected neighborhood of the set* $\mathfrak{S}$ *at* $t \in \mathfrak{T}_0$ *such that*

$$\xi > 0, \ \mathfrak{V}_\xi(t;t_0) = \{\mathbf{S} : \mathbf{S} \in \mathfrak{R}^n, \ V(t,\mathbf{S}) < \xi\}, \ \ \forall t \in \mathfrak{T}_0, \tag{11.30}$$

and

1) $V(t,\mathbf{S})$ *is strictly less than* ξ *on* $\mathfrak{V}_\xi(t;t_0)$ *at every instant* $t \in \mathfrak{T}_0$,

$$\xi > 0, \ \mathbf{S} \in \mathfrak{V}_\xi(t;t_0) \Longrightarrow V(t,\mathbf{S}) < \xi, \ \ \forall t \in \mathfrak{T}_0. \tag{11.31}$$

2) If $\xi \in \mathfrak{R}^+$, *then the boundary* $\partial\mathfrak{V}_\xi(t;t_0)$ *of* $\mathfrak{V}_\xi(t;t_0)$ *is the hypersurface* $V(t,\mathbf{S}) \equiv \xi$,

$$\xi \in \mathfrak{R}^+ \Longrightarrow V(t,\mathbf{S}) = \xi \Longleftrightarrow \mathbf{S} \in \partial\mathfrak{V}_\xi(t;t_0), \ \forall t \in \mathfrak{T}_0. \tag{11.32}$$

3) There are $\xi_S \in \mathfrak{R}^+ \cup \{\infty\}$ and a connected constant neighborhood $\mathcal{N}(\mathfrak{S})$ of the set $\mathfrak{S}$ such that the union $\mathfrak{V}_\xi(\mathfrak{T}_0)$ of all $\mathfrak{V}_\xi(t;t_0)$ over $\mathfrak{T}_0$,

$$\mathfrak{V}_\xi(\mathfrak{T}_0) = \mathfrak{V}_{\xi M}(\mathfrak{T}_0;\mathfrak{S}) = \cup\left[\mathfrak{V}_\xi(t;t_0) : t \in \mathfrak{T}_0\right], \tag{11.33}$$

is the nonempty bounded connected neighborhood of the set $\mathfrak{S}$ for every $(\xi \leq \xi_S) \in \mathfrak{R}^+\cup\{\infty\}$,

$$\exists \xi_S \in \mathfrak{R}^+ \cup \{\infty\},\ \exists \mathcal{N}(\mathfrak{S}) \subseteq \mathfrak{R}^n \Longrightarrow$$
$$\forall(\xi \leq \xi_S) \in \mathfrak{R}^+ \cup \{\infty\},\ \ \mathfrak{V}_\xi(\mathfrak{T}_0) \subseteq \mathcal{N}(\mathfrak{S}). \tag{11.34}$$

The function $V(.)$ ***is the generating function of the set $\mathfrak{V}_\xi(t)$ relative to the set $\mathfrak{S}$ on $\mathfrak{T}_0$.***

*The expression "**on** $\mathfrak{T}_0$" and "t_0" are to be omitted if and only if $\mathfrak{T}_0 = \mathfrak{T}$. Then and only then, $\mathfrak{V}_\xi(t;t_0) = \mathfrak{V}_\xi(t)$,*

$$\mathfrak{T}_0 = \mathfrak{T} \Longrightarrow \mathfrak{V}_\xi(t;t_0) = \mathfrak{V}_\xi(t). \tag{11.35}$$

*The expression "**relative to** $\mathfrak{S}$" is to be omitted if and only if $\mathfrak{S} = \mathfrak{O} = \{\mathbf{0}_n\}$.*

Note 267 *The conditions 1) and 2) of Definition 266 are equivalent to the condition a-2) of Definition 263 (Section 11.4). They are inherent for the time-varying function $V(.)$ to be positive definite.*

The conditions 1) and 2) of Definition 266 reflect an important property of the set $\mathfrak{V}_\xi(t;t_0)$ associated with the function $V(.)$ and its values less than, or equal to, ξ_S.

Example 268 *Let $\mathfrak{S} = \mathfrak{O}$ and $V(t,\mathbf{S}) = e^{-t}d(\mathbf{S},\mathfrak{O})$. It is globally positive semidefinite. It is not positive definite. There is not a time-invariant positive definite function $W(.)$ such that $V(.)$ obeys the condition a-2) of Definition 263 because*

$$t \longrightarrow \infty \Longrightarrow V(t,\mathbf{S}) = e^{-t}d(\mathbf{S},\mathfrak{O}) \longrightarrow 0,\ \forall \mathbf{S} \in \mathfrak{R}^n \backslash \mathfrak{O}.$$

For every $\xi \in \mathfrak{R}^+$ the set $\mathfrak{V}_\xi(t)$ diverges to $\mathfrak{R}^n$ as $t \longrightarrow \infty$,

$$\mathfrak{V}_\xi(t) = \left\{\mathbf{S} : e^{-t}d(\mathbf{S},\mathfrak{O}) < \xi\right\} = \left\{\mathbf{S} : d(\mathbf{S},\mathfrak{O}) < \xi e^t\right\} \longrightarrow \mathfrak{R}^n \text{ as } t \longrightarrow \infty$$

(see Figure 11.2).

Example 269 *Let $V(t,\mathbf{S}) = \left(1+t^2\right)d(\mathbf{S},\mathfrak{S})$. It is globally positive definite relative to $\mathfrak{S}$. It obeys the condition a-2) of Definition 263 for the globally positive definite time-invariant function $W(\mathbf{S}) = d(\mathbf{S},\mathfrak{S})$ relative to $\mathfrak{S}$. For every $\xi \in \mathfrak{R}^+$, the set $\mathfrak{V}_\xi(t)$ is a subset of the time-invariant connected neighborhood $\mathcal{N}_\xi(\mathfrak{S})$ of $\mathfrak{S}$ on $\mathfrak{T}$,*

$$\mathfrak{V}_\xi(t) = \left\{\mathbf{S} : \left(1+t^2\right)d(\mathbf{S},\mathfrak{S}) < \xi\right\} =$$
$$= \left\{\mathbf{S} : d(\mathbf{S},\mathfrak{S}) < \xi\left(1+t^2\right)^{-1}\right\} \subseteq \mathcal{N}_\xi(\mathfrak{S}),\ \forall t \in \mathfrak{T},$$

which implies

$$\mathfrak{V}_\xi(\mathfrak{T}_0) \subseteq \mathcal{N}_\xi(\mathfrak{S}).$$

Notice that the set $\mathfrak{V}_\xi(t)$ is asymptotically contractive to $\mathfrak{S}$ (Definition 108), which means that the closure $Cl\mathfrak{V}_\xi(t)$ of the set $\mathfrak{V}_\xi(t)$ contracts asymptotically to the closure $Cl\mathfrak{S}$ of the set $\mathfrak{S}$ as $t \longrightarrow \infty$,

$$t \longrightarrow \infty \Longrightarrow Cl\mathfrak{V}_\xi(t) = \left\{\mathbf{S} : d(\mathbf{S},\mathfrak{S}) \leq \xi\left(1+t^2\right)^{-1}\right\} \longrightarrow Cl\mathfrak{S}.$$

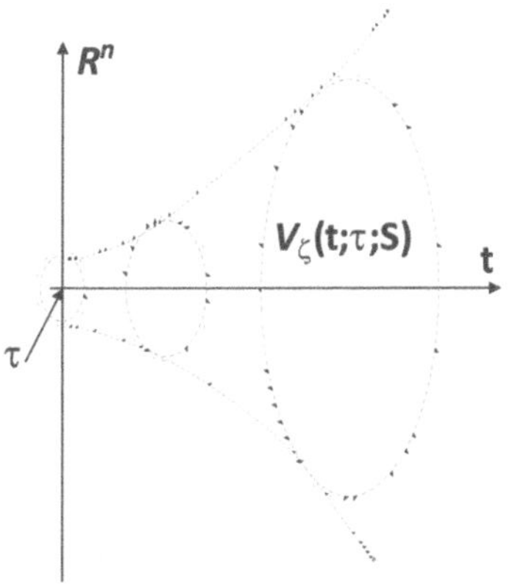

Figure 11.2: The set $\mathfrak{V}_{\varsigma}(t;\tau)$, related to the *time*-varying semidefinite function $V(t,\mathbf{S}) = e^{-t}d(\mathbf{S},\mathfrak{O})$, is unbounded.

This means that there is not a lower open time-invariant limit set of the set $\mathfrak{V}_{\xi}(t)$ *as* $t \longrightarrow \infty$. *For arbitrarily small* $\varepsilon \in \mathfrak{R}^{+}$, *there is* $\tau \in \mathfrak{T}$ *such that* $\mathfrak{V}_{\xi}(t)$ *is the proper subset of the open neighborhood* $\mathcal{N}_{\varepsilon}(\mathfrak{S})$ *of* $\mathfrak{S}$ *at every* $(t \geq \tau) \in \mathfrak{T}$,

$$\mathfrak{V}_{\xi}(t) \subset \mathcal{N}_{\varepsilon}(\mathfrak{S}),\ \forall (t \geq \tau) \in \mathfrak{T}_0.$$

Property 270 ***The property of the boundaries of the sets*** $\mathfrak{V}_{\xi}(t;t_0)$

The boundaries $\partial\mathfrak{V}_{\xi_i}(t;t_0)$ *of the sets* $\mathfrak{V}_{\xi_i}(t;t_0)$ *induced by the positive definite function function* $V(.)$ *relative to the set* $\mathfrak{S}$, $i = 1,2$, $\xi_1 \neq \xi_2$, *which are subsets of the neighborhood* $\mathcal{N}(\mathfrak{S})$ *introduced in Definition 263, do not ever intersect on* $\mathfrak{T}_0$*:*

$$\partial\mathfrak{V}_{\xi_1}(t;t_0) \cap \partial\mathfrak{V}_{\xi_2}(t;t_0) = \phi,\ \forall t \in \mathfrak{T}_0. \tag{11.36}$$

Proof. Let a function $V(.) : \mathfrak{T}_0 \times \mathfrak{R}^n \longrightarrow \mathfrak{R}$ be positive definite relative to $\mathfrak{S}$. It satisfies Definition 263. Let $\mathcal{N}(\mathfrak{S})$ be the neighborhood introduced in Definition 263. Let $\xi_i \in \mathfrak{R}^{+}$ be such that $\mathfrak{V}_{\xi_i}(t;t_0) \subseteq \mathcal{N}(\mathfrak{S})$, $\forall t \in \mathfrak{T}_0$, $i = 1,2$, $\xi_1 \neq \xi_2$. Let us assume that the statement of Property 270 is wrong, i.e., that there is $\sigma \in \mathfrak{T}_0$ such that $\partial\mathfrak{V}_{\xi_1}(\sigma;t_0) \cap \partial\mathfrak{V}_{\xi_2}(\sigma;t_0) \neq \phi$. Let $\mathbf{S} \in \partial\mathfrak{V}_{\xi_1}(\sigma;t_0) \cap \partial\mathfrak{V}_{\xi_2}(\sigma;t_0)$. Because $\mathbf{S} \in \partial\mathfrak{V}_{\xi_1}(\sigma;t_0)$ then $V(t,\mathbf{S}) = \xi_1$, But $\mathbf{S} \in \partial\mathfrak{V}_{\xi_2}(\sigma;t_0)$ so that $V(t,\mathbf{S}) = \xi_2 \neq \xi_1$. This contradicts the continuity of the positive definite function $V(.)$ relative to $\mathfrak{S}$ on $\mathfrak{T}_0 \times \mathcal{N}(\mathfrak{S})$, (11.19). The contradiction is the consequence of the assumption that the statement of Property 270 is wrong, i.e., that (11.36) is wrong. Hence, the statement of Property 270 is correct; i.e., (11.36) is valid. ■

The following proposition discovers another property of the sets $\mathfrak{V}_{\xi}(t;t_0)$ induced by the positive definite function $V(.)$ relative to $\mathfrak{S}$.

Lemma 271 ***Positive definiteness relative to*** $\mathfrak{S}$***, comparison function, and the set*** $\mathfrak{V}_{\xi}(t;t_0)$ $\surd$

If the set $\mathfrak{S}$ *is time-invariant, connected, and bounded, and if a function* $V(.)$ *is positive definite relative to* $\mathfrak{S}$ *on* $\mathfrak{T}_0 \times \mathcal{N}(\mathfrak{S})$, *where* $\mathcal{N}(\mathfrak{S}) \subseteq \mathfrak{R}^n$ *is a time-invariant connected neighborhood of the set* $\mathfrak{S}$, *then there is* $\eta \in \mathfrak{R}^{+}$ *such that for every* $\varsigma \in]0,\eta[$ *the set* $\mathfrak{V}_{\varsigma}(t;t_0)$ *is bounded on* $\mathfrak{T}_0$ *and* $\mathfrak{V}_{\varsigma}(t;t_0) \subseteq \mathcal{N}(\mathfrak{S})$, $\forall t \in \mathfrak{T}_0$.

Proof. Let the set $\mathfrak{S}$ be *time*-invariant, connected, and bounded. Let $\mathcal{N}(\mathfrak{S})$ be a bounded neighborhood of the set $\mathfrak{S}$ such that $V(.)$ is positive definite relative to $\mathfrak{S}$ on $\mathfrak{T}_0 \times \mathcal{N}(\mathfrak{S})$. Let

$$\mu = min\left[d(\mathbf{S},\mathfrak{S}) : \mathbf{S} \in \partial\mathcal{N}(\mathfrak{S})\right]. \tag{11.37}$$

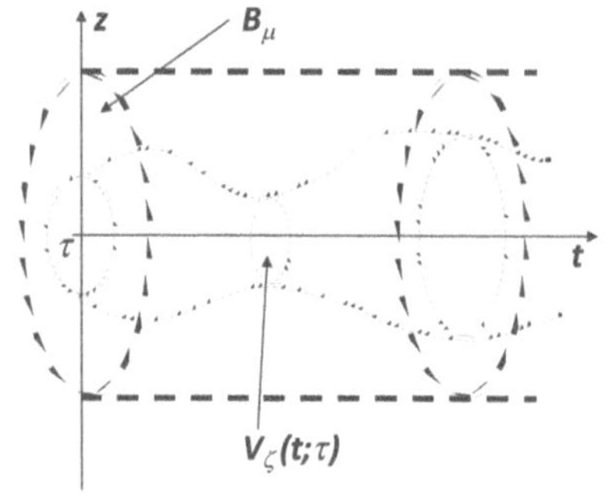

Figure 11.3: The set $\mathfrak{V}_\varsigma(t;t_0;\Upsilon)$ is bounded. It is always a proper subset of the hyperball $\mathfrak{N}_\mu(\Upsilon)$ for $\Upsilon = \mathfrak{O}$.

Let $\varphi(.) : \mathfrak{R}_+ \longrightarrow \mathfrak{R}_+$ be a comparison function of the class $\mathcal{K}_{[0,\mu[}$, which obeys

$$V(t,\mathbf{S}) \geq \varphi[d(\mathbf{S},\mathfrak{S})], \ \forall (t,\mathbf{S}) \in \mathfrak{T}_0 \mathrm{x} \mathcal{N}_\mu(\mathfrak{S})$$

due to (11.28). We continue the proof by contradiction. Let all the preceding conditions hold and let the set $\mathfrak{V}_\varsigma(t;t_0)$, $\varsigma \in]0,\eta[$ for $\eta = \varphi(\mu)$, be unbounded on $\mathfrak{T}_0$. Then there is a moment $\sigma \in \mathfrak{T}_0$ such that $\mathfrak{V}_\varsigma(\sigma;t_0) \setminus \mathcal{N}_\mu(\mathfrak{S}) \neq \phi$. Let $\mathbf{S}^* \in \mathfrak{V}_\varsigma(\sigma;t_0) \setminus \mathcal{N}_\mu(\mathfrak{S})$, which implies both $d(\mathbf{S}^*,\mathfrak{S}) \geq \mu$ and $V(\sigma,\mathbf{S}^*) \geq \varphi(\mu) = \eta$ due to (11.28). Hence, $\eta > \varsigma$ and $\mathbf{S}^* \notin \mathfrak{V}_\varsigma(\sigma;t_0) \setminus \mathcal{N}_\mu(\mathfrak{S})$ (due to Definition 266), which contradicts $\mathbf{S}^* \in \mathfrak{V}_\varsigma(\sigma;t_0) \setminus \mathcal{N}_\mu(\mathfrak{S})$. This and $\mathfrak{V}_\varsigma(\sigma;t_0) \cap \mathcal{N}_\mu(\mathfrak{S}) \neq \phi$, $\forall t \in \mathfrak{T}_0$ prove that the set $\mathfrak{V}_\varsigma(t;t_0)$, $\varsigma \in]0,\eta[$, is bounded on $\mathfrak{T}_0$, and that $\mathfrak{V}_\varsigma(t;t_0)$ is a subset of the neighborhood $\mathcal{N}_\mu(\mathfrak{S})$ on $\mathfrak{T}_0$, i.e., $\mathfrak{V}_\xi(\mathfrak{T}_0) \subseteq \mathcal{N}_\mu(\mathfrak{S})$ (see Figure 11.3). ■

Conclusion 272 *Example 268 and Example 269 illustrate the important and useful property of the positive definiteness of a function $V(.)$ relative to $\mathfrak{S}$ on $\mathfrak{T}_0$, which is the boundedness of the associated set $\mathfrak{V}_\xi(t;t_0)$ on $\mathfrak{T}_0$. This need not hold for the set $\mathfrak{V}_\xi(t;t_0)$ if the function $V(.)$ is only positive semidefinite relative to $\mathfrak{S}$ on $\mathfrak{T}_0$.*

Lyapunov classical theory concerns nonlinear systems with stability properties valid on the *time*-invariant sets. It deals with Lyapunov functions $V(.)$ for which the associated sets $\mathfrak{V}_\xi(t;t_0)$ are not asymptotically contractive. The works [128], [139], [115], [223], [224] extend it to nonlinear systems with stability properties valid on the sets that are asymptotically contractive, hence, they extend it to the use of Lyapunov functions $V(.)$ for which the associated sets $\mathfrak{V}_\xi(t;t_0)$ are asymptotically contractive.

Definition 273 ***Set increasing***

Let the function $V(.)$, $V(t,\mathbf{S}) \in \mathfrak{C}(\mathfrak{T}_0 \times \mathfrak{R}^n)$, induce sets $\mathfrak{V}_{\varsigma_i}(t;t_0)$, (Definition 266). The sets $\mathfrak{V}_{\varsigma_i}(t;t_0)$ are ***set increasing relative to $\mathfrak{S}$ on the neighborhood $\mathcal{N}(\mathfrak{S})$ of $\mathfrak{S}$*** *if and only if*

$$\begin{aligned} &Cl\mathfrak{V}_{\xi_1}(t;t_0) \subset Cl\mathfrak{V}_{\xi_2}(t;t_0), \ \forall t \in \mathfrak{T}_0, \\ &\partial\mathfrak{V}_{\xi_1}(t;t_0) \cap \partial\mathfrak{V}_{\xi_2}(t;t_0) = \phi, \forall t \in \mathfrak{T}_0, \\ &if \ and \ only \ if \ \xi_1 < \xi_2, \ \forall t \in \mathfrak{T}_0, \end{aligned} \tag{11.38}$$

for $\xi_i \in \mathfrak{R}$ such that $\mathfrak{V}_{\xi_i}(t;t_0) \subseteq \mathcal{N}(\mathfrak{S})$, $i = 1,2$, $\forall t \in \mathfrak{T}_0$.

Lemma 274 *Let* $V(.)$, $V(t,\mathbf{S}) \in \mathfrak{C}(\mathfrak{T} \times \mathfrak{R}^n)$, *be a positive definite function relative to* $\mathfrak{S}$. *Then the sets* $\mathfrak{V}_\xi(t;t_0)$, $\mathfrak{V}_{\xi_j}(t;t_0) \subseteq \mathcal{N}(\mathfrak{S})$, $\forall t \in \mathfrak{T}_0$, *are set increasing.*

Proof. Let the conditions of Lemma 274 statement hold. Let us assume that the sets $\mathfrak{V}_\xi(t;t_0)$, $\mathfrak{V}_{\xi_j}(t;t_0) \subseteq \mathcal{N}(\mathfrak{S})$, $\forall t \in \mathfrak{T}_0$, are not set increasing. This means that there are $\sigma \in \mathfrak{T}_0$ and $\xi_i \in \mathfrak{R}^+$, $i = 1,2$, for which $\mathfrak{V}_{\xi_i}(t;t_0) \subseteq \mathcal{N}(\mathfrak{S}), \forall t \in \mathfrak{T}_0$, $\xi_1 < \xi_2$, such that $\mathfrak{V}_{\xi_2}(\sigma;t_0) \subseteq \mathfrak{V}_{\xi_1}(\sigma;t_0)$. If the sets $\mathfrak{V}_{\xi_2}(\sigma;t_0)$ and $\mathfrak{V}_{\xi_1}(\sigma;t_0)$ were equal, $\mathfrak{V}_{\xi_2}(\sigma;t_0) = \mathfrak{V}_{\xi_1}(\sigma;t_0)$, then their boundaries would be equal; $\partial\mathfrak{V}_{\xi_2}(\sigma;t_0) = \partial\mathfrak{V}_{\xi_1}(\sigma;t_0)$, which is impossible due to Property 270. If $\mathfrak{V}_{\xi_2}(\sigma;t_0) \subset \mathfrak{V}_{\xi_1}(\sigma;t_0)$ were valid, then there would be $\mathbf{S} \in \mathfrak{V}_{\xi_2}(\sigma;t_0)$ such that both $\mathbf{S} \in \mathfrak{V}_{\xi_1}(\sigma;t_0)$ and $V(\sigma,\mathbf{S}) > \xi_1$ were valid. However, $V(\sigma,\mathbf{S}) > \xi_1$ is impossible for $\mathbf{S} \in \mathfrak{V}_{\xi_1}(\sigma;t_0)$ due to (11.31). The assumption that the sets $\mathfrak{V}_\xi(t;t_0)$, $\mathfrak{V}_{\xi_j}(t;t_0) \subseteq \mathcal{N}(\mathfrak{S})$, $\forall t \in \mathfrak{T}_0$, are not set increasing is invalid. This implies that the sets $\mathfrak{V}_\xi(t;t_0)$ are set increasing. ■

11.6 Decrescent functions and constant sets

The following functional property is indispensable for every *time*-dependent function $V(.)$ to assure that the associated set $\mathfrak{V}_\xi(t;t_0)$ is not asymptotically contractive. If the function $V(.)$ is *time*-independent, then it is automatically decrescent, i.e., its associated set $\mathfrak{V}_\xi(\mathfrak{S})$ is *time*-invariant and cannot be asymptotically contractive.

Definition 275 ***Time-varying decrescent function relative to a constant set*** $\mathfrak{S}$ ***on*** $\mathfrak{T}_0$

A function $V(.) : \mathfrak{T}_0 \times \mathfrak{R}^n \longrightarrow \mathfrak{R}$ *is* ***decrescent relative to a constant set*** $\mathfrak{S}$ ***on*** $\mathfrak{T}_0$ *if and only if*

(i) $V(.)$ *is defined and continuous on* $\mathfrak{T}_0 \times \mathcal{N}(\mathfrak{S})$,

$$V(t,\mathbf{S}) \in \mathfrak{C}[\mathfrak{T}_0 \times \mathcal{N}(\mathfrak{S})], \tag{11.39}$$

and

(ii) there is a time-invariant connected neighborhood $\mathcal{N}(\mathfrak{S})$ *of* $S = \mathbf{0}_n$ *and a time-invariant positive definite function* $W(.)$ *relative to* $\mathfrak{S}$ *on* $\mathcal{N}(\mathfrak{S})$ *such that*

$$W(\mathbf{S}) \geq V(t,\mathbf{S}),\ \forall (t,\mathbf{S}) \in \mathfrak{T}_0 \times \mathcal{N}(\mathfrak{S}). \tag{11.40}$$

It is ***globally decrescent relative to*** $\mathfrak{S}$ ***on*** $\mathfrak{T}_0$ *if and only if the function* $W(.)$ *is globally positive definite relative to* $\mathfrak{S}$ *and* $\mathcal{N} = \mathfrak{R}^n$.

*The expression "**on*** $\mathfrak{T}_0$*" is to be omitted if and only if* $\mathfrak{T}_0 = \mathfrak{T}$.

*The expression "**relative to*** $\mathfrak{S}$*" is to be omitted if and only if* $\mathfrak{S} = \mathfrak{O} = \{\mathbf{0}_n\}$.

Lemma 262 and Definition 275 imply:

Lemma 276 ***Time-varying decrescent functions relative to a constant set*** $\mathfrak{S}$ ***on*** $\mathfrak{T}_0$ ***and comparison functions***

Necessary and sufficient for a time-varying function $V(.) : \mathfrak{T}_0 \times \mathfrak{R}^n \longrightarrow \mathfrak{R}$ *to be* ***decrescent relative to a constant set*** $\mathfrak{S}$ ***on*** $\mathfrak{T}_0 \times \mathcal{N}(\mathfrak{S})$, *where* $\mathcal{N}(\mathfrak{S})$ *is a time-invariant connected constant neighborhood of* $\mathfrak{S}$, *is that both (i) and (ii) hold:*

(i) $V(.)$ *is defined and continuous on* $\mathfrak{T}_0 \times \mathcal{N}(\mathfrak{S})$,

$$V(t,\mathbf{S}) \in \mathfrak{C}[\mathfrak{T}_0 \times \mathcal{N}(\mathfrak{S})]. \tag{11.41}$$

(ii)There exists a comparison function $\varphi(.) \in \mathcal{K}_{[0,\alpha[}$,

$$\alpha = sup\ \{d(\mathbf{S},\mathfrak{S}):\ \mathbf{S} \in \mathcal{N}(\mathfrak{S})\},$$

such that

$$\varphi[d(\mathbf{S},\mathfrak{S})] \geq V(t,\mathbf{S}),\ \forall (t,\mathbf{S}) \in \mathfrak{T}_0 \times \mathcal{N}(\mathfrak{S}). \tag{11.42}$$

The following functional properties are also useful. They simplify stability tests.

Definition 277 ***Time-varying radially unbounded function relative to a constant set*** $\mathfrak{S}$ ***on*** $\mathfrak{T}_0$

A function $V(.) : \mathfrak{T}_0 \times \mathfrak{R}^n \longrightarrow \mathfrak{R}$ *is* ***radially unbounded relative to a constant set*** $\mathfrak{S}$ ***on*** $\mathfrak{T}_0$ *if and only if it diverges to infinity as* $d(\mathbf{S},\mathfrak{S})$ *diverges to infinity at every instant* $t \in \mathfrak{T}_0$,

$$d(\mathbf{S},\mathfrak{S}) \longrightarrow \infty \Longrightarrow V(t,\mathbf{S}) \longrightarrow \infty,\quad \forall t \in \mathfrak{T}_0.$$

*The expression "****on*** $\mathfrak{T}_0$*" is to be omitted if and only if* $\mathfrak{T}_0 = \mathfrak{T}$.
*The expression "****relative to*** $\mathfrak{S}$*" is to be omitted if and only if* $\mathfrak{S} = \mathfrak{O} = \{\mathbf{0}_n\}$.

Example 278 *[232]The function* $V(t,\mathbf{S}) = \left[(1+t^2)(2+t^2)^{-1}\right] d(\mathbf{S},\mathfrak{S})$ *is radially unbounded,*

$$d(\mathbf{S},\mathfrak{S}) \longrightarrow \infty \Longrightarrow V(t,\mathbf{S}) = \frac{1+t^2}{2+t^2} d(\mathbf{S},\mathfrak{S}) \longrightarrow \infty,\quad \forall t \in \mathfrak{T}.$$

However, globally positive definite functions defined by

$$V_1(t,\mathbf{S}) = \frac{1+t^2}{2+t^2}\frac{d(\mathbf{S},\mathfrak{S})}{1+d(\mathbf{S},\mathfrak{S})},\quad V_2(t,\mathbf{S}) = (1+t^2)\frac{d(\mathbf{S},\mathfrak{S})}{1+d(\mathbf{S},\mathfrak{S})}$$

are not radially unbounded,

$$d(\mathbf{S},\mathfrak{S}) \longrightarrow \infty \Longrightarrow V_1(t,\mathbf{S}) = \frac{1+t^2}{2+t^2}\frac{d(\mathbf{S},\mathfrak{S})}{1+d(\mathbf{S},\mathfrak{S})} \longrightarrow \frac{1+t^2}{2+t^2} < \infty,\ \forall t \in \mathfrak{T}$$
$$d(\mathbf{S},\mathfrak{S}) \longrightarrow \infty \Longrightarrow V_2(t,\mathbf{S}) = (1+t^2)\frac{d(\mathbf{S},\mathfrak{S})}{1+d(\mathbf{S},\mathfrak{S})} \longrightarrow (1+t^2) < \infty,\ \forall t \in \mathfrak{T}.$$

Definition 279 ***Time-varying radially increasing function relative to a constant set*** $\mathfrak{S}$ ***on*** $\mathfrak{T}_0$

A function $V(.) : \mathfrak{T}_0 \times \mathfrak{R}^n \longrightarrow \mathfrak{R}$ *is* ***radially increasing relative to a constant set*** $\mathfrak{S}$ ***on*** $\mathfrak{T}_0$ *if and only if*

$$\begin{array}{rcl} \forall \lambda_i & \in & \mathfrak{R}^+,\ i=1,2,\ \lambda_2 > \lambda_1 \Longrightarrow \\ V(t,\lambda_2\mathbf{S}) & > & V(t,\lambda_1\mathbf{S}),\ \forall (t,\mathbf{S}) \in \mathfrak{T}_0 \times \mathfrak{R}^n \backslash Cl\mathfrak{S}. \end{array} \tag{11.43}$$

*The expression "****on*** $\mathfrak{T}_0$*" is to be omitted if and only if* $\mathfrak{T}_0 = \mathfrak{T}$.
*The expression "****relative to*** $\mathfrak{S}$*" is to be omitted if and only if* $\mathfrak{S} = \mathfrak{O} = \{\mathbf{0}_n\}$.

Example 280 *[232] The globally positive definite functions defined by*

$$V_1(t,\mathbf{S}) = \frac{1+t^2}{2+t^2}\left(1 - e^{-d(\mathbf{S},\mathfrak{S})}\right),$$
$$V_2(t,\mathbf{S}) = (1+t^2)\left(1 - e^{-d(\mathbf{S},\mathfrak{S})}\right),$$
$$V_3(t,\mathbf{S}) = (1+t^2)\, d(\mathbf{S},\mathfrak{S})\left(1 + e^{d(\mathbf{S},\mathfrak{S})}\right),$$

are radially increasing relative to $\mathfrak{S}$ *on* $\mathfrak{T}_0$*, but* $V_1(t,\mathbf{S})$ *and* $V_2(t,\mathbf{S})$ *are not radially unbounded, whereas* $V_3(t,\mathbf{S})$ *is both radially increasing relative to* $\mathfrak{S}$ *on* $\mathfrak{T}_0$ *and radially unbounded relative to* $\mathfrak{S}$ *on* $\mathfrak{T}_0$*. However, the globally positive definite functions defined by*

$$\begin{aligned} V_4(t,\mathbf{S}) &= \frac{1+t^2}{2+t^2}\frac{d(\mathbf{S},\mathfrak{S})}{1+d^2(\mathbf{S},\mathfrak{S})}, \\ V_5(t,\mathbf{S}) &= \left(1+t^2\right)\frac{d(\mathbf{S},\mathfrak{S})}{1+d^2(\mathbf{S},\mathfrak{S})}, \\ V_6(t,\mathbf{S}) &= \left(1+t^2\right)\frac{\left(1-e^{-d(\mathbf{S},\mathfrak{S})}\right)}{1+e^{d(\mathbf{S},\mathfrak{S})}}, \end{aligned}$$

are neither radially increasing relative to $\mathfrak{S}$ *on* $\mathfrak{T}_0$ *nor radially unbounded relative to* $\mathfrak{S}$ *on* $\mathfrak{T}_0$*.*

The preceding definitions permit us to present fundamental Lyapunov results for the nonlinear systems (4.142), (4.143) and (5.22), (5.23). The books [217] and [230] contain Lyapunov's original results for *time*-invariant ISO linear systems and their extensions to two other classes of *time*-invariant linear systems.

Lemma 281 ***Decrescent function relative to*** $\mathfrak{S}$***, comparison function and the set*** $\mathfrak{V}_\xi(t)$***.***

If a function $V(.)$ *is decrescent relative to* $\mathfrak{S}$ *on* $\mathfrak{T}_0 \times \mathcal{N}(\mathfrak{S})$*, where* $\mathcal{N}(\mathfrak{S})$ *is a time-invariant connected neighborhood of the set* $\mathfrak{S}$*, then there is* $\xi \in \mathfrak{R}^+$ *such that for every* $\varsigma \in]0,\xi[$ *there is a neighborhood* $\mathcal{N}_\eta(\mathfrak{S})$ *of* $\mathfrak{S}$*, which is a subset of the set* $\mathfrak{V}_\varsigma(t;t_0)$ *for all* $t \in \mathfrak{T}_0$*,*

$$\begin{aligned} \exists\xi &\in \mathfrak{R}^+ \Longrightarrow \forall\varsigma \in]0,\xi[,\ \exists\eta \in \mathfrak{R}^+,\ \eta = \eta(\varsigma), \\ &\Longrightarrow \mathcal{N}_\eta(\mathfrak{S}) \subset \mathfrak{V}_\varsigma(t;t_0),\ \forall t \in \mathfrak{T}_0. \end{aligned}$$

Proof. Let $\mathcal{N}(\mathfrak{S})$ be a connected neighborhood of the set $\mathfrak{S}$ such that $V(.)$ is decrescent relative to $\mathfrak{S}$ on $\mathfrak{T}_0 \times \mathcal{N}(\mathfrak{S})$. Let μ be defined by Equation (11.37). Let $\varphi(.) : \mathfrak{R}_+ \longrightarrow \mathfrak{R}_+$ be a comparison function of the class $\mathcal{K}_{[0,\mu[}$, which obeys

$$\varphi[d(\mathbf{S},\mathfrak{S})] \geq V(t,\mathbf{S}),\ \forall(t,\mathbf{S}) \in \mathfrak{T}_0 \times \mathcal{N}_\mu(\mathfrak{S})$$

due to (11.42). We continue the proof by contradiction. Let all the preceding conditions hold and let the set $\mathfrak{V}_\zeta(t;t_0)$, $\zeta \in]0,\xi[$ for $\xi = \varphi(\mu)$, be such that for every $\eta \leq \varphi^I(\zeta)$, $\eta \in \mathfrak{R}^+$, there is a moment $\sigma \in \mathfrak{T}_0$ at which $\mathfrak{V}_\zeta(\sigma;t_0)$ is a subset of $\mathcal{N}_\eta(\mathfrak{S})$, $\mathfrak{V}_\zeta(\sigma;t_0) \subset \mathcal{N}_\eta(\mathfrak{S})$. Let $\mathbf{S}^* \in \mathcal{N}_\eta(\mathfrak{S}) \backslash \mathfrak{V}_\zeta(\sigma;t_0)$, which implies both $d(\mathbf{S}^*,\mathfrak{S}) < \eta$ and $V(\sigma,\mathbf{S}^*) \geq \zeta$. Hence, $\varphi[d(\mathbf{S}^*,\mathfrak{S})] < \varphi(\eta) \leq \zeta$. This and (11.42) imply

$$\zeta \geq \varphi(\eta) > \varphi[d(\mathbf{S}^*,\mathfrak{S})] \geq V(\sigma,\mathbf{S}^*) \geq \zeta,$$

which shows the contradiction $\zeta > \zeta$. The contradiction proves that for every $\zeta \in]0,\xi[$ there is $\eta \subset \mathfrak{R}^+$ such that the neighborhood $\mathcal{N}_\eta(\mathfrak{S})$ is subset of the set $\mathfrak{V}_\zeta(t;t_0)$ on $\mathfrak{T}_0$,

$$\begin{aligned} \exists\xi \in \mathfrak{R}^+ \Longrightarrow \forall\zeta \in]0,\xi[,\ \exists\eta \in \mathfrak{R}^+,\ \eta = \eta(\varsigma) \Longrightarrow \\ \mathfrak{V}_\zeta(t;t_0) \supset \mathcal{N}_\eta(\mathfrak{S}),\ \forall t \in \mathfrak{T}_0 \end{aligned}$$

(see Figure 11.4). ∎

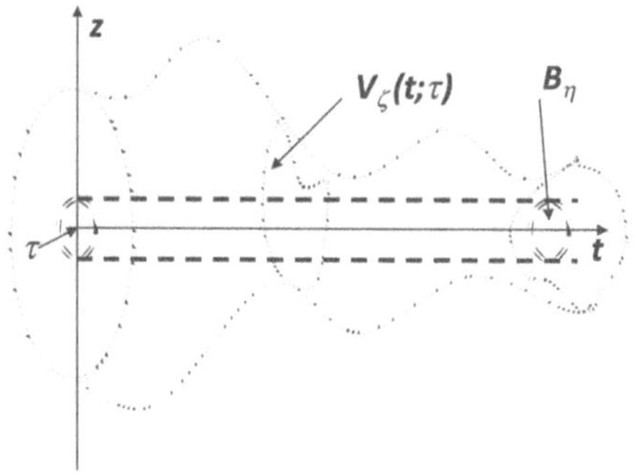

Figure 11.4: The neighborhood $\mathfrak{N}_\eta(\mathfrak{S})$ of $\mathfrak{S} = \mathfrak{O}$ is the hyperball $\mathfrak{B}_\eta$ that is always the subset of the set $\mathfrak{V}_\varsigma(t;\tau)$ induced by the *time*-varying decrescent function $V(.)$.

Example 282 *The function $V(t,\mathbf{S}) = (1+t^2)\, d(\mathbf{S},\mathfrak{S})$ is not a decrescent function relative to $\mathfrak{S}$. There is not a time-invariant positive definite function $W(.)$ such that $V(.)$ obeys Definition 275, i.e., (11.40), because*

$$t \longrightarrow \infty \Longrightarrow V(t,\mathbf{S}) = (1+t^2)\, d(\mathbf{S},\mathfrak{S}) \longrightarrow \infty,\ \forall (\mathbf{S} \neq \mathbf{0}_n) \in \mathfrak{R}^n.$$

For every $\xi \in \mathfrak{R}^+$ the set $\mathfrak{V}_\xi(t)$ is asymptotically contractive relative to $\mathfrak{S}$ (Example 269),

$$\mathfrak{V}_\xi(t) = \left\{\mathbf{S} : d(\mathbf{S},\mathfrak{S}) < \xi\left(1+t^2\right)^{-1}\right\} \longrightarrow \mathfrak{S} \text{ as } t \longrightarrow \infty.$$

For every positive definite function $W(.)$ there is $t_0 \in \mathfrak{T}$ such that

$$V(t,\mathbf{S}) > W(\mathbf{S}),\ \forall (t > t_0) \in \mathfrak{T}.$$

For every $\zeta \in \mathfrak{R}^+$there is $t_0 \in \mathfrak{T}$ such that $\mathfrak{V}_\xi(t)$ is the proper subset of $\mathcal{N}_\zeta(\mathfrak{S})$ at every $t \in \mathfrak{T}_0$,

$$\mathfrak{V}_\xi(t) \subset \mathcal{N}_\zeta(\mathfrak{S}),\ \forall t \in \mathfrak{T}_0.$$

Example 283 *The function $V(t,\mathbf{S}) = (1+t^2)\left(2+t^2\right)^{-1} d(\mathbf{S},\mathfrak{S})$ is a decrescent function relative to $\mathfrak{S}$ because it satisfies Definition 275 for the (globally) positive definite function $W(\mathbf{S}) = d(\mathbf{S},\mathfrak{S})$ relative to $\mathfrak{S}$,*

$$V(t,\mathbf{S}) = \frac{1+t^2}{2+t^2} d(\mathbf{S},\mathfrak{S}) \leq d(\mathbf{S},\mathfrak{S}),\ \forall (t,\mathbf{S}) \in \mathfrak{T} \times \mathfrak{R}^n.$$

For every $\xi \in \mathfrak{R}^+$there is $\zeta \in \mathfrak{R}^+$ such that the set $\mathfrak{W}_\zeta(\mathfrak{S})$,

$$\mathfrak{W}_\zeta(\mathfrak{S}) = \{\mathbf{S} : W(\mathbf{S}) < \zeta\} = \{\mathbf{S} : d(\mathbf{S},\mathfrak{S}) < \zeta\},$$

is always a subset of the set $\mathfrak{V}_\xi(t)$,

$$\mathfrak{W}_\zeta(\mathfrak{S}) \subset \mathfrak{V}_\xi(t),\ \forall t \in \mathfrak{T}.$$

11.7 Set invariance relative to a function

The concept of the set positive invariance with respect to a function was introduced in [115, Definition 8, p. 542], [170, Definition 1, p. 45] and used in [195, Definition19, p. 34].

Definition 284 ***Set positive invariance with respect to a function that is positive definite relative to the constant set*** $\mathfrak{A}$

A time-invariant set $\mathfrak{S}$ *is:*

a) ***Positively invariant with respect to a function*** $V(.) : \mathfrak{T} \times \mathfrak{R}^n \longrightarrow \mathfrak{R}_+$, ***which is positive definite relative to the constant set*** $\mathfrak{A}$ ***on*** $\mathfrak{T}_0$, *if and only if for* $t_0 \in \mathfrak{T}$ *there is* $\xi_0 = \xi(t_0) \in \mathfrak{R}$ *such that*

$$Cl\mathfrak{A} \subset \mathfrak{V}_{\xi_0}(t) \subseteq \mathfrak{S},\ \forall t \in \mathfrak{T}_0. \tag{11.44}$$

The expression "***on*** $\mathfrak{T}_0$" *is to be omitted if and only if* $\mathfrak{T}$ *replaces* $\mathfrak{T}_0$.

b) ***Invariant with respect to the function*** $V(.)$, ***which is positive definite relative to the set*** $\mathfrak{A}$, *if and only if there is* $\xi \in \mathfrak{R}$ *such that*

$$Cl\mathfrak{A} \subset \mathfrak{V}_{\xi}(t) \subseteq \mathfrak{S}(t), \forall t \in \mathfrak{T}. \tag{11.45}$$

The expression "***relative to the set*** $\mathfrak{A}$" *is to be omitted if and only if* $\mathfrak{A} = \mathfrak{O} = \{\mathbf{0}_n\}$.

A useful relationship between the function positive definiteness and sets follows.

Lemma 285 *[115, Proposition 6, p. 542], [195, Proposition 14, p. 35]* ***The (positive) invariance of a neighborhood*** $\mathcal{N}(\mathfrak{S})$ ***of the constant set*** $\mathfrak{S}$ ***with respect to a function*** $V(.) : \mathfrak{T} \times \mathfrak{R}^n \longrightarrow \mathfrak{R}$, *(*$V(.) : \mathfrak{T}_0 \times \mathfrak{R}^n \longrightarrow \mathfrak{R}$*)*, ***which is positive definite relative to the set*** $\mathfrak{A}$

Every time-invariant, connected, and bounded neighborhood $\mathcal{N}(\mathfrak{A})$ *of* $\mathfrak{A}$, *is (positive) invariant (on* $\mathfrak{T}_0$*) with respect to every function* $V(.)$ *(on* $\mathfrak{T}_0$*), which is positive definite relative to* $\mathfrak{A}$, *respectively.*

Proof. The statements follow directly from Lemma 271 (Section 11.5). ∎

Example 286 *Let* $\mathfrak{A} = \mathfrak{O}$. *The set*

$$\mathfrak{S} = \left\{\mathbf{S} : \mathbf{S} \in \mathfrak{R}^2,\ (2|S_1| + |S_2|) \leq 1\right\}$$

is invariant relative to the function $V(.)$,

$$V(t, \mathbf{S}) = \left(1 + t^2\right)(2|S_1| + |S_2|),\ \mathbf{S} = \begin{bmatrix} S_1 \\ S_2 \end{bmatrix},$$

because the set $\mathfrak{V}_{\xi}(t)$,

$$\mathfrak{V}_{\xi}(t) = \left\{\mathbf{S} : \mathbf{S} \in \mathfrak{R}^2,\ V(t, \mathbf{S}) = \left(1 + t^2\right)(2|S_1| + |S_2|) < \xi\right\}$$

obeys the conditions a) and b) of Definition 284 for every $\xi \in]0, 1]$.

If a set $\mathfrak{S}$ is bounded, then there can exist a function $\omega(.)$ such that its constant value $\xi \in \mathfrak{R}$, $\omega(\mathbf{S}) = \xi$, determines a hypersurface representing the boundary $\partial\mathfrak{S}$ of the set $\mathfrak{S}$. More precisely:

Definition 287 ***The boundary function of a set***

A function $\omega(.) : \mathfrak{R}^n \longrightarrow \mathfrak{R}$ *is* ***the boundary function of a connected set*** $\mathfrak{S}$, $\mathfrak{S} \subset \mathfrak{R}^n$, *if and only if the set* $\mathfrak{S}$ *has the boundary* $\partial\mathfrak{S}$ *and there is* $\xi \in \mathfrak{R}$ *such that*

$$\omega(\mathbf{S}) = \xi \Longleftrightarrow \mathbf{S} \in \partial\mathfrak{S}.$$

Example 288 *The boundary function of the bounded set:*

$$\mathfrak{S} = \left\{ \mathbf{S} : \mathbf{S} \in \mathfrak{R}^2, \ (2\,|S_1| + |S_2|) \leq 1 \right\}$$

is

$$\omega(\mathbf{S}) = 2\,|S_1| + |S_2|\,.$$

Its boundary is

$$\partial\mathfrak{S} = \left\{ \mathbf{S} : \mathbf{S} \in \mathfrak{R}^2, \, \omega(\mathbf{S}) = (2\,|S_1| + |S_2|) = 1 \right\} \Longrightarrow \xi = 1.$$

Definition 289 ***Differentiability of bounded connected sets***

A bounded connected set $\mathfrak{S}$ with the boundary function $\omega(.) : \mathfrak{R}^n \longrightarrow \mathfrak{R}$ is ***differentiable*** *if and only if its boundary function is differentiable on $\mathfrak{R}^n$.*

Definition 290 ***Radially increasing boundary function on the set boundary***

The boundary function $\omega(.) : \mathfrak{R}^n \longrightarrow \mathfrak{R}$ of a bounded set $\mathfrak{S}$, $\mathfrak{S} \subset \mathfrak{R}^n$, is ***radially increasing on its boundary*** *$\partial\mathfrak{S}$ if and only if there is $\zeta \in]0,1[$ such that $\mathbf{S} \in \partial\mathfrak{S}$ implies*

$$\omega\left[(1-\lambda)\,\mathbf{S}\right] < \omega(\mathbf{S}) < \omega\left[(1+\lambda)\,\mathbf{S}\right], \ \forall\lambda \in]0,\zeta], \ \forall\mathbf{S} \in \partial\mathfrak{S}.$$

Lemma 291 ***Positive invariance of time-invariant sets relative to the constant set $\mathfrak{S}$***

Let $\mathcal{N}(\mathfrak{S})$ be a connected nonempty neighborhood of the set $\mathfrak{S}$. Let both the neighborhood $\mathcal{N}(\mathfrak{S})$ of $\mathfrak{S}$ and the system motions $\mathcal{S}(.;t_0;\mathbf{S}_0)$ be time-continuous on $\mathfrak{T}_0$ for every $\mathbf{S}_0 \in Cl\mathcal{N}(\mathfrak{S})$, $t_0 \in \mathfrak{T}_i$. If both

1) the boundary function $\omega(.) : \mathfrak{R}^n \longrightarrow \mathfrak{R}$ of $\mathcal{N}(\mathfrak{S})$ is radially increasing on $\partial\mathcal{N}(\mathfrak{S})$ and

2) $D^+\omega\left[\mathbf{S}(t;t_0;\mathbf{S}_0)\right] < 0, \ \forall(t,\mathbf{S}_0) \in \mathfrak{T}_0 \times Cl\mathcal{N}(\mathfrak{S})$,

then the closure $Cl\mathcal{N}(\mathfrak{S})$ of $\mathcal{N}(\mathfrak{S})$ is positive invariant with respect to the system motions $\mathcal{S}(.;t_0;\mathbf{S}_0)$ on $\mathfrak{T}_0$.

Proof. Let the conditions of Lemma 291 hold. Let us assume that the closure $Cl\mathcal{N}(\mathfrak{S})$ of $\mathcal{N}(\mathfrak{S})$ is not positive invariant relative to the system motions $\mathcal{S}(.;t_0;\mathbf{S}_0)$. This means, due to Definition 106, that there is a pair $(\mathbf{S}^*, t^*) \in Cl\mathcal{N}(\mathfrak{S}) \times \mathfrak{T}_0$ and $\widetilde{t} \in \mathfrak{R}^+$ such that both $\mathbf{S}(t_0;t_0;\mathbf{S}_0) = \mathbf{S}_0 \in Cl\mathcal{N}(\mathfrak{S})$ and $\mathbf{S}(t;t^*;\mathbf{S}^*) \notin Cl\mathcal{N}(\mathfrak{S})$, $\forall t \in]t^*, \widetilde{t}[$. Thus, Definition 287, and Definition 290 imply that both $\omega(\mathbf{S}^*) = \xi$ and $\omega\left[\mathbf{S}(t;t^*;\mathbf{S}^*)\right] > \xi$, $\forall t \in]t^*, \overline{t}[$, where such $\overline{t} \in]t^*, \widetilde{t}[$ exists due to both the *time* invariance of $\mathfrak{S}$ and continuity on $\mathfrak{T}_0$ of the motions $\mathcal{S}(.;t_0;\mathbf{S}_0)$ for every $\mathbf{S}_0 \in Cl\mathcal{N}(\mathfrak{S})$. The consequence is that

$$\left\langle \omega\left[\mathbf{S}(t+\theta;t^*;\mathbf{S}^*)\right] - \omega(\mathbf{S}^*) \right\rangle \theta^{-1} > 0, \quad \forall\theta \in]0, \overline{t} - t^*[.$$

This implies in the limit as $\theta \longrightarrow 0^+$:

$$D^+\omega\left[\mathbf{S}(t;t^*;\mathbf{S}^*)\right] =$$
$$= \lim\left\{ \left\langle \omega\left[\mathbf{S}(t+\theta;t^*;\mathbf{S}^*)\right] - \omega(t^*, \mathbf{S}^*) \right\rangle \theta^{-1}; \theta \longrightarrow 0^+ \right\} > 0.$$

This result contradicts condition 2) of the lemma statement. The contradiction is the consequence of the assumption that the closure $Cl\mathcal{N}(\mathfrak{S})$ of $\mathcal{N}(\mathfrak{S})$ is not positive invariant with respect to the system motions $\mathcal{S}(.;t_0;\mathbf{S}_0)$. The assumption fails, which proves that the closure $Cl\mathcal{N}(\mathfrak{S})$ of $\mathcal{N}(\mathfrak{S})$ is positive invariant with respect to the system motions $\mathcal{S}(.;t_0;\mathbf{S}_0)$ on $\mathfrak{T}_0$. ■

Chapter 12

Consistent stability methodology basis

12.1 Features of consistent methodology

The first crucial problem of the Lyapunov method implementation is the problem of the selection or of the determination of an appropriate function $V(.)$ for a given nonlinear system. It is called the system **Lyapunov function**. Lyapunov's original results do not give any guideline for solving this problem. They leave us to use the trial-error procedure that can be endless if we test a stability property of the system. Consistent Lyapunov Methodology theoretically solves (and practically by using power and energy approach) this problem.

The physical approach uses the fact that energy / matter $E(.)$ and power $P(.)$ create motions and control of physical systems. It suggests accepting the system energy (energy and/or matter) $E(.)$ for the function $V(.)$, the system power for the function $-W(.)$ so that $V^{(1)}(.) = E^{(1)}(.) = P(.) = -W(.)$.

If we apply Lyapunov's original, i.e., the inverse, methodology for nonlinear systems, then we should ensure first the demanded characteristics of the function $V(.)$ and then continue to test the properties of $V^{(1)}(.)$. For the system to have the required stability property, it is then sufficient, but not necessary, that $V^{(1)}(.)$ possesses the specified characteristics. If $V^{(1)}(.)$ does not possess them, the problem is not solved. The procedure should be repeated with another choice of the function $V(.)$. The procedure can be endless, which is its crucial disadvantage.

This approach linked with the system energy $E(.)$ and power $P(.)$, which are interrelated by $E^{(1)}(.) = P(.)$, is applicable only if the energy has the demanded features. To solve the problem, the system power should also possess the necessary characteristics. If it does not, then the problem is not solved.

If we apply Consistent Lyapunov Methodology, then we first select $W(.)$ with the characteristics demanded for $V^{(1)}(.)$. Then we test the properties of the function $V(.)$ obtained as the solution of the differential equation $V^{(1)}(.) = -W(.)$. For the system to have the required stability property it is both necessary and sufficient that the solution function $V(.)$ possesses the specified characteristics. We get the exact complete solution to the problem in a single application of the procedure. It results in the direct construction of the system Lyapunov function $V(.)$ if the solution is affirmative. If the solution is not affirmative, then

DOI: 10.1201/9781003744528-18

the system does not have the Lyapunov function and it does not possess the requested property.

The subsequent parts of the book will show how we can use the Lyapunov method and methodologies to solve stability problems of *time*-varying nonlinear systems.

In order to mathematically explain the characteristics of **the Consistent Lyapunov methodology** (CLM) for the nonlinear systems we remind ourselves of Lyapunov's original methodology for the *time*-invariant linear systems.

For the *time*-invariant linear system

$$\frac{d\mathbf{S}}{dt} = A\mathbf{S} \tag{12.1}$$

the Lyapunov function $V(.)$ is the unique solution of the following differential equation for **an arbitrary time-invariant positive definite function** $W(.)$:

$$\frac{dV(\mathbf{S})}{dt} = -W(\mathbf{S}),\ \mathbf{S} = \mathbf{S}(t;\mathbf{S}_0)\,. \tag{12.2}$$

This means that the Eulerian derivative $V^{(1)}(\mathbf{S}) = dV(\mathbf{S})/dt$ is taken along the solutions $\mathcal{S}(.;\mathbf{S}_0)$ of the *time*-invariant linear system (12.1).

The new *consistent Lyapunov methodology (CLM)* for the nonlinear systems starts with the generalization of (12.2) in order to correspond to the stability study of the *time*-varying nonlinear system **(5.22), (5.23)**:

$$\begin{gathered} W(t,\mathbf{S}) \in \mathfrak{C}^1(\mathfrak{T}_0 \times \mathfrak{R}^n) \Longrightarrow \\ V^{(1)}(t,\mathbf{S}) = \frac{dV(t,\mathbf{S})}{dt} = -W(t,\mathbf{S}),\quad \mathbf{S} = \mathbf{S}(t;t_0;\mathbf{S}_0)\,, \end{gathered} \tag{12.3}$$

or, in general,

$$W(t,\mathbf{S}) \in \mathfrak{C}(\mathfrak{T}_0 \times \mathfrak{R}^n) \Longrightarrow D^+V(t,\mathbf{S}) = -W(t,\mathbf{S}),\ \mathbf{S} = \mathbf{S}(t;t_0;\mathbf{S}_0)\,. \tag{12.4}$$

The Dini derivative $D^+V(t,\mathbf{S})$ of the function $V(.)$ is taken along the system motions $\mathcal{S}(.;t_0;\mathbf{S}_0)$. The function $W(.)$ is **the generating function**. The following two classes of the generating functions enable the determination of the conditions that are both necessary and sufficient for all three fundamental tasks unsolved in the original Lyapunov stability theory of the nonlinear systems:

1. The establishment of the stability conditions that are not expressed in terms of the existence of a Lyapunov function $V(.)$.

2. The establishment of the stability conditions that are expressed in terms of the properties of the generating function $W(.)$ and, by it induced, the solution Lyapunov function $V(.)$ of (12.4).

3. The determination of both the necessary and sufficient conditions for asymptotic stability of the positive invariant set $\mathfrak{I}^+$ of the *time*-varying nonlinear system **(5.22), (5.23)**.

4. The establishment of the necessary and sufficient conditions of, and the procedure for, the direct generation of the system Lyapunov function $V(.)$ so that the function $V(.)$ is obtained after the single application of the conditions and the procedure. The obtained result is then final.

5. The specification of the necessary and sufficient conditions for a set to be the asymptotic stability domain.

The functions $V(.)$ and $W(.)$ induce the sets $\mathfrak{V}_\varsigma(t,t_0)$ and $\mathfrak{W}_\varsigma(t,t_0)$, respectively, Definition 266, Section 11.5.

12.2 System smoothness properties

What follows establishes the Consistent Lyapunov methodology (*CLM*) for the constant set $\mathcal{I}^+$ of the *time-varying* nonlinear systems **(5.22), (5.23)**.

Condition 292 *The time-invariant set* $\mathcal{I}^+$ *satisfies Property 197, Section 7.2.*

Let $\mathbb{N}(t;t_0;\mathcal{I}^+)$ be an open, continuous, connected *time*-varying neighborhood of the set $\mathcal{I}^+$,

$$\mathbb{N}\left(t;t_0;\mathcal{I}^+\right) \in \mathfrak{C}\left(\mathfrak{T}_0 \times \mathfrak{T}_i\right),$$
$$Cl\mathcal{I}^+ \subset In\mathbb{N}\left(t;t_0;\mathcal{I}^+\right), \ \forall (t;t_0) \in \mathfrak{T}_0 \times \mathfrak{T}_i, \tag{12.5}$$

such that $\mathbb{N}(\mathfrak{T}_0;\mathfrak{T}_i;\mathcal{I}^+)$ is also an open connected neighborhood of the set $\mathcal{I}^+$:

$$Cl\mathcal{I}^+ \subset In\mathbb{N}\left(\mathfrak{T}_0;\mathfrak{T}_i;\mathcal{I}^+\right). \tag{12.6}$$

where $\mathbb{N}(\mathfrak{T}_0;\mathfrak{T}_i;\mathcal{I}^+)$ is the lower set limit $\mathbb{N}_m(\mathfrak{T}_0;\mathfrak{T}_i;\mathcal{I}^+)$ of the set $\mathbb{N}(t;t_0;\mathcal{I}^+)$ on $\mathfrak{T}_0 \times \mathfrak{T}_i$ defined as follows:

$$\mathbb{N}\left(\mathfrak{T}_0;\mathfrak{T}_i;\mathcal{I}^+\right) = \mathbb{N}_m\left(\mathfrak{T}_0;\mathfrak{T}_i;\mathcal{I}^+\right) = \cap\left\{\mathbb{N}\left(t;t_0;\mathcal{I}^+\right) : (t,t_0) \in \mathfrak{T}_0 \times \mathfrak{T}_i\right\}. \tag{12.7}$$

The following definitions determine the advanced system properties additional to Property 147 and Property 148 when the system positive invariant set is constant $\mathcal{I}^+$ (see Note 149 and Note 155, all in Section 4.11).

Definition 293 ***Weak smoothness property***
Let Property 197 (Section 7.2) hold.

(i) There is an open, continuous, connected neighborhood $\mathbb{N}(t;t_0;\mathcal{I}^+)$, $\mathbb{N}(t;t_0;\mathcal{I}^+) \subseteq \mathfrak{R}^n$, *of the set* $\mathcal{I}^+$ *for every* $(t,t_0) \in \mathfrak{T}_0 \times \mathfrak{T}_i$ *such that* $\mathbb{N}(\mathfrak{T}_0;\mathfrak{T}_i;\mathcal{I}^+)$ *(12.7) is open connected neighborhood of the set* $\mathcal{I}^+$, *and*

$$\forall (t,t_0,\mathbf{S}_0) \in \mathfrak{T}_0 \times \mathfrak{T}_i \times \mathbb{N}(t_0;t_0;\mathcal{I}^+)$$

the following conditions are valid:

a) The system ***(5.22), (5.23)*** *has a unique motion* $\mathcal{S}(.;t_0;\mathbf{S}_0)$ *through* $\mathbf{S}_0$ *at* t_0 *on* $\mathfrak{T}_0$, *and*

b) the system motion $\mathcal{S}(.;t_0;\mathbf{S}_0)$ *is defined, continuous in* $(t,t_0,\mathbf{S}_0) \in \mathfrak{T}_0 \times \mathfrak{T}_i \times \mathbb{N}(t_0;t_0;\mathcal{I}^+)$ *and possesses Property P4 117, Section 4.2.*

(ii) For every $(t_0,\mathbf{S}_0) \in \mathfrak{T}_i \times [\mathfrak{R}^n \backslash Cl\mathbb{N}(t_0;t_0;\mathcal{I}^+)]$ *every motion of the system is continuous in* $t \in \mathfrak{T}_0$.

(iii) There is not a positive invariant set of the system ***(5.22), (5.23)*** *in the set difference*

$$\mathbb{N}\left(t;t_0;\mathcal{I}^+\right)\backslash\mathcal{I}^+, \ \forall (t,t_0) \in \mathfrak{T}_0 \times \mathfrak{T}_i. \tag{12.8}$$

If and only if the preceding conditions hold for $\mathbb{N}(t;t_0;\mathcal{I}^+) \equiv \mathfrak{R}^n$, *then the weak smoothness property holds on* $\mathfrak{R}^n$.

Definition 294 ***The first-order weak smoothness property***
(i) There is an open continuous connected neighborhood $\mathbb{N}(t;t_0;\mathcal{I}^+)$,

$$\mathbb{N}(t;t_0;\mathcal{I}^+) \subseteq \mathfrak{R}^n, \ \mathbb{N}(t;t_0;\mathcal{I}^+) \in \mathfrak{C}\left(\mathfrak{T}_i\right),$$

of the set $\mathcal{I}^+$ for every $(t, t_0) \in \mathfrak{T}_0 \times \mathfrak{T}_i$ such that

$$\mathbb{N}(\mathfrak{T}_0; \mathfrak{T}_i; \mathcal{I}^+) = \mathbb{N}_m(\mathfrak{T}_0; \mathfrak{T}_i; \mathcal{I}^+) = \cap \left[\mathbb{N}(t; t_0; \mathcal{I}^+) : (t, t_0) \in \mathfrak{T}_0 \times \mathfrak{T}_i\right]$$

is open connected neighborhood of the set $\mathcal{I}^+$, and

$$\forall (t, t_0, \mathbf{S}_0) \in \mathfrak{T}_0 \times \mathfrak{T}_i \times \mathbb{N}(t_0; t_0; \mathcal{I}^+)$$

the following conditions hold:

*a) The system **(5.22), (5.23)** has a unique motion $\mathcal{S}(.; t_0; \mathbf{S}_0)$ through $\mathbf{S}_0$ at t_0 on $\mathfrak{T}_0$.*

b) $\mathbf{S}(t; t_0; \mathbf{S}_0)$ is defined, continuous, and continuously and differentiable for all triplets $(t, t_0, \mathbf{S}_0) \in \mathfrak{T}_0 \times \mathfrak{T}_i \times \mathbb{N}(t_0; t_0; \mathcal{I}^+)$ and possesses Property P4 117.

(ii) For every $(t_0, \mathbf{S}_0) \in \mathfrak{T}_i \times [\mathfrak{R}^n \backslash Cl\mathbb{N}(t_0; t_0; \mathcal{I}^+)]$, every motion of the system is continuous in $t \in \mathfrak{T}_0$.

*(iii) There is not a positive invariant set of the system **(5.22), (5.23)** in the set difference (12.8).*

If and only if the preceding conditions hold for $\mathbb{N}(t; t_0; \mathcal{I}^+) \equiv \mathfrak{R}^n$, then the first-order weak smoothness property holds on $\mathfrak{R}^n$.

Definition 295 ***Strong smoothness property***

*(i) The system **(5.22), (5.23)** possesses the weak smoothness property 293.*

(ii) If the boundary $\partial\mathbb{N}(t; t_0; \mathcal{I}^+)$ of $\mathbb{N}(t; t_0; \mathcal{I}^+)$ is nonempty at any moment $t \in \mathfrak{T}_0$, and for any $t_0 \in \mathfrak{T}_i$, then every system motion passing through $\mathbf{S}_0 \in \partial\mathbb{N}(t_0; t_0; \mathcal{I}^+)$ at any $t_0 \in \mathfrak{T}_i$ satisfies the following:

$$inf \left\{d\left[\mathbf{S}(t; t_0; \mathbf{S}_0), \mathcal{I}^+\right] : t \in \mathfrak{T}_0\right\} > 0, \ \forall (t_0, \mathbf{S}_0) \in \mathfrak{T}_i \times \partial\mathbb{N}(t_0; t_0; \mathcal{I}^+).$$

If and only if the preceding conditions hold for $\mathbb{N}(t; t_0; \mathcal{I}^+) \equiv \mathfrak{R}^n$, then the strong smoothness property holds on $\mathfrak{R}^n$.

Definition 296 ***The first-order strong smoothness property***

*(i) The system **(5.22), (5.23)** possesses the first-order weak smoothness property 294.*

(ii) If the boundary $\partial\mathbb{N}(t; t_0; \mathcal{I}^+)$ of $\mathbb{N}(t; t_0; \mathcal{I}^+)$ is nonempty at any moment $t \in \mathfrak{T}_0$, and for any $t_0 \in \mathfrak{T}_i$, then every system motion passing through $\mathbf{S}_0 \in \partial\mathbb{N}(t_0; t_0; \mathcal{I}^+)$ at every $t_0 \in \mathfrak{T}_i$ satisfies the following:

$$inf \left\{d\left[\mathbf{S}(t; t_0; \mathbf{S}_0), \mathcal{I}^+\right] : t \in \mathfrak{T}_0\right\} > 0, \ \forall (t_0, \mathbf{S}_0) \in \mathfrak{T}_i \times \partial\mathbb{N}(t_0; t_0; \mathcal{I}^+).$$

If and only if the preceding conditions hold for $\mathbb{N}(t; t_0; \mathcal{I}^+) \equiv \mathfrak{R}^n$, then the first-order strong smoothness property holds on $\mathfrak{R}^n$.

Comment 297 ***On smoothness properties***

Both the weak smoothness property and the strong smoothness property allow nondifferentiability of the system motions $\mathbf{S}(t; t_0; \mathbf{S}_0)$ with respect to $(t, t_0, \mathbf{S}_0) \in \mathfrak{T}_0 \times \mathfrak{T}_i \times \mathbb{N}(t_0; t_0; \mathcal{I}^+)$.

Both the first-order weak smoothness property and the first-order strong smoothness property concern systems with differentiable motions $\mathbf{S}(t; t_0; \mathbf{S}_0)$ with respect to $(t, t_0, \mathbf{S}_0) \in \mathfrak{T}_0 \times \mathfrak{T}_i \times \mathbb{N}(t_0; t_0; \mathcal{I}^+)$.

When we treat a physical system, we can often draw conclusions about the smoothness of the system motions from the system physical properties.

12.3 Systems and generating functions

What follows introduces two families of the generating functions $W(.)$ that induce the sets $\mathfrak{W}_{\zeta}(t,t_0;\mathcal{I}^+)$, Definition 266, Section 11.5.

$D^+V(t,\mathbf{S})$ denotes the right-hand upper Dini derivative of $V(t,\mathbf{S})$ along system motions at a moment t (Appendix B).

Let

$$\mathbb{N}_M(\mathfrak{T}_i;\mathfrak{T}_i;\mathcal{I}^+) = \cup\left[\mathbb{N}(t_0;t_0;\mathcal{I}^+) : (t,t_0) \in \mathfrak{T}_0 \times \mathfrak{T}_i\right]. \tag{12.9}$$

Definition 298 ***Family*** $L(\mathfrak{T}_0,\mathfrak{T}_i;\mathbb{N};\mathcal{I}^+)$ ***of the generating functions*** $W(.)$

Let $\mathbb{N}(t;t_0;\mathcal{I}^+)$ *be an open, continuous, connected neighborhood of the set* $\mathcal{I}^+$ *such that* $\mathbb{N}_M(\mathfrak{T}_i;\mathfrak{T}_i;\mathcal{I}^+)$ *(12.9) is also an open connected neighborhood of the set* $\mathcal{I}^+$. *A function* $W(.) : \mathfrak{T}\times\mathfrak{R}^n \longrightarrow \mathfrak{R}_+$ *belongs to the family* $L(\mathfrak{T}_0,\mathfrak{T}_i;\mathbb{N};\mathcal{I}^+)$ *if and only if:*

1) $W(.)$ *is continuous on* $\mathfrak{T}\times\mathbb{N}_M(\mathfrak{T}_0;\mathfrak{T}_i;\mathcal{I}^+)$:

$$W(t,\mathbf{S}) \in \mathfrak{C}\left[\mathfrak{T}\times\mathbb{N}_M(\mathfrak{T}_0;\mathfrak{T}_i;\mathcal{I}^+)\right]. \tag{12.10}$$

2) The equations (12.11) and (12.12):

$$D^+V(t,\mathbf{S}) = -W(t,\mathbf{S}),\ \forall (t,\mathbf{S}) \in \mathfrak{T}\times\mathbb{N}(t;t_0;\mathcal{I}^+),\ \forall t_0 \in \mathfrak{T}_i, \tag{12.11}$$

$$V(t,\mathbf{S}) = 0 \iff \mathbf{S}\in\partial\mathcal{I}^+,\ \forall (t,t_0)\in\mathfrak{T}_0\times\mathfrak{T}_i, \tag{12.12}$$

have a solution $V(.) : \mathfrak{T}\times\mathfrak{R}^n \longrightarrow \mathfrak{R}_+$, *which is continuous in* $(t,\mathbf{S}) \in \mathfrak{T}\times Cl\mathbb{N}_{\mu}(\mathcal{I}^+)$ *for an arbitrarily small* $\mu\in\mathfrak{R}^+$, $\mu = \mu(W)$, *and which obeys (12.13) for some* $\omega_{\mu}(\mathbf{S}) \in \mathfrak{C}[Cl\mathbb{N}_{\mu}(\mathcal{I}^+)]$:

$$V(t,\mathbf{S}) \le \omega_{\mu}(\mathbf{S};\mathcal{I}^+),\ \forall (t,t_0,\mathbf{S}) \in \mathfrak{T}_0\times\mathfrak{T}_i\times Cl\mathbb{N}_{\mu}(\mathcal{I}^+). \tag{12.13}$$

3) The following holds for any $\zeta\in\mathfrak{R}^+$ *that satisfies* $Cl\mathfrak{W}_{\zeta}(t,t_0;\mathcal{I}^+) \subset \mathbb{N}(t;t_0;\mathcal{I}^+)$ *for every* $(t,t_0)\in\mathfrak{T}_0\times\mathfrak{T}_i$:

$$\begin{aligned} min\left\{W(t,\mathbf{S}) : (t,t_0,\mathbf{S}) \in \mathfrak{T}_0\times\mathfrak{T}_i\times\left[\mathbb{N}(t;t_0;\mathcal{I}^+)\setminus\mathfrak{W}_{\zeta}(t,t_0;\mathcal{I}^+)\right]\right\} = \\ = a = a(\zeta;W)\in\mathfrak{R}^+. \end{aligned} \tag{12.14}$$

Definition 299 ***Family*** $L^1(\mathfrak{T}_0,\mathfrak{T}_i;\mathbb{N};\mathcal{I}^+)$ ***of the generating functions*** $W(.)$

Let $\mathbb{N}(t;t_0;\mathcal{I}^+)$ *be an open, continuous connected neighborhood of the set* $\mathcal{I}^+$ *such that* $\mathbb{N}_M(\mathfrak{T}_i;\mathfrak{T}_i;\mathcal{I}^+)$ *(12.9) is an open connected neighborhood of the set* $\mathcal{I}^+$. *A function* $W(.) : \mathfrak{T}\times\mathfrak{R}^n \longrightarrow \mathfrak{R}_+$ *belongs to the family* $L^1(\mathfrak{T}_0,\mathfrak{T}_i;\mathbb{N};\mathcal{I}^+)$ *if and only if:*

1) $W(.)$ *is differentiable on* $\mathfrak{T}\times\mathbb{N}_M(\mathfrak{T}_0;\mathfrak{T}_i;\mathcal{I}^+)$:

$$W(t,\mathbf{S}) \in \mathfrak{C}^1\left[\mathfrak{T}\times\mathbb{N}_M(\mathfrak{T}_0;\mathfrak{T}_i;\mathcal{I}^+)\right]. \tag{12.15}$$

2) The equations (12.16) and (12.17):

$$V^{(1)}(t,\mathbf{S};\mathcal{I}^+) = -W(t,\mathbf{S}),\ \forall (t,t_0,\mathbf{S}) \in \mathfrak{T}_0\times\mathfrak{T}_i\times\mathbb{N}(t_0;t_0;\mathcal{I}^+), \tag{12.16}$$

$$V(t,\mathbf{S}) = 0 \iff \mathbf{S}\in\partial\mathcal{I}^+,\ \forall (t,t_0)\in\mathfrak{T}_0\times\mathfrak{T}_i, \tag{12.17}$$

have a solution $V(.) : \mathfrak{T}\times\mathfrak{R}^n \longrightarrow \mathfrak{R}_+$ *that is differentiable in* $(t,\mathbf{S}) \in \mathfrak{T}\times\mathbb{N}(t;t_0;\mathcal{I}^+)$ *for an arbitrarily small* $\mu\in\mathfrak{R}^+$, $\mu=\mu(W)$, *and which obeys (12.13) for some* $\omega_{\mu}(\mathbf{S};\mathcal{I}^+) \in \mathfrak{C}[Cl\mathbb{N}_{\mu}(\mathcal{I}^+)]$,

3) The condition (12.14) holds for any $\zeta\in\mathfrak{R}^+$ *satisfying* $Cl\mathfrak{W}_{\zeta}(t,t_0;\mathcal{I}^+) \subset \mathbb{N}(t;t_0;\mathcal{I}^+)$ *for every* $(t,t_0)\in\mathfrak{T}_0\times\mathfrak{T}_i$.

The family $E\left[\mathfrak{T}_0,\mathfrak{T}_i,\mathbb{N}\left(.\right);\mathcal{I}^+\right]$ of the generating functions $W\left(.;\mathcal{I}^+\right)$ is defined by:

Definition 300 ***Family*** $E\left(\mathfrak{T}_0,\mathfrak{T}_i;\mathbb{N};\mathcal{I}^+\right)$ ***of the generating functions*** $W(.)$

Let $\mathbb{N}\left(t;t_0;\mathcal{I}^+\right)$ *be an open, continuous, connected neighborhood of the set* $\mathcal{I}^+$ *such that* $\mathbb{N}_M\left(\mathfrak{T}_i;\mathfrak{T}_i;\mathcal{I}^+\right)$ *(12.9) is also an open connected neighborhood of the set* $\mathcal{I}^+$. *A function* $W(.):\mathfrak{T}\times\mathfrak{R}^n\longrightarrow\mathfrak{R}_+$ *belongs to the family* $E\left(\mathfrak{T}_0,\mathfrak{T}_i;\mathbb{N};\mathcal{I}^+\right)$ *if and only if:*

1) $W(.)$ *is continuous on* $\mathfrak{T}\times\mathbb{N}_M\left(\mathfrak{T}_0;\mathfrak{T}_i;\mathcal{I}^+\right)$:

$$W(t,\mathbf{S})\in\mathfrak{C}\left[\mathfrak{T}\times\mathbb{N}_M\left(\mathfrak{T}_0;\mathfrak{T}_i;\mathcal{I}^+\right)\right], \tag{12.18}$$

2) The equations (12.19) and (12.20):

$$\begin{aligned} &D^+U(t,\mathbf{S})=-\left[1-U(t,\mathbf{S})\right]W(t,\mathbf{S}),\\ &\forall(t,t_0,\mathbf{S})\in\mathfrak{T}_0\times\mathfrak{T}_i\times\mathbb{N}\left(t;t_0;\mathcal{I}^+\right), \end{aligned} \tag{12.19}$$

$$U(t,\mathbf{S})=0\Longleftrightarrow\mathbf{S}\in\partial\mathcal{I}^+,\ \forall(t,t_0)\in\mathfrak{T}_0\times\mathfrak{T}_i, \tag{12.20}$$

have a solution $U\left(.;\mathcal{I}^+\right):\mathfrak{T}\times\mathfrak{R}^n\longrightarrow\mathfrak{R}_+$ *that is continuous in* $(t,\mathbf{S})\in\mathfrak{T}\times\mathbb{N}\left(t;t_0;\mathcal{I}^+\right)$ *for an arbitrarily small* $\mu\in\mathfrak{R}^+$, $\mu=\mu\left(W\right)$, *and which obeys (12.21) for some* $\omega_\mu\left(\mathbf{S}\right)\in\mathfrak{C}\left[Cl\mathbb{N}_\mu\left(\mathcal{I}^+\right)\right]$, *or* $\omega_\mu\left(\mathbf{S};\mathcal{I}^+\right)\in\mathfrak{C}\left[Cl\mathbb{N}_\mu\left(\mathcal{I}^+\right)\right]$:

$$U(t,\mathbf{S})\leq\omega_\mu\left(\mathbf{S};\mathcal{I}^+\right),\ \forall\left(t,t_0,\mathbf{S}\right)\in\mathfrak{T}_0\times\mathfrak{T}_i\times Cl\mathbb{N}_\mu\left(\mathcal{I}^+\right). \tag{12.21}$$

3) The following holds for any $\zeta\in\mathfrak{R}^+$ *satisfying* $Cl\mathfrak{W}_\zeta\left(t,t_0;\mathcal{I}^+\right)\subset\mathbb{N}\left(t;t_0;\mathcal{I}^+\right)$ *for every* $(t,t_0)\in\mathfrak{T}_0\times\mathfrak{T}_i$:

$$min\left\{\begin{array}{c} W(t,\mathbf{S}):(t,t_0,\mathbf{S})\in\\ \in\mathfrak{T}_0\times\mathfrak{T}_i\times\left[\mathbb{N}\left(t;t_0;\mathcal{I}^+\right)\backslash\mathfrak{W}_\zeta\left(t,t_0;\mathcal{I}^+\right)\right]\end{array}\right\}=a=a\left(\zeta;W\right)\in\mathfrak{R}^+. \tag{12.22}$$

Definition 301 ***Family*** $E^1\left(\mathfrak{T}_0,\mathfrak{T}_i;\mathbb{N};\mathcal{I}^+\right)$ ***of the generating functions*** $W(.)$

Let $\mathbb{N}\left(t;t_0;\mathcal{I}^+\right)$ *be an open, continuous, connected neighborhood of the set* $\mathcal{I}^+$ *such that* $\mathbb{N}_M\left(\mathfrak{T}_i;\mathfrak{T}_i;\mathcal{I}^+\right)$ *(12.9) is also an open connected neighborhood of the set* $\mathcal{I}^+$. *A function* $W(.):\mathfrak{T}\times\mathfrak{R}^n\longrightarrow\mathfrak{R}_+$ *belongs to the family* $E^1\left(\mathfrak{T}_0,\mathfrak{T}_i;\mathbb{N};\mathcal{I}^+\right)$ *if and only if:*

1) $W(.)$ *is differentiable on* $\mathfrak{T}\times\mathbb{N}_M\left(\mathfrak{T}_0;\mathfrak{T}_i;\mathcal{I}^+\right)$:

$$W(t,\mathbf{S})\in\mathfrak{C}^1\left[\mathfrak{T}\times\mathbb{N}_M\left(\mathfrak{T}_0;\mathfrak{T}_i;\mathcal{I}^+\right)\right]. \tag{12.23}$$

2) The equations (12.24) and (12.25):

$$U^{(1)}(t,\mathbf{S})=-\left[1-U(t,\mathbf{S})\right]W(t,\mathbf{S}),\ \forall(t,t_0,\mathbf{S})\in\mathfrak{T}_0\times\mathfrak{T}_i\times\mathbb{N}\left(t;t_0;\mathcal{I}^+\right), \tag{12.24}$$

$$U(t,\mathbf{S})=0\Longleftrightarrow\mathbf{S}\in\partial\mathcal{I}^+,\ \forall(t,t_0)\in\mathfrak{T}_0\times\mathfrak{T}_i, \tag{12.25}$$

have a solution $U(.):\mathfrak{T}\times\mathfrak{R}^n\longrightarrow\mathfrak{R}_+$ *that is differentiable in* $(t,\mathbf{S})\in\mathfrak{T}\times Cl\mathbb{N}_\mu\left(\mathcal{I}^+\right)$ *for an arbitrarily small* $\mu\in\mathfrak{R}^+$, $\mu=\mu\left(W\right)$, *and which obeys (12.21) for some* $\omega_\mu\left(\mathbf{S};\mathcal{I}^+\right)\in\mathfrak{C}\left[Cl\mathbb{N}_\mu\left(\mathcal{I}^+\right)\right]$.

3) The condition (12.22) holds for any $\zeta\in\mathfrak{R}^+$ *satisfying* $Cl\mathfrak{W}_\zeta\left(t,t_0;\mathcal{I}^+\right)\subset\mathbb{N}\left(t;t_0;\mathcal{I}^+\right)$ *for every* $(t,t_0)\in\mathfrak{T}_0\times\mathfrak{T}_i$.

Note 302 ***Stability and functional families*** $L\left(\mathfrak{T}_0,\mathfrak{T}_i;\mathbb{N};\mathcal{I}^+\right)$, $L^1\left(\mathfrak{T}_0,\mathfrak{T}_i;\mathbb{N};\mathcal{I}^+\right)$, ***as well as*** $E\left(\mathfrak{T}_0,\mathfrak{T}_i;\mathbb{N};\mathcal{I}^+\right)$, ***and*** $E^1\left(\mathfrak{T}_0,\mathfrak{T}_i;\mathbb{N};\mathcal{I}^+\right)$

There is not any condition imposed on the system ***(5.22), (5.23)*** *in Definition 298-Definition 301. The families* $L\left(\mathfrak{T}_0,\mathfrak{T}_i;\mathbb{N};\mathcal{I}^+\right)$, $L^1\left(\mathfrak{T}_0,\mathfrak{T}_i;\mathbb{N};\mathcal{I}^+\right)$, $E\left(\mathfrak{T}_0,\mathfrak{T}_i;\mathbb{N};\mathcal{I}^+\right)$, *and* $E^1\left(\mathfrak{T}_0,\mathfrak{T}_i;\mathbb{N};\mathcal{I}^+\right)$ *are independent of stability properties of the system.*

Comment 303 ***The relationship between the functions $V(.)$ and $U(.)$ for the same generating function $W(.)$***

Let the generating function $W(.)$ be the same in Definition 298 and Definition 300. The functions $V()$ generated by $W(.) \in L(\mathfrak{T}_0, \mathfrak{T}_i; \mathbb{N}; \mathcal{I}^+)$ and the function $U()$ generated by $W(.) \in E(\mathfrak{T}_0, \mathfrak{T}_i; \mathbb{N}; \mathcal{I}^+)$ are interrelated, due to (12.11) and (12.19), in the differential form, and in the explicit form [495], by:

$$\begin{gathered} D^+U(t,\mathbf{S}) = [1 - U(t,\mathbf{S})]\, D^+V(t,\mathbf{S}),\ or \\ U^{(1)}(t,\mathbf{S}) = [1 - U(t,\mathbf{S})]\, V^{(1)}(t,\mathbf{S};\mathcal{I}^+), \\ \forall (t,t_0,\mathbf{S}) \in \mathfrak{T}_0 \times \mathfrak{T}_i \times \mathbb{N}\left(t;t_0;\mathcal{I}^+\right), \end{gathered} \tag{12.26}$$

$$U(t,\mathbf{S}) = 1 - exp\left[-V(t,\mathbf{S})\right],\ \forall (t,t_0,\mathbf{S}) \in \mathfrak{T}_0 \times \mathfrak{T}_i \times \mathbb{N}\left(t;t_0;\mathcal{I}^+\right), \tag{12.27}$$

$$U(t,\mathbf{S}) = 0 \iff V(t,\mathbf{S}) = 0,\ and\ U(t,\mathbf{S}) \longrightarrow 1 \iff V(t,\mathbf{S}) \longrightarrow \infty. \tag{12.28}$$

12.4 Lemma A

Lemma 304 ***Lemma A***

*Let the system **(5.22)**, **(5.23)** possess the weak smoothness property, Definition **293**. Let the set $\mathcal{I}^+$ be uniformly attractive on $\mathfrak{T}_i$ with the instantaneous attraction domain $\mathcal{D}_A(t_0;\mathcal{I}^+)$ at $t_0 \in \mathfrak{T}_i$ obeying $\mathcal{D}_A(t_0;\mathcal{I}^+) \subseteq \mathbb{N}(t_0;t_0;\mathcal{I}^+)$ for all $t_0 \in \mathfrak{T}_i$ and with the domain $\mathcal{D}_{UA}(\mathfrak{T}_i;\mathcal{I}^+)$ of uniform attraction on $\mathfrak{T}_i$.*

a) If $\mathfrak{T}_i \subset \mathfrak{T}$, then:

1) $(t_0, \mathbf{S}_0) \in \mathfrak{T}_i \times \mathcal{D}_A(t_0;\mathcal{I}^+)$ implies $\mathbf{S}(t;t_0;\mathbf{S}_0) \in \mathcal{D}_A(t;\mathcal{I}^+)$ for all $t \in \mathfrak{T}_0$, for every $t_0 \in \mathfrak{T}$, which means that $\mathcal{D}_A(t;\mathcal{I}^+)$ is a positively invariant set of the system on $\mathfrak{T}_0 \times \mathfrak{T}_i$,

2) $\mathcal{D}_A(t;\mathcal{I}^+)$ is an open continuous connected neighborhood of the set $\mathcal{I}^+$ at any $(t;t_0) \in \mathfrak{T}_0 \times \mathfrak{T}_i$:

$$\mathcal{D}_A\left(t;\mathcal{I}^+\right) \in \mathfrak{C}\left(\mathfrak{T}_0 \times \mathfrak{T}_i\right),\ \mathcal{D}_A\left(t;\mathcal{I}^+\right) \equiv In\mathcal{D}_A\left(t;\mathcal{I}^+\right),$$

3) $\mathcal{D}_{UA}(\mathfrak{T}_i;\mathcal{I}^+)$ is a connected neighborhood of $\mathcal{I}^+$. If

$$\mathcal{D}_A\left(t;\mathcal{I}^+\right) = \mathcal{D}_{UA}\left(\mathfrak{T}_i;\mathcal{I}^+\right)\ \forall (t;t_0) \in \mathfrak{T}_0 \times \mathfrak{T}_i$$

then $\mathcal{D}_{UA}(\mathfrak{T}_i;\mathcal{I}^+)$ is also a positively invariant set of the system on $\mathfrak{T}_0 \times \mathfrak{T}_i$.

b) If $\mathfrak{T}_i = \mathfrak{T}$, then:

1) $\mathcal{D}_A(t;\mathcal{I}^+)$ is a positively invariant set of the system on $\mathfrak{T}_0 \times \mathfrak{T}$, i.e., $(t_0, \mathbf{S}_0) \in \mathfrak{T} \times \mathcal{D}_A(t_0;\mathcal{I}^+)$ implies $\mathbf{S}(t;t_0;\mathbf{S}_0) \in \mathcal{D}_A(t;\mathcal{I}^+)$ for all $t \in \mathfrak{T}_0$, for every $t_0 \in \mathfrak{T}$.

2) $\mathcal{D}_A(t;\mathcal{I}^+)$ is an open continuous connected neighborhood of $\mathcal{I}^+$ at any $t \in \mathfrak{T}$:

$$\begin{gathered} \mathcal{D}_A\left(t;\mathcal{I}^+\right) \in \mathfrak{C}\left(\mathfrak{T}\right), \\ \mathcal{D}_A\left(t;\mathcal{I}^+\right) \equiv In\mathcal{D}_A\left(t;\mathcal{I}^+\right) \supset Cl\mathcal{I}^+,\ \forall (t;t_0) \in \mathfrak{T}_0 \times \mathfrak{T}_i. \end{gathered}$$

3) $\mathcal{D}_A(\mathfrak{T};\mathcal{I}^+)$ is a connected neighborhood of $\mathcal{I}^+$. If

$$\mathcal{D}_A\left(t;\mathcal{I}^+\right) = \mathcal{D}_A\left(\mathfrak{T};\mathcal{I}^+\right)\ \forall (t;t_0) \in \mathfrak{T} \times \mathfrak{T},$$

then $\mathcal{D}_A(\mathfrak{T};\mathcal{I}^+)$ is also a positively invariant set of the system on $\mathfrak{T} \times \mathfrak{T}$.

Proof. of Lemma A

Let the system **(5.22)**, **(5.23)** possess a Weak smoothness property, Definition **293**. Let the set $\mathcal{I}^+$ be uniformly attractive on $\mathfrak{T}_i$ with the instantaneous domain $\mathcal{D}_A(t;\mathcal{I}^+)$ of attraction at $t \in \mathfrak{T}_i$, which obeys $\mathcal{D}_A(t;\mathcal{I}^+) \subseteq \mathbb{N}(t;t;\mathcal{I}^+)$ for all $t \in \mathfrak{T}_i$, and with the domain $\mathcal{D}_{UA}(\mathfrak{T}_i;\mathcal{I}^+)$ of the uniform attraction on $\mathfrak{T}_i$, Definition 8.6.

a) Let $t_0 \in \mathfrak{T}_i$ and $t^* \in \mathfrak{T}_i$, $t^* \neq t_0$. Let $\mathbf{S}^* = \mathbf{S}(t^*;t_0;\mathbf{S}_0)$ for any $\mathbf{S}_0 \in \mathcal{D}_A(t_0;\mathcal{I}^+)$. Then, $\mathbf{S}(t;t_0;\mathbf{S}_0) \longrightarrow \mathcal{I}^+$ as $t \longrightarrow \infty$. Because

$$\mathbf{S}(t;t^*;\mathbf{S}^*) \equiv \mathbf{S}[t;t^*;\mathbf{S}(t^*;t_0;\mathbf{S}_0)],$$

which is true due to (i) of the Weak smoothness property, and

$$\mathcal{D}_A(t_0;\mathcal{I}^+) \subseteq \mathbb{N}(t_0;t_0;\mathcal{I}^+)$$

for all $t_0 \in \mathfrak{T}_i$, then $\mathbf{S}(t;t^*;\mathbf{S}^*) \longrightarrow \mathcal{I}^+$ as $t \longrightarrow \infty$. Hence, $\mathbf{S}^* = \mathbf{S}(t^*;t_0;\mathbf{S}_0) \in \mathcal{D}_A(t^*;\mathcal{I}^+)$, which proves the statement under a-1). Notice that the statement under a-1) is the same as the statement under a) of Lemma 227, Section 8.4.

Let $\zeta \in \mathfrak{R}^+$ be such that

$$\mathfrak{N}_{2\zeta}(\mathcal{I}^+) \subset \mathcal{D}_A(t;\mathcal{I}^+), \ \forall t \in \mathfrak{T}_i.$$

It exists (Definition 225). Let it be assumed that the domain of attraction $\mathcal{D}_A(t;\mathcal{I}^+)$ is not open for all $t_0 \in \mathfrak{T}_i$. Let there exist $t_0' \in \mathfrak{T}_i$ and

$$\mathbf{S}_0' \in \partial\mathcal{D}_A(t_0';\mathcal{I}^+) \cap \mathcal{D}_A(t_0';\mathcal{I}^+).$$

Let $\varepsilon \in]0, \zeta/2[$. Then, (i) of the Weak smoothness property and

$$\mathcal{D}_A(t;\mathcal{I}^+) \subseteq \mathbb{N}(t;t_0;\mathcal{I}^+)$$

for all $t \in \mathfrak{T}_0$ and for every $t_0 \in \mathfrak{T}_i$, imply the existence of $\theta \in \mathfrak{R}^+$, $\theta = \theta(t_0', \mathbf{S}_0', \varepsilon)$, such that $\|\mathbf{S}_0 - \mathbf{S}_0'\| < \theta$ ensures

$$\|\mathbf{S}(t_0' + 2\sigma'; t_0'; \mathbf{S}_0) - \mathbf{S}(t_0' + 2\sigma'; t_0'; \mathbf{S}_0')\| < \varepsilon,$$

where $\sigma' = \tau(t_0, \varsigma, \mathbf{S}_0)$, Definition 225. Because $\varepsilon < \zeta/2$ and

$$d[\mathbf{S}(t_0' + 2\sigma'; t_0'; \mathbf{S}_0'), \mathcal{I}^+] < \zeta,$$

then $\mathbf{S}(t_0' + 2\sigma'; t_0'; \mathbf{S}_0) \in \mathfrak{N}_{2\zeta}[\mathcal{I}^+] \subset \mathcal{D}_A(t_0';\mathcal{I}^+)$. Hence,

$$\mathbf{S}_0 \in \mathcal{D}_A(t_0';\mathcal{I}^+).$$

Any $\mathbf{S}_0$ obeying $\|\mathbf{S}_0 - \mathbf{S}_0'\| < \theta$ may be selected in a θ-neighborhood of $\mathbf{S}_0'$ out of the domain $\mathcal{D}_A(t_0';\mathcal{I}^+)$, which is contradicted by the obtained $\mathbf{S}_0 \in \mathcal{D}_A(t_0';\mathcal{I}^+)$. The former is true and the latter is wrong, showing that there are not $t_0' \in \mathfrak{T}_i$ and $\mathbf{S}_0' \in \partial\mathcal{D}_A(t_0';\mathcal{I}^+) \cap \mathcal{D}_A(t_0';\mathcal{I}^+)$. If $\mathbf{S}_0' \in \partial\mathcal{D}_A(t_0';\mathcal{I}^+)$, then

$$\mathbf{S}_0' \notin In\mathcal{D}_A(t_0';\mathcal{I}^+).$$

The set $\mathcal{D}_A(t;\mathcal{I}^+)$ is open for all $(t, t_0) \in \mathfrak{T}_0 \times \mathfrak{T}_i$, and it is a neighborhood of $\mathcal{I}^+$ due to Definition 225 for every $t_0 \in \mathfrak{T}_i$. Therefore, $\mathcal{D}_A(t;\mathcal{I}^+) \equiv In\mathcal{D}_A(t;\mathcal{I}^+)$ and it is a neighborhood of $\mathcal{I}^+$ on $\mathfrak{T}_i$. Altogether, the domain $\mathcal{D}_A(t;\mathcal{I}^+)$ is an open neighborhood of $\mathcal{I}^+$ on $\mathfrak{T}_i$. In order to prove $\mathcal{D}_A(t;\mathcal{I}^+) \in \mathfrak{C}(\mathfrak{T}_0 \times \mathfrak{T}_i)$ we use a contradiction. Let there exist $t^* \in \mathfrak{T}_0$ such that $\mathcal{D}_A(t;\mathcal{I}^+)$ is discontinuous at $(t^*, t_0^*) \in \mathfrak{T}_0 \times \mathfrak{T}_i$. As a consequence, there are

$\varepsilon^* \in \mathfrak{R}^+$ and sequences $J^* \subseteq \{1, 2, ..., n, ...\}$ and $K^* \subseteq \{1, 2, ..., n, ...\}$ such that $t_j \longrightarrow t^*$as $j \longrightarrow \infty$, $j \in J^*$, and $t_k \longrightarrow t_0^*$ as $k \longrightarrow \infty$, $k \in K^*$, and that there is $\mathbf{S}^* \in \mathcal{D}_A(t_0^*; \mathcal{I}^+)$ for which $d[\mathbf{S}^*, \mathcal{D}_A(t_j; \mathcal{I}^+)] \geq \varepsilon^*$, $\forall (j,k) \in J^* \times K^*$, and/or there is $\mathbf{W}^* \in \mathcal{D}_A(t_j; \mathcal{I}^+)$, $\forall (j,k) \in J^* \times K^*$, for which $d[\mathbf{W}^*, \mathcal{D}_A(t_0^*; \mathcal{I}^+)] \geq \varepsilon^*$. Let $0 < \xi < \varepsilon/2$ and $\mathfrak{N}_\xi(\mathcal{I}^+) \subseteq \mathcal{D}_A(t; \mathcal{I}^+)$, $\forall (t, t_0) \in \mathfrak{T}_0 \times \mathfrak{T}_i$, which is possible due to the uniform attraction of $\mathcal{I}^+$ on $\mathfrak{T}_i$, (Definition 228). Let $m \in K^*$ be such that $t_m > t_0^* + \tau(t_0^*, \xi/2, \mathbf{S}^*)$, $t_m \in \mathfrak{T}_i$. This guarantees (Definition 225): $\mathbf{S}(t_m; t_0; \mathbf{S}^*) \in \mathfrak{N}_{\xi/2}(\mathcal{I}^+)$. Let $\delta = \delta(t_0^*, \mathbf{S}^*, m, \xi/2) \in \mathfrak{R}^+$, $\delta < \xi/2$ and $\psi = \psi(t_0^*, \mathbf{S}^*, m, \xi/2) \in \mathfrak{R}^+$ obey that

$$|t_l - t_0^*| < \psi \textit{ and } \|\mathbf{S}_0 - \mathbf{S}^*\| < \delta,\ l \in K^* \Longrightarrow$$
$$\|\mathbf{S}(t_m; t_l; \mathbf{S}_0) - \mathbf{S}(t_m; t_0^*; \mathbf{S}^*)\| < \xi/2,$$

which is possible due to continuity of the system motions (Weak smoothness property). Hence, $\mathbf{S}(t_m; t_0^*; \mathbf{S}^*) \in \mathfrak{N}_{\xi/2}(\mathcal{I}^+)$ implies

$$\mathbf{S}(t_m; t_l; \mathbf{S}_0) \in \mathfrak{N}_\xi(\mathcal{I}^+).$$

This further yields

$$\mathbf{S}(t_m; t_l; \mathbf{S}_0) \in \mathcal{D}_A(t_m; \mathcal{I}^+) \textit{ and } \mathbf{S}_0 \in \mathcal{D}_A(t_0; \mathcal{I}^+).$$

Also,

$$\|\mathbf{S}_0 - \mathbf{S}^*\| < \delta < \xi/2 < \varepsilon/4 \textit{ and } \mathbf{S}_0 \in \mathcal{D}_A(t_l; \mathcal{I}^+)$$

imply $d[\mathbf{S}^*, \mathcal{D}_A(t_l; \mathcal{I}^+)] < \varepsilon^*$, which contradicts

$$d[\mathbf{S}^*, \mathcal{D}_A(t_k; \mathcal{I}^+)] \geq \varepsilon^*, \forall k \in K^*,$$

and disproves the existence of $(t_0^*, \mathbf{S}^*) \in \mathfrak{T}_i \times \mathcal{D}_A(t_0^*; \mathcal{I}^+)$ for which

$$d[\mathbf{S}^*, \mathcal{D}_A(t_j; \mathcal{I}^+)] \geq \varepsilon^*, \forall (j,k) \in J^* \times K^*.$$

Analogously, we show that there are not $\mathbf{W}^*$ and t_0^* as defined above. This proves the continuity of $\mathcal{D}_A(t_0; \mathcal{I}^+)$ on $\mathfrak{T}_i$. The statement under a-2) is correct.

Furthermore, $\mathcal{D}_A(t; \mathcal{I}^+)$ is a neighborhood of $\mathcal{I}^+$ by definition (Definition 225). Its connectedness is proved by contradiction as follows. Let us assume that it is not connected. Then, there are, for example, N disjoint sets $\mathcal{D}_{Ak}(t)$, $k = 1, 2, \cdots, N$, such that $\mathcal{D}_A(t; \mathcal{I}^+) = \cup[\mathcal{D}_{Ak}(t) : k = 1, 2, ..., N]$. At least one of the sets $\mathcal{D}_{Ak}(t)$ is not a neighborhood of $\mathcal{I}^+$, e.g., $\mathcal{D}_{A1}(t)$ and let $\mathcal{D}_{Am}(t)$, $\mathcal{D}_{Am}(t) \subset \mathcal{D}_A(t; \mathcal{I}^+)$, $m \in \{2, 3, \cdots, N\}$, be a connected neighborhood of $\mathcal{I}^+$, which is possible because $\mathcal{I}^+$ is a compact connected set. Then $\mathbf{S}_0 \in \mathcal{D}_{A1} \subset \mathcal{D}_A(t; \mathcal{I}^+)$ implies $\mathbf{S}(t; t_0; \mathbf{S}_0) \longrightarrow \mathcal{I}^+$ as $t \longrightarrow \infty$, $\forall t_0 \in \mathfrak{T}_i$. There is $t_1 \in \mathfrak{T}_0$ such that $\mathbf{S}(t_1; t_0; \mathbf{S}_0) \notin \mathcal{D}_A[t_1; \mathcal{I}^+(t_1; t_0)]$ because of the continuity of $\mathbf{S}(t; t_0; \mathbf{S}_0)$ in $t \in \mathfrak{T}_0$, $\forall t_0 \in \mathfrak{T}_i$, and because $\mathcal{D}_{A1}(t)$ is a disjoint subset of $\mathcal{D}_A(t; \mathcal{I}^+)$, which is not a neighborhood of $\mathcal{I}^+$. However, this is impossible due to the fact that $\mathbf{S}[t; t_1; \mathbf{S}(t_1; t_0; \mathbf{S}_0)] \equiv \mathbf{S}(t; t_0; \mathbf{S}_0) \longrightarrow \mathcal{I}^+$ as $t \longrightarrow \infty$, $\forall t_0 \in \mathfrak{T}_i$, which implies $\mathbf{S}(t_1; t_0; \mathbf{S}_0) \in \mathcal{D}_A[t_1; \mathcal{I}^+(t_1; t_0)]$, which is correct and contradicts $\mathbf{S}(t_1; t_0; \mathbf{S}_0) \notin \mathcal{D}_A[t_1; \mathcal{I}^+(t_1; t_0)]$. Hence, the assumption on disconnectedness of $\mathcal{D}_A(t; \mathcal{I}^+)$ is incorrect, which completes the proof of all statements under a) by noting that the positive invariance of $\mathcal{D}_A(t; \mathcal{I}^+)$ under a-3) results directly from 1) in the case $\mathcal{D}_{UA}(\mathfrak{T}_i; \mathcal{I}^+) = \mathcal{D}_A(t; \mathcal{I}^+)$ for all $t \in \mathfrak{T}_i$. Since, by definition, $\mathcal{D}_A(t_0; \mathcal{I}^+)$ is a neighborhood of $\mathcal{I}^+$ for every $t_0 \in \mathfrak{T}_i$, then it is neighborhood of $\mathcal{I}^+$ on $\mathfrak{T}_i$. This completes the proof of all statements under a).

b) The assertions under b) follow directly from those under a) in the case $\mathfrak{T}_i = \mathfrak{T}$. ∎

12.5 Lemma B

Lemma 305 ***Lemma B***

a) Let the system ***(5.22), (5.23)*** *possess a Weak smoothness property. Let the set* $\mathcal{I}^+$ *be asymptotically stable at* $t_0 \in \mathfrak{T}_i$ *with the instantaneous attraction domain* $\mathcal{D}_A(t_0;\mathcal{I}^+)$ *at every* $t_0 \in \mathfrak{T}_i$*, which obeys*

$$\mathcal{D}_A(t_0;\mathcal{I}^+) \subseteq \mathbb{N}\left[t_0;t_0;\mathcal{I}^+\right], \ \forall t_0 \in \mathfrak{T}_i. \tag{12.29}$$

Then the domains $\mathcal{D}_A(t_0;\mathcal{I}^+)$*,* $\mathcal{D}_S(t_0;\mathcal{I}^+)$*,* $\mathcal{D}_{AS}(t_0;\mathcal{I}^+)$ *are interrelated by*

$$\begin{aligned} \mathcal{D}_A(t_0;\mathcal{I}^+) &\subseteq \mathcal{D}_S(t_0;\mathcal{I}^+), \ \forall t_0 \in \mathfrak{T}_i, \\ \mathcal{D}_{AS}(t_0;\mathcal{I}^+) &= \mathcal{D}_A(t_0;\mathcal{I}^+), \ \forall t_0 \in \mathfrak{T}_i. \end{aligned} \tag{12.30}$$

b) Let the set $\mathcal{I}^+$ *of the system* ***(5.22), (5.23)*** *be uniformly asymptotically stable on* $\mathfrak{T}_i$*. Let its domain* $\mathcal{D}_{UA}(\mathfrak{T}_i;\mathcal{I}^+)$ *of uniform attraction on* $\mathfrak{T}_i$ *obey*

$$\mathcal{D}_{UA}(\mathfrak{T}_i;\mathcal{I}^+) \subseteq \mathbb{N}(\mathfrak{T}_0;\mathfrak{T}_i;\mathcal{I}^+), \tag{12.31}$$

then its domains $\mathcal{D}_{UA}(\mathfrak{T}_i;\mathcal{I}^+)$*,* $\mathcal{D}_{US}(\mathfrak{T}_i;\mathcal{I}^+)$ *and* $\mathcal{D}_{UAS}(\mathfrak{T}_i;\mathcal{I}^+)$ *are interrelated by*

$$\begin{aligned} \mathcal{D}_{UA}(\mathfrak{T}_i;\mathcal{I}^+) &\subseteq \mathcal{D}_{US}(\mathfrak{T}_i;\mathcal{I}^+), \\ \mathcal{D}_{UAS}(\mathfrak{T}_i;\mathcal{I}^+) &= \mathcal{D}_{UA}(\mathfrak{T}_i;\mathcal{I}^+). \end{aligned} \tag{12.32}$$

Proof. of Lemma B

a) Let the conditions under a) hold. Let $\mathbf{S}_0 \in \mathcal{D}_A(t_0;\mathcal{I}^+)$ be arbitrary. The constancy of the set $\mathcal{I}^+$ and the continuity of the system state $\mathbf{S}(t;t_0;\mathbf{S}_0)$ in $(t,t_0,\mathbf{S}_0) \in \mathfrak{T}_0 \times \mathfrak{T}_i \times \mathbb{N}(t_0;t_0;\mathcal{I}^+)$, the condition 12.29 and $\mathbf{S}_0 \in \mathcal{D}_A(t_0;\mathcal{I}^+)$ imply

$$max\left\{d\left[\mathbf{S}(t;t_0;\mathbf{S}_0),\mathcal{I}^+\right]: \ t \in \mathfrak{T}_0\right\} < \infty.$$

Let

$$\varepsilon = 2max\left\{d\left[\mathbf{S}(t;t_0;\mathbf{S}_0),\mathcal{I}^+\right]: \ t \in \mathfrak{T}_0\right\} \in \mathfrak{R}^+.$$

This ensures $\mathbf{S}_0 \in \mathcal{D}_S(t_0;\varepsilon;\mathcal{I}^+)$ in view of Definition 201 and Definition 204, which imply $\mathbf{S}_0 \in \mathcal{D}_S(t_0;\mathcal{I}^+)$. Altogether, $\mathbf{S}_0 \in \mathcal{D}_A(t_0;\mathcal{I}^+)$ yields $\mathbf{S}_0 \in \mathcal{D}_S(t_0;\mathcal{I}^+)$, which proves $\mathcal{D}_A(t_0;\mathcal{I}^+) \subseteq \mathcal{D}_S(t_0;\mathcal{I}^+)$ for every $t_0 \in \mathfrak{T}_i$. This result and Definition 235, i.e., its first equation (9.1), complete the proof of the statement under a).

b) Let the set $\mathcal{I}^+$ of the system **(5.22), (5.23)** be uniformly asymptotically stable on $\mathfrak{T}_i$ so that the condition (12.31) is satisfied.. Let $\mathbf{S}_0 \in \mathcal{D}_{UA}(\mathfrak{T}_i;\mathcal{I}^+)$ be arbitrary. Hence

$$max\left\{d\left[\mathbf{S}(t;t_0;\mathbf{S}_0),\mathcal{I}^+\right]: (t;t_0) \in \mathfrak{T}_0 \times \mathfrak{T}_i\right\} < \infty$$

due to the constancy of the set $\mathcal{I}^+$ and the continuity of the system state $\mathbf{S}(t;t_0;\mathbf{S}_0)$ in $(t,t_0,\mathbf{S}_0) \in \mathfrak{T}_0 \times \mathfrak{T}_i \times \mathcal{D}_{UA}(\mathfrak{T}_i;\mathcal{I}^+)$. Let

$$\varepsilon = 2max\left\{d\left[\mathbf{S}(t;t_0;\mathbf{S}_0),\mathcal{I}^+\right]: (t;t_0) \in \mathfrak{T}_0 \times \mathfrak{T}_i\right\} \in \mathfrak{R}^+$$

so that

$$\mathbf{S}_0 \in \mathcal{D}_S(\mathfrak{T}_i;\varepsilon;\mathcal{I}^+) = \cap\left[\mathcal{D}_S(t_0;\varepsilon;\mathcal{I}^+) : t_0 \in \mathfrak{T}_i\right] \neq \phi.$$

Therefore, $\mathbf{S}_0 \in \mathcal{D}_S(\mathfrak{T}_i;\mathcal{I}^+)$ (Definition 212). The result that $\mathbf{S}_0 \in \mathcal{D}_{UA}(\mathfrak{T}_i;\mathcal{I}^+)$ implies $\mathbf{S}_0 \in \mathcal{D}_S(\mathfrak{T}_i;\mathcal{I}^+)$ proves

$$\mathcal{D}_{UA}(\mathfrak{T}_i;\mathcal{I}^+) \subseteq \mathcal{D}_{US}(\mathfrak{T}_i;\mathcal{I}^+).$$

This and Definition 238 prove

$$\mathcal{D}_{UAS}\left(\mathfrak{T}_i;\mathcal{I}^+\right) = \mathcal{D}_{UA}\left(\mathfrak{T}_i;\mathcal{I}^+\right)$$

and complete the proof of both the statement under b) and of Lemma B.

Let $\varepsilon = 2max\left\{d\left[\mathbf{S}\left(t;t_0;\mathbf{S}_0\right),\mathcal{I}^+\right]:\ t \in \mathfrak{T}_0\right\}$. This ensures $\mathbf{S}_0 \in \mathcal{D}_S\left(t_0;\varepsilon;\mathcal{I}^+\right)$ in view of Definition 201 and Definition 204, which imply $\mathbf{S}_0 \in \mathcal{D}_S\left(t_0;\mathcal{I}^+\right)$. Altogether, $\mathbf{S}_0 \in \mathcal{D}_A\left(t_0;\mathcal{I}^+\right)$ yields $\mathbf{S}_0 \in \mathcal{D}_S\left(t_0;\mathcal{I}^+\right)$, which proves $\mathcal{D}_A\left(t_0;\mathcal{I}^+\right) \subseteq \mathcal{D}_S\left(t_0;\mathcal{I}^+\right)$ for every $t_0 \in \mathfrak{T}_i$. This result and Definition 235, i.e., its first equation (9.1), complete the proof of both the statement under a) and Lemma B 305. ■

Chapter 13

Asymptotic stability criteria

13.1 Systems with continuous motions

The form of the uniform asymptotic stability conditions depend on the smoothness property of the system (Definition 293 or Definition 295) and on the family of the generating functions (Definition 298 or Definition 300).

13.1.1 System with strong smoothness property

Theorem 306 ***Uniform asymptotic stability of the set $\mathcal{I}^{+}$ on $\mathfrak{T}_0 \times \mathfrak{T}_i$ of the system (5.22), (5.23): Strong smoothness property***

Let the system **(5.22)**, **(5.23)** *possess Strong smoothness property* ***295***.

In order for the set $\mathcal{I}^{+}$ to be uniformly asymptotically stable on $\mathfrak{T}_i$, for a set $\mathcal{N}(t_0;\mathcal{I}^{+})$ to be the asymptotic stability domain $\mathcal{D}_{AS}(t_0;\mathcal{I}^{+})$ of $\mathcal{I}^{+}$ at $t_0 \in \mathfrak{T}_i$:

$$\mathcal{N}(t_0;\mathcal{I}^{+}) = \mathcal{D}_{AS}(t_0;\mathcal{I}^{+}),\ t_0 \in \mathfrak{T}_i,$$

and for the set $\mathcal{N}(\mathfrak{T}_i;\mathcal{I}^{+})$ to be the domain $\mathcal{D}_{UAS}(\mathfrak{T}_i;\mathcal{I}^{+})$ of the uniform asymptotic stability of $\mathcal{I}^{+}$ on $\mathfrak{T}_i$:

$$\mathcal{N}(\mathfrak{T}_i;\mathcal{I}^{+}) = \mathcal{D}_{UAS}(\mathfrak{T}_i;\mathcal{I}^{+})$$

it is both necessary and sufficient that:

1) the set $\mathcal{N}(t_0;\mathcal{I}^{+})$ is an open, continuous, connected neighborhood of $\mathcal{I}^{+}$ and $\mathcal{N}(t_0;\mathcal{I}^{+}) \subseteq \mathbb{N}(t_0;t_0;\mathcal{I}^{+})$ for every $t_0 \in \mathfrak{T}_i$,

2) the set $\mathcal{N}(\mathfrak{T}_i;\mathcal{I}^{+}) = \cap\{\mathcal{N}(t_0;\mathcal{I}^{+}) : t_0 \in \mathfrak{T}_i\}$ is a connected neighborhood of $\mathcal{I}^{+}$ on $\mathfrak{T}_i$ such that

$$\mathcal{N}(\mathfrak{T}_i;\mathcal{I}^{+}) \subseteq \mathbb{N}(\mathfrak{T}_i;\mathfrak{T}_i;\ \mathcal{I}^{+}), \tag{13.1}$$

3) $\mathbf{f}(t,\mathbf{S}) = \mathbf{0}_n$ for all $(t,t_0) \in \mathfrak{T}_0 \times \mathfrak{T}_i$, is possible only for $\mathbf{S} \notin \{\mathcal{N}[t;\mathcal{I}^{+}(t;t)] \setminus \mathcal{I}^{+}(t;t)\}$,

4) for any decrescent positive definite function $W(.)$ relative to $\mathcal{I}^{+}$ on $\mathfrak{T}_0 \times \mathfrak{T}_i \times \mathbb{N}(t;t_0;\mathcal{I}^{+})$, which obeys:

a) $W(.) \in L(\mathfrak{T}_0;\mathfrak{T}_i,\mathbb{N},\mathcal{I}^{+})$, the equations (12.11) and (12.12) have the unique solution function $V(.)$ with the following properties:

(i) $V(.)$ is a decrescent positive definite function relative to $\mathcal{I}^{+}$ on $\mathfrak{T}_0 \times \mathfrak{T}_i \times \mathcal{N}(t;t_0;\mathcal{I}^{+})$,

and

(ii) if the boundary $\partial\mathcal{N}(t_0;\mathcal{I}^{+})$ of $\mathcal{N}(t_0;\mathcal{I}^{+})$ is nonempty, then $\mathbf{S} \longrightarrow \partial\mathcal{N}(t;\mathcal{I}^{+})$, $\mathbf{S} \in \mathcal{N}(t;\mathcal{I}^{+})$, implies $V(t,\mathbf{S}) \longrightarrow \infty$, $\forall t \in \mathfrak{T}_i$,

or obeys:

DOI: 10.1201/9781003744528-19

b) $W(.) \in E(\mathfrak{T}_0;\mathfrak{T}_i;\mathbb{N},\mathcal{I}^+)$, the equations (12.19) and (12.20) have the unique solution function $U(.)$ with the following properties:

(i) $U(.)$ is a decrescent positive definite function relative to $\mathcal{I}^+$ on $\mathfrak{T}_0 \times \mathfrak{T}_i \times \mathcal{N}(t;t_0;\mathcal{I}^+)$,

and

(ii) if the boundary $\partial\mathcal{N}(t_0;\mathcal{I}^+)$ of $\mathcal{N}(t_0;\mathcal{I}^+)$ is nonempty, then $\mathbf{S} \longrightarrow \partial\mathcal{N}(t;\mathcal{I}^+)$, $\mathbf{S} \in \mathcal{N}(t;\mathcal{I}^+)$, implies $U(t,\mathbf{S}) \longrightarrow 1$, $\forall t \in \mathfrak{T}_i$.

Proof. The functions $W(.)$ and $V(.)$ are positive definite relative to the set $\mathcal{I}^+$. They induce the sets $\mathfrak{W}_\zeta(t,t_0)$ and $\mathfrak{V}_\xi(t,t_0)$, respectively, Definition 266, Section 11.5.

Necessity. Let the system **(5.22), (5.23)** possess Strong smoothness property **295**. Let the set $\mathcal{I}^+$ of the system **(5.22), (5.23)** be uniformly asymptotically stable on $\mathfrak{T}_i$. Definition 234, Definition 235, Definition 237 and Definition 238 hold. They imply that the set $\mathcal{I}^+$ has the domain $\mathcal{D}_{AS}(t_0;\mathcal{I}^+)$ of the asymptotic stability at every $t_0 \in \mathfrak{T}_i$ and the domain $\mathcal{D}_{UAS}(\mathfrak{T}_i;\mathcal{I}^+)$ of the uniform asymptotic stability on $\mathfrak{T}_i$. The preceding definitions, Definition 201, Definition 204, Definition 211, Definition 212, Definition 225, Definition 226, Definition 228, and Definition 229 show that the system also has the stability domain $\mathcal{D}_S(t_0;\mathcal{I}^+)$ and the attraction domain $\mathcal{D}_A(t_0;\mathcal{I}^+)$ at every $t_0 \in \mathfrak{T}_i$, the uniform stability domain $\mathcal{D}_{US}(\mathfrak{T}_i;\mathcal{I}^+)$, and the uniform attraction domain $\mathcal{D}_{UA}(\mathfrak{T}_i;\mathcal{I}^+)$ on $\mathfrak{T}_i$ so that (13.2) holds:

$$\mathcal{D}_{AS}(t_0;\mathcal{I}^+) \subseteq \mathcal{D}_S(t_0;\mathcal{I}^+),\ \mathcal{D}_{AS}(t_0;\mathcal{I}^+) \subseteq \mathcal{D}_A(t_0;\mathcal{I}^+),\ \forall t_0 \in \mathfrak{T}_i,$$
$$\mathcal{D}_{UAS}(\mathfrak{T}_i;\mathcal{I}^+) \subseteq \mathcal{D}_{US}(\mathfrak{T}_i;\mathcal{I}^+),\ \mathcal{D}_{UAS}(\mathfrak{T}_i;\mathcal{I}^+) \subseteq \mathcal{D}_{UA}(\mathfrak{T}_i;\mathcal{I}^+). \tag{13.2}$$

Also, $\mathcal{D}_A(t_0;\mathcal{I}^+)$ is a connected neighborhood of $\mathcal{I}^+$ for every $t_0 \in \mathfrak{T}_i$, and $\mathcal{D}_{UA}(\mathfrak{T}_i;\mathcal{I}^+)$ is a connected neighborhood of $\mathcal{I}^+$ (Definition 228 and Definition 229).

The set $\mathbb{N}(t;t_0;\mathcal{I}^+)$ is a connected neighborhood of $\mathcal{I}^+$ at every $t_0 \in \mathfrak{T}_i$ and

$$\mathbb{N}(\mathfrak{T}_i;\mathfrak{T}_i;\mathcal{I}^+) = \cap\left[\mathbb{N}(t_0;t_0;\mathcal{I}^+) : t_0 \in \mathfrak{T}_i\right]$$

is also a neighborhood of $\mathcal{I}^+$ (Weak smoothness property **293** induced by Strong smoothness property **295**). Hence,

$$\mathcal{D}_A(t_0;\mathcal{I}^+) \cap \mathbb{N}(t_0;t_0;\mathcal{I}^+) \neq \phi\ \forall t_0 \in \mathfrak{T}_i,$$

and

$$\mathcal{D}_{UA}(\mathfrak{T}_i;\mathcal{I}^+) \cap \mathbb{N}(\mathfrak{T}_i;\mathcal{I}^+) \neq \phi.$$

Let us prove

$$\mathcal{D}_{AS}(t_0;\mathcal{I}^+) \subseteq \mathbb{N}(t_0;t_0;\mathcal{I}^+),\ \forall t_0 \in \mathfrak{T}_i. \tag{13.3}$$

If this were not true, then there would exist

$$(t_0,\mathbf{S}_0) \in \mathfrak{T}_i \times \left[\mathcal{D}_A(t_0;\mathcal{I}^+) \setminus Cl\mathbb{N}(t_0;t_0;\mathcal{I}^+)\right],$$

which would mean

$$\mathbf{S}_0 \in \mathcal{D}_A(t_0;\mathcal{I}^+) \cap \partial\mathbb{N}(t_0;t_0;\mathcal{I}^+)\ \ or\ \ \mathbf{S}_0 \in \mathcal{D}_A(t_0;\mathcal{I}^+) \setminus Cl\mathbb{N}(t_0;t_0;\mathcal{I}^+)$$

due to $\mathcal{D}_A(t_0;\mathcal{I}^+) \cap \mathbb{N}(t_0;t_0;\mathcal{I}^+) \neq \phi$ and the fact that $\mathbb{N}(t_0;t_0;\mathcal{I}^+)$ is an open set (due to 1) of Weak smoothness property **293** induced by 1) of Strong smoothness property **295**). If $\mathbf{S}_0 \in \mathcal{D}_A(t_0;\mathcal{I}^+) \cap \partial\mathbb{N}(t_0;t_0;\mathcal{I}^+)$, then

$$inf\left\{d\left[\mathbf{S}(t;t_0;\mathbf{S}),\mathcal{I}^+\right] : t \in \mathfrak{T}_0\right\} > 0$$

due to (ii) of Strong smoothness property **295**), which would mean $\mathbf{S}_0 \notin \mathcal{D}_A(t_0;\mathcal{I}^+)$ and would contradict $\mathbf{S}_0 \in \mathcal{D}_A(t_0;\mathcal{I}^+) \cap \partial\mathbb{N}(t;t_0;\mathcal{I}^+)$. Hence, $\mathbf{S}_0 \notin \mathcal{D}_A(t_0;\mathcal{I}^+) \cap \partial\mathbb{N}(t;t_0;\mathcal{I}^+)$ and $\mathcal{D}_A(t_0;\mathcal{I}^+) \cap \partial\mathbb{N}(t;t_0;\mathcal{I}^+) = \phi$. If

$$\mathbf{S}_0 \in \mathcal{D}_A(t_0;\mathcal{I}^+) \backslash Cl\mathbb{N}(t;t_0;\mathcal{I}^+),$$

then

$$lim\left[d\left[\mathbf{S}(t;t_0;\mathbf{S}_0),\mathcal{I}^+\right] : t \longrightarrow \infty\right] = 0, \forall t_0 \in \mathfrak{T}_i,$$

which together with (i) of Strong smoothness property **295**, (ii) of Weak smoothness property **293**, and $\mathbb{N}(t;t_0;\mathcal{I}^+) \in \mathfrak{C}(\mathfrak{T}_i)$ (due to (i) of Weak smoothness property **293**) would imply the existence of $(t^*,t_0^*) \in (In\mathfrak{T}_0) \times \mathfrak{T}_i$ such that $\mathbf{S}(t^*;t_0^*;\mathbf{S}) \in \partial\mathbb{N}(t^*,t_0^*;\mathcal{I}^+)$. This is impossible as shown above. The assumed triplet $(t^*,t_0^*) \in (In\mathfrak{T}_0) \times \mathfrak{T}_i$ does not exist. Hence, $\mathcal{D}_A(t_0;\mathcal{I}^+) \backslash Cl\mathbb{N}(t;t_0;\mathcal{I}^+) = \phi$, which, with $\mathcal{D}_A(t_0;\mathcal{I}^+) \cap \partial\mathbb{N}(t;t_0;\mathcal{I}^+) = \phi$ and $\mathcal{D}_A(t_0;\mathcal{I}^+) \cap \mathbb{N}(t;t_0;\mathcal{I}^+) \neq \phi$, implies $\mathbb{N}(t;t_0;\mathcal{I}^+) \supseteq \mathcal{D}_A(t_0;\mathcal{I}^+) \equiv \mathcal{D}_{AS}(t_0;\mathcal{I}^+)$ by having in mind that both $\mathcal{D}_A(t_0;\mathcal{I}^+)$ and $\mathbb{N}(t;t_0;\mathcal{I}^+)$ are open neighborhoods of $\mathcal{I}^+$ on $\mathfrak{T}_i$, in view of Lemma A 12.4, (i) of Weak smoothness property **293** and (i) of Strong smoothness property **295**, and that $\mathcal{D}_A(t_0;\mathcal{I}^+) \equiv \mathcal{D}_{AS}(t_0;\mathcal{I}^+)$ due to Lemma B 12.5 This completes the proof of (13.3).

Let

$$\mathcal{N}(t_0;\mathcal{I}^+) \equiv \mathcal{D}_{AS}(t_0;\mathcal{I}^+) \tag{13.4}$$

so that $\mathbb{N}(t_0;t_0;\mathcal{I}^+) \supseteq \mathcal{N}(t_0;\mathcal{I}^+)$. The set $\mathcal{N}(t_0;\mathcal{I}^+)$ is an open continuous connected neighborhood of $\mathcal{I}^+$ on $\mathfrak{T}_i$, (a-2 of Lemma A 12.4), and $\mathcal{N}(\mathfrak{T}_i;\mathcal{I}^+) = \mathcal{D}_{UAS}(\mathfrak{T}_i;\mathcal{I}^+)$ is a connected neighborhood of $\mathcal{I}^+$ [a-3) of Lemma A, Section 12.4,]. Also,

$$\mathcal{N}(\mathfrak{T}_i;\mathcal{I}^+) = \cap\left[\mathcal{N}(t_0;\mathcal{I}^+) : t_0 \in \mathfrak{T}_i\right]$$

because of $\mathcal{N}(t_0;\mathcal{I}^+) \equiv \mathcal{D}_A(t_0;\mathcal{I}^+)$, and

$$\mathcal{N}(\mathfrak{T}_i;\mathcal{I}^+) = \cap\left[\mathcal{N}(t_0;\mathcal{I}^+) : t_0 \in \mathfrak{T}_i\right] = \cap\left[\mathcal{D}_A(t_0;\mathcal{I}^+) : t_0 \in \mathfrak{T}_i\right] = \\ = \mathcal{D}_{UA}(\mathfrak{T}_i;\mathcal{I}^+).$$

These results prove the necessity of the conditions 1) and 2).

From $\mathcal{D}_S(t_0;\mathcal{I}^+) \supseteq \mathcal{D}_A(t_0;\mathcal{I}^+)$, $\mathcal{D}_A(t_0;\mathcal{I}^+) \equiv \mathcal{D}_{AS}(t_0;\mathcal{I}^+)$ due to a) of Lemma B 12.5 and from $\mathcal{D}_{AS}(t_0;\mathcal{I}^+) = \mathcal{N}(t_0;\mathcal{I}^+)$, Equation 13.4, it follows that there is not an equilibrium state of the system in $[\mathcal{N}(t;\mathcal{I}^+) \backslash \mathcal{I}^+]$. It implies that $\mathbf{f}(t,\mathbf{S}) = \mathbf{0}_n$ for all $(t,t_0) \in \mathfrak{T}_0 \times \mathfrak{T}_i$, is possible only for $\mathbf{S} \notin [\mathcal{N}(t;\mathcal{I}^+) \backslash \mathcal{I}^+]$. This proves the necessity of the condition 3).

From $\mathcal{N}(t_0;\mathcal{I}^+) \equiv \mathcal{D}_{AS}(t_0;\mathcal{I}^+)$ it follows that $\mathfrak{T}_0$ is the interval of the existence of the system motions $\mathbf{S}(t;t_0;\mathbf{S}_0)$ for every $(t,t_0,\mathbf{S}_0) \in \mathfrak{T}_0 \times \mathfrak{T}_i \times \mathcal{N}(t_0;\mathcal{I}^+)$, due to Definition 225 and Definition 234. Let $W(.) \in L(\mathfrak{T}_0,\mathfrak{T}_i,\mathbb{N},\mathcal{I}^+)$, Definition 298, be an arbitrarily selected positive definite decrescent function with respect to $\mathcal{I}^+$ on $\mathfrak{T}_i \times \mathbb{N}(t;t_0;\mathcal{I}^+)$. Therefore, there is $\mu \in \mathfrak{R}^+$, $\mu = \mu(W)$, due to Weak smoothness property **293** induced by Strong smoothness property **295**, such that there exists a solution $V(.)$ to the equations (12.11) and (12.12) that is continuous in $(t,\mathbf{S}) \in \mathfrak{T}_i \times \mathcal{N}_\mu(\mathcal{I}^+)$ and satisfies (12.13). Hence,

$$|V(t,\mathbf{S})| < \infty, \ \forall (t,\mathbf{S}) \in \mathfrak{T}_i \times Cl\mathcal{N}_\mu(\mathcal{I}^+). \tag{13.5}$$

Let $\beta \in]1,\infty[$ and $\zeta \in \mathfrak{R}^+$ be such that

$$Cl\mathcal{N}_\beta(\mathcal{I}^+) \cap Cl\mathcal{N}_\mu(\mathcal{I}^+) \cap \mathbb{N}(t;t;\mathcal{I}^+) \supset \mathfrak{W}_\zeta(t;\mathcal{I}^+), \ \forall t \in \mathfrak{T}_i. \tag{13.6}$$

Existence of such β and ζ is guaranteed by positive definiteness of $W(.)$ on $\mathfrak{T}_i \times \mathbb{N}(t;t;\mathcal{I}^+)$ and by the fact that $\mathbb{N}(t;t;\mathcal{I}^+)$ is a neighborhood of $\mathcal{I}^+$ for all $t \in \mathfrak{T}_i$. Let $t_0 \in \mathfrak{T}_i$ be arbitrary and $\tau \in \mathfrak{R}^+$, $\tau = \tau(t_0, \mathbf{S}_0, \zeta; W)$, be such that for any $\mathbf{S}_0 \in \mathcal{N}(t_0;\mathcal{I}^+)$ the following condition holds:

$$\mathbf{S}(t;t_0;\mathbf{S}_0) \in Cl\mathfrak{W}_\zeta(t;\mathcal{I}^+), \ \forall t \in [t_0+\tau, \infty[. \tag{13.7}$$

Such τ exists due to Definition 225, Definition 235, $\mathbf{S}_0 \in \mathcal{N}(t_0;\mathcal{I}^+)$ and the fact that

$$\mathcal{D}_A(t_0;\mathcal{I}^+) \equiv \mathcal{D}_{AS}(t_0;\mathcal{I}^+) \equiv \mathcal{N}(t_0;\mathcal{I}^+).$$

Notice that $\mathbf{S}_0 \in \mathcal{N}(t_0;\mathcal{I}^+)$ also yields

$$d\left[\mathbf{S}(\infty;t_0;\mathbf{S}_0), \mathcal{I}^+\right] = 0, \quad \mathbf{S}(\infty;t_0;\mathbf{S}_0) \in Cl\mathcal{I}^+. \tag{13.8}$$

Let (12.11) be integrated from $t \in \mathfrak{T}_0$ to ∞,

$$V[\infty, \mathbf{S}(\infty;t_0;\mathbf{S}_0)] - V[t, \mathbf{S}(t;t_0;\mathbf{S}_0)] =$$
$$= -\int_t^\infty W[\sigma, \mathbf{S}(\sigma;t_0;\mathbf{S}_0)]\,d\sigma, \ \forall (t,\mathbf{S}) \in \mathfrak{T}_0 \times \mathcal{N}(t_0;\mathcal{I}^+). \tag{13.9}$$

Now, (12.12) and (13.8) reduce (13.9) to

$$V[t, \mathbf{S}(t;t_0;\mathbf{S}_0)] = \int_t^\infty W[\sigma, \mathbf{S}(\sigma;t_0;\mathbf{S}_0)]\,d\sigma =$$
$$= \int_t^{t_0+\tau} W[\sigma, \mathbf{S}(\sigma;t_0;\mathbf{S}_0)]\,d\sigma + \int_{t_0+\tau}^\infty W[\sigma, \mathbf{S}(\sigma;t_0;\mathbf{S}_0)]\,d\sigma,$$
$$\forall (t,\mathbf{S}_0) \in \mathfrak{T}_0 \times \mathcal{N}(t_0;\mathcal{I}^+). \tag{13.10}$$

Positive invariance of $\mathcal{D}_A(t_0;\mathcal{I}^+)$ on $\mathfrak{T}_i$ with respect to the system motions (Lemma A, Section 12.4,), $\mathbb{N}(t;t_0;\mathcal{I}^+) \supseteq \mathcal{D}_{AS}(t_0;\mathcal{I}^+) \equiv \mathcal{D}_A(t_0;\mathcal{I}^+) \equiv \mathcal{N}(t_0;\mathcal{I}^+)$ (13.4), continuity of $\mathbf{S}(t;t_0;\mathbf{S}_0)$ in $(t;t_0;\mathbf{S}_0) \in \mathfrak{T}_0 \times \mathfrak{T}_i \times \mathbb{N}(t;t_0;\mathcal{I}^+)$, [due to (i-b) of Weak smoothness property **293**], continuity, positive definiteness, and the decrescent property of $W(.)$ relative to $\mathcal{I}^+$ on $\mathfrak{T}_0 \times \mathfrak{T}_i \times \mathbb{N}(t;t_0;\mathcal{I}^+)$, the definition of τ, (13.7), and compactness of $[t, t_0+\gamma]$ for any $\gamma \in \mathfrak{R}_+$ imply:

$$\left| \int_t^{t_0+\gamma} W[\sigma, \mathbf{S}(\sigma;t_0;\mathbf{S}_0)]\,d\sigma \right| < \infty, \ \forall (\gamma, t_0, \mathbf{S}_0) \in \mathfrak{R}_+ \times \mathfrak{T}_i \times \mathcal{N}(t_0;\mathcal{I}^+). \tag{13.11}$$

Now, (13.5)-(13.7), (13.10), and (13.11) for $\gamma = \tau$ yield

$$|V[t, \mathbf{S}(t;t_0;\mathbf{S}_0)]| < \infty, \ \forall (t, t_0, \mathbf{S}_0) \in \mathfrak{T}_0 \times \mathfrak{T}_i \times \mathcal{N}(t_0;\mathcal{I}^+). \tag{13.12}$$

Let us replace t by t_0 and $\mathbf{S}_0$ by $\mathbf{S}$ in (13.12). Then,

$$|V(t,\mathbf{S})| < \infty, \ \forall (t,\mathbf{S}) \in \mathfrak{T}_i \times \mathcal{N}(t;\mathcal{I}^+). \tag{13.13}$$

Continuity of $W(.)$ on $\mathfrak{T}_i \times \mathbb{N}((t;t_0;\mathcal{I}^+)$, $W(.) \in L(\mathfrak{T}_0, \mathfrak{T}_i, \mathbb{N}, \mathcal{I}^+)$, Definition 298, $\mathbb{N}(t;t_0;\mathcal{I}^+) \supseteq \mathcal{N}(t_0;\mathcal{I}^+)$, (13.10), and (13.13) prove

$$V(t,\mathbf{S}) \in \mathfrak{C}\left[\mathfrak{T}_i \times \mathcal{N}(t;\mathcal{I}^+)\right] = \mathfrak{C}\left[\mathfrak{T}_i \times \mathcal{D}_{AS}(t;\mathcal{I}^+)\right]. \tag{13.14}$$

Positive invariance of $\mathcal{D}_A(t_0;\mathcal{I}^+)$, (a-1) of Lemma A, Section 12.4), $\mathcal{D}_A(t_0;\mathcal{I}^+) \equiv \mathcal{D}_{AS}(t_0;\mathcal{I}^+) \equiv \mathcal{N}(t_0;\mathcal{I}^+)$, (13.4), continuity of $\mathbf{S}(t;t_0;\mathbf{S}_0)$ in $(t,t_0,\mathbf{S}_0) \in \mathfrak{T}_0 \times \mathfrak{T}_i \times \mathcal{D}_{AS}(t_0;\mathcal{I}^+)$, positive definiteness, and the decrescent property of $W(.)$ relative to the set $\mathcal{I}^+$ on $\mathfrak{T}_0 \times \mathfrak{T}_i \times \mathcal{N}(t;t_0;\mathcal{I}^+)$, $W(.) \in L(\mathfrak{T}_0,\mathfrak{T}_i,\mathbb{N},\mathcal{I}^+)$, (12.13), Definition 298, the definition of τ (13.7), and compactness of $[t,t_0+\tau]$ guarantee the existence of $\zeta_k(.):\mathfrak{R}^n \longrightarrow \mathfrak{R}$, $k=1,2$, $\zeta_1(\mathbf{S}) \in \mathfrak{C}\{\mathcal{N}_M(\mathfrak{T}_i;\mathcal{I}^+)\}$, and $\zeta_2(\mathbf{S}) \in \mathfrak{C}\{\mathcal{N}_m(\mathfrak{T}_i;\mathcal{I}^+)\}$, where

$$\mathcal{N}_M(\mathfrak{T}_i;\mathcal{I}^+) = \cup\left[\mathcal{N}(t;\mathcal{I}^+) : t \in \mathfrak{T}_i\right],$$
$$\mathcal{N}_m(\mathfrak{T}_i;\mathcal{I}^+) = \cap\left[\mathcal{N}(t;\mathcal{I}^+) : t \in \mathfrak{T}_i\right],$$

and $\psi_k(.):\mathfrak{R}^n \longrightarrow \mathfrak{R}$, $k=1,2$, such that

$$0 < \zeta_1(\mathbf{S}_0) \leq \int_t^{t_0+\tau} \psi_1\left[\mathbf{S}(\sigma;t_0;\mathbf{S}_0)\right] d\boldsymbol{\sigma},$$
$$\forall(t,t_0,\mathbf{S}) \in \mathfrak{T}_0 \times \mathfrak{T}_i \times \left\{\mathcal{N}(t;\mathcal{I}^+) \setminus Cl\mathcal{N}_\mu(t;\mathcal{I}^+)\right\}, \tag{13.15}$$

$$\infty > \zeta_2(\mathbf{S}_0) \geq \int_t^{t_0+\tau} \psi_2\left[\mathbf{S}(\sigma;t_0;\mathbf{S}_0)\right] d\boldsymbol{\sigma},$$
$$\forall(t,t_0,\mathbf{S}) \in \mathfrak{T}_0 \times \mathfrak{T}_i \times \left\{\mathcal{N}_m(\mathfrak{T}_i;\mathcal{I}^+) \setminus Cl\mathcal{N}_\mu(\mathfrak{T}_i;\mathcal{I}^+)\right\}, \tag{13.16}$$

and

$$\psi_1(\mathbf{S}) \in \mathfrak{C}\left[\mathcal{N}_M(\mathfrak{T}_i;\mathcal{I}^+)\right],$$
$$\psi_2(\mathbf{S}) \in \mathfrak{C}\left[\mathcal{N}_m(\mathfrak{T}_i;\mathcal{I}^+)\right], \tag{13.17}$$

$$\psi_k(\mathbf{S}) = 0,\ \forall \mathbf{S} \in \partial\mathcal{I}^+,\ \forall t \in \mathfrak{T}_i,\ k=1,2, \tag{13.18}$$

$$\psi_1(\mathbf{S}) > 0,\ \forall \mathbf{S} \in \left\{\mathcal{N}_M(\mathfrak{T}_i;\mathcal{I}^+) \setminus \mathcal{I}^+\right\},\ \forall t \in \mathfrak{T}_i,$$
$$\psi_2(\mathbf{S}) > 0,\ \forall \mathbf{S} \in \left[\mathcal{N}_m(\mathfrak{T}_i;\mathcal{I}^+) \setminus \mathcal{I}^+\right], \tag{13.19}$$

$$\psi_1(\mathbf{S}) \leq W(t,\mathbf{S}),\ \forall(t,\mathbf{S}) \in \mathfrak{T}_i \times \left\{\mathcal{N}_M(\mathfrak{T}_i;\mathcal{I}^+) \setminus \mathcal{I}^+\right\}, \tag{13.20}$$
$$\psi_2(\mathbf{S}) \geq W(t,\mathbf{S}),\ \forall(t,\mathbf{S}) \in \mathfrak{T}_i \times \mathcal{N}_m\left[\mathfrak{T}_i;\mathcal{I}^+ \setminus \mathcal{I}^+\right]. \tag{13.21}$$

Such functions $\psi_k(.)$, $k=1,2$, exist due to the decrescent property and positive definiteness of $W(.)$ relative to the set $\mathcal{I}^+$ on the set product $\mathfrak{T}_0 \times \mathfrak{T}_i \times \mathbb{N}(t;t_0;\mathcal{I}^+)$,

$$W(.) \in L(\mathfrak{T}_0;\mathfrak{T}_i;\mathbb{N},\mathcal{I}^+),$$

and $\mathbb{N}(t;t;\mathcal{I}^+) \supseteq \mathcal{N}(t;\mathcal{I}^+)$, $\forall t \in \mathfrak{T}_i$. They can be of the form

$$\psi_k(\mathbf{S}) = g_k\left\{d\left[\mathbf{S},\mathcal{I}^+\right]\right\},\ k=1,2,$$

together with $g_k(.)$ in the class $\mathcal{K}$ of the comparison functions (Definition 250, Section 11.2), $g_k(.) \in \mathcal{K}$. Let $w_k(.):\mathfrak{R}^n \longrightarrow \mathfrak{R}$, $k=1,2$, obey (13.22) through (13.24):

$$w_k(\mathbf{S}) \in \mathfrak{C}(\mathfrak{R}^n) \text{ and } w_k(\mathbf{S}) = 0\ \forall \mathbf{S} \in \partial\mathcal{I}^+,\ \forall k=1,2, \tag{13.22}$$

$$0 < w_1(\mathbf{S}) \leq$$
$$\left\{\begin{array}{c} \zeta_1(\mathbf{S}),\ \forall \mathbf{S} \in \{\mathcal{N}_M(\mathfrak{T}_i;\mathcal{I}^+) \setminus Cl\mathcal{N}_\mu([\mathcal{I}^+])\}, \\ \omega_\mu(\mathbf{S}),\ \forall \mathbf{S} \in \{Cl\mathcal{N}_\mu[\mathcal{I}^+] \setminus \mathcal{I}^+\}, \end{array}\right\} \tag{13.23}$$

$$w_2(\mathbf{S}) \geq \left\{ \begin{array}{c} \left\langle \begin{array}{c} \zeta_2(\mathbf{S}) + \omega_\mu(\mathbf{S}),\ \forall \mathbf{S} \in \\ \in \{\mathcal{N}_m(\mathfrak{T}_i;\mathcal{I}^+) \backslash Cl\mathcal{N}_\mu[\mathcal{I}^+]\} \end{array} \right\rangle, \\ \omega_\mu(\mathbf{S}),\ \forall \mathbf{S} \in \{Cl\mathcal{N}_\mu[\mathcal{I}^+] \backslash \mathcal{I}^+\}, \end{array} \right\} \tag{13.24}$$

where $\omega_\mu(.)$ is defined by (12.13). Now, (12.13), (13.10), positive definiteness of $W(.)$ relative to $\mathcal{I}^+$ on $\mathfrak{T}_0 \times \mathfrak{T}_i \times \mathbb{N}(t;t_0;\mathcal{I}^+)$, the positive invariance of the set $\mathcal{I}^+$ on $\mathfrak{T}_0 \times \mathfrak{T}_i$, and (13.15) through (13.24) yield the following for $(t_0, \mathbf{S}_0) = (t, \mathbf{S})$:

$$w_1(\mathbf{S}) \leq V(t,\mathbf{S}), \forall (t,\mathbf{S}) \in \mathfrak{T}_i \times \{\mathcal{N}(t;\mathcal{I}^+) \backslash In\mathcal{I}^+\}, \tag{13.25}$$

$$V(t,\mathbf{S}) \leq w_2(\mathbf{S}),\ \forall (t,\mathbf{S}) \in \mathfrak{T}_i \times \{\mathcal{N}_m[\mathfrak{T}_i;\mathcal{I}^+ \backslash \mathcal{I}^+]\}, \tag{13.26}$$

$$V(t,\mathbf{S}) \leq 0,\ \forall (t,\mathbf{S}) \in \mathfrak{T}_i \times \mathcal{I}^+, \tag{13.27}$$

$$V(t,\mathbf{S}) = 0,\ \ (t,\mathbf{S}) \in \mathfrak{T}_i \times \partial\mathcal{I}^+. \tag{13.28}$$

From $W(.) \in L(\mathfrak{T}_0, \mathfrak{T}_i, \mathbb{N}, \mathcal{I}^+)$, (12.12), (13.14), and (13.25) through (13.28) it follows that a solution function $V(.)$ to (12.11), (12.12) is decrescent, positive definite, hence continuous, relative to $\mathcal{I}^+$ on $\mathfrak{T}_0 \times \mathfrak{T}_i \times \mathcal{N}(t;\mathcal{I}^+)$. Let it be assumed that there exist two such solutions $V_1(.)$ and $V_2(.)$ to (12.11), (12.12). Hence,

$$\begin{gathered} V_1(t_0,\mathbf{S}_0) - V_2(t_0,\mathbf{S}_0) = \\ = \int_{t_0}^{\infty} \{W[\sigma, \mathbf{S}_1(\sigma;t_0;\mathbf{S}_0)] - W[\sigma, \mathbf{S}_2(\sigma;t_0;\mathbf{S}_0)]\}\, d\sigma, \\ \forall (t,t_0,\mathbf{S}_0) \in \mathfrak{T}_0 \times \mathfrak{T}_i \times \mathcal{N}(t_0;\mathcal{I}^+). \end{gathered} \tag{13.29}$$

Uniqueness of the motions $\mathcal{S}(.;t_0;\mathbf{S}_0)$, $\forall (t_0,\mathbf{S}_0) \in \mathfrak{T}_i \times \mathcal{N}(t_0;\mathcal{I}^+)$ (Weak smoothness property **293**), $\mathbb{N}(t_0;t_0;\mathcal{I}^+) \supseteq \mathcal{N}(t_0;\mathcal{I}^+)$, and uniqueness of $W(t,\mathbf{S})$ for every $(t,t_0,\mathbf{S}) \in \mathfrak{T}_0 \times \mathfrak{T}_i \times \mathbb{N}(t;t_0;\mathcal{I}^+)$, [due to the positive definiteness of $W(.)$ on $\mathbb{N}(t;t_0;\mathcal{I}^+)$, imply

$$\begin{gathered} \int_{t_0}^{\infty} \{W[\sigma, \mathbf{S}_1(\sigma;t_0;\mathbf{S}_0)] - W[\sigma, \mathbf{S}_2(\sigma;t_0;\mathbf{S}_0)]\}\, d\sigma \\ = \int_{t_0}^{\infty} \{W[\sigma, \mathbf{S}(\sigma;t_0;\mathbf{S}_0)] - W[\sigma, \mathbf{S}(\sigma;t_0;\mathbf{S}_0)]\}\, d\sigma = 0, \\ \forall (t,t_0,\mathbf{S}_0) \in \mathfrak{T}_0 \times \mathfrak{T}_i \times \mathcal{N}(t_0;\mathcal{I}^+). \end{gathered}$$

This and (13.29) prove

$$V_1(t_0,\mathbf{S}_0) \equiv V_2(t_0,\mathbf{S}_0).$$

The function $V(.)$ is the unique solution to (12.11), (12.12). This completes the proof of the necessity of the condition 4-a-i). If $\partial\mathcal{N}(t_0;\mathcal{I}^+) \neq \phi$, then let $t_0 \in \mathfrak{T}_i$ be arbitrary and $\mathbf{S}_k$, $k = 1, 2, ...$, be a sequence converging to U, $U \in \partial\mathcal{N}(t_0;\mathcal{I}^+)$, $\mathbf{S}_k \longrightarrow U$ as $k \longrightarrow \infty$, $\mathbf{S}_k \in \mathcal{N}(t_0;\mathcal{I}^+)$, for all $k = 1, 2, ...$ Let $\zeta \in \mathfrak{R}^+$ be arbitrarily chosen so that $\mathcal{N}(\mathfrak{T}_i;\mathcal{I}^+) \supset Cl\mathfrak{W}_\zeta(t;\mathcal{I}^+)$ for all $t \in \mathfrak{T}_i$, defines $Cl\mathfrak{W}_\zeta(t;\mathcal{I}^+)$. Such ζ exists because $W(.)$ is positive definite relative to $\mathcal{I}^+$ on $\mathfrak{T}_0 \times \mathfrak{T}_i \times \mathbb{N}(t;t_0;\mathcal{I}^+)$, $\mathcal{N}(\mathfrak{T}_i;\mathcal{I}^+) \subseteq \mathbb{N}(\mathfrak{T}_i;\mathcal{I}^+)$ and because $\mathcal{N}(\mathfrak{T}_i;\mathcal{I}^+)$ is a neighborhood of $\mathcal{I}^+$. Let τ_k, $\tau_k = \tau_k\ (t_0;\mathbf{S}_k;\zeta;\mathcal{I}^+) \in \mathfrak{R}_+$, be the first instant satisfying (13.30):

$$\mathbf{S}(t;t_0;\mathbf{S}_k) \in Cl\mathfrak{W}_\zeta(t;\mathcal{I}^+),\ \forall (t,t_0) \in [t_0 + \tau_k, \infty[\times \mathfrak{T}_i. \tag{13.30}$$

The existence of such τ_k is ensured by $\mathbf{S}_k \in \mathcal{N}(t_0;\mathcal{I}^+)$, $\mathcal{N}(t_0;\mathcal{I}^+) \equiv \mathcal{D}_{AS}(t_0;\mathcal{I}^+)$ and by the fact that $\cap [\mathfrak{W}_\varsigma(t;\mathcal{I}^+) : t \in \mathfrak{T}_i]$ is a neighborhood of $\mathcal{I}^+$ due to the decrescent property of $W(.)$ relative to $\mathcal{I}^+$ on $\mathfrak{T}_0 \times \mathfrak{T}_i \times \mathcal{N}(t;t_0;\mathcal{I}^+)$ (Lemma 276, Section 11.6). Continuity of $\mathbf{S}(t;t_0;\mathbf{S}_0)$ in $(t;t_0;\mathbf{S}_0) \in \mathfrak{T}_0 \times \mathfrak{T}_i \times \mathbb{N}(t_0;t_0;\mathcal{I}^+)$, (Weak smoothness property **293**), $\mathbb{N}(t_0;t_0;\mathcal{I}^+) \supseteq \mathcal{D}_A(t_0;\mathcal{I}^+) = \mathcal{N}(t_0;\mathcal{I}^+)$, the positive invariance of $\mathcal{D}_A(t;\mathcal{I}^+)$, [a-1) of Lemma A, Section 12.4], the fact that $\cap[\mathcal{D}_A(t;\mathcal{I}^+) : t \in \mathfrak{T}_i] = \mathcal{D}_{UA}(\mathfrak{T}_i;\mathcal{I}^+)$ (Definition 228 and Definition 229) is a neighborhood of $\mathcal{I}^+$, Definition 229), and $\mathbf{S}_k \in \mathcal{N}(t_0;\mathcal{I}^+)$ imply

$$\tau_k \longrightarrow \infty \text{ as } k \longrightarrow \infty. \tag{13.31}$$

Let $m \in \{1,2,...\}$ be such that $\mathbf{S}_k \in \{\mathcal{N}(t_0;\mathcal{I}^+) \backslash Cl\mathfrak{W}_\zeta(t_0;\mathcal{I}^+)\}$ for all $k = m, m+1, ...$, and $\mathbf{S}_k \longrightarrow \partial\mathcal{N}(t_0;\mathcal{I}^+)$ as $k \longrightarrow \infty$. Such $\mathbf{S}_k$ exists because $\mathcal{N}(t_0;\mathcal{I}^+) \equiv \mathcal{D}_A(t_0;\mathcal{I}^+)$ is open [a-2) of Lemma A, Section 12.4,] and $\mathcal{N}(t_0;\mathcal{I}^+) \supset Cl\mathfrak{W}_\zeta(t_0;\mathcal{I}^+)$.

Let α be defined by Equation (12.14) in condition 3) of Definition 298 that determines the family $L(\mathfrak{T}_0;\mathfrak{T}_i,\mathbb{N},\mathcal{I}^+)$, Section 12.3. Then

$$\chi = min\left\{W(t,\mathbf{S}) : (t,\mathbf{S}) \in \mathfrak{T}_i \times \left[\mathcal{N}(t;t;\mathcal{I}^+) \backslash \mathfrak{W}_\zeta(t;\mathcal{I}^+)\right]\right\} \in \mathfrak{R}^+, \ \chi \leq \alpha, \tag{13.32}$$

due to $W(.) \in L(\mathfrak{T}_0;\mathfrak{T}_i,\mathbb{N},\mathcal{I}^+)$ and

$$\mathfrak{N}(t;t_0;\mathcal{I}^+) \subseteq \mathbb{N}(t;t_0;\mathcal{I}^+), \ \forall (t;t_0) \in \mathfrak{T}_0 \times \mathfrak{T}_i.$$

Therefore, (13.10), (13.30), (13.32), and the definitions of χ and τ_k yield $V(t_0,\mathbf{S}_k) \geq \chi\tau_k$, $\forall t_0 \in \mathfrak{T}_i$, which together with (13.31) prove the necessity of the condition 4-a-ii). The conditions under 4-b) follow from 4-a) due to (12.11), (12.12), (12.19), (12.20), and (12.26)-(12.28). This completes the proof of the necessity part.

Sufficiency. Let all the conditions of the theorem statement hold. The function $V(.)$ is the solution to (12.11), (12.12), [or, $V(.)$ is the solution to (12.19), (12.20)], and it is positive definite and decrescent relative to $\mathcal{I}^+$ on $\mathfrak{T}_0 \times \mathfrak{T}_i \times \mathbb{N}(t;t_0;\mathcal{I}^+)$. The function $W(.) \in L(\mathfrak{T}_0,\mathfrak{T}_i,\mathbb{N},\mathcal{I}^+)$, and/or, $W(.) \in E(\mathfrak{T}_0,\mathfrak{T}_i,\mathbb{N})$ is also positive definite and decrescent relative to $\mathcal{I}^+$ on $\mathfrak{T}_0 \times \mathfrak{T}_i \times \mathbb{N}(t;t_0;\mathcal{I}^+)$. They satisfy Definition 263, Section 11.4, and Definition 275, Section 11.6.

Let us prove that the set $\mathcal{I}^+$ of the system **(5.22), (5.23)** is asymptotically stable on $\mathfrak{T}_i$.

We refer now to Definition 263, Definition 266, Property 270, Section 11.5. Let $\varepsilon \in \mathfrak{R}^+$ be arbitrary. Let $t_0 \in \mathfrak{T}_i$ be also arbitrary. Let $\zeta(\varepsilon) \in \mathfrak{R}^+$ be such that

$$\mathfrak{V}_{\zeta(\varepsilon)}(t;t_0) \subset N_\varepsilon(\mathcal{I}^+), \ \forall (t;t_0) \in \mathfrak{T}_0 \times \mathfrak{T}_i. \tag{13.33}$$

This, (12.11), (12.12), positive definiteness of both functions $W(.)$, and $V(.)$ relative to $\mathcal{I}^+$ on $\mathfrak{T}_0 \times \mathfrak{T}_i \times \mathcal{N}(t;t_0;\mathcal{I}^+)$ due to

$$\mathcal{N}(t_0;\mathcal{I}^+) \subseteq \mathbb{N}(t_0;t_0;\mathcal{I}^+),$$

guarantee that $V[t,\mathbf{S}(t;t_0;\mathbf{S}_0)]$ is strictly decreasing on $\mathfrak{T}_0$ for every $\mathbf{S}_0 \in \mathfrak{V}_{\zeta(\varepsilon)}(t_0;t_0)$, i.e.,

$$\begin{gathered} V[t_1,\mathbf{S}(t_1;t_0;\mathbf{S}_0)] > V[t_2,\mathbf{S}(t_2;t_0;\mathbf{S}_0)], \\ \forall (t_1,t_2) \in \mathfrak{T}_0, \ t_1 > t_2, \ \forall (t_0,\mathbf{S}_0) \in \mathfrak{T}_i \times \mathfrak{V}_{\zeta(\varepsilon)}(t_0;t_0), \end{gathered} \tag{13.34}$$

and

$$\begin{gathered} N_\varepsilon(\mathcal{I}^+) \supset \mathfrak{V}_{\zeta(\varepsilon)}(t_1;t_0) \supset \mathfrak{V}_{\zeta(\varepsilon)}(t_2;t_0), \\ \forall (t_1,t_2) \in \mathfrak{T}_0, \ t_1 > t_2, \ \forall t_0 \in \mathfrak{T}_i. \end{gathered} \tag{13.35}$$

The results (13.34) and (13.35) show that the set $\mathfrak{V}_{\zeta(\varepsilon)}(t;t_0)$ satisfies the conditions (7.6), (7.7) of stability Definition 201 for every $t_0 \in \mathfrak{T}_i$:

$$\forall \varepsilon \in \mathfrak{R}^+,\ \mathcal{N}\left(\varepsilon;t_0;\mathcal{I}^+\right) = \mathfrak{V}_{\zeta(\varepsilon)}(t;t_0) \Longrightarrow$$
$$\mathbf{S}(t;t_0;\mathbf{S}_0) \in \mathcal{N}_\varepsilon\left(t;\mathcal{I}^+\right),\ \forall (t,t_0,\mathbf{S}_0) \in \mathfrak{T}_0 \times \mathfrak{T}_i \times \mathcal{N}\left(\varepsilon;t_0;\mathcal{I}^+\right). \tag{13.36}$$

The set $\mathcal{I}^+$ is stable on $\mathfrak{T}_i$. It has the stability (strict stability) domain on $\mathfrak{T}_i$, Definition 204 (Definition 206). In order to prove attraction and asymptotic stability of the set $\mathcal{I}^+$ on $\mathfrak{T}_i$, let us assume the existence of $\psi \in \mathfrak{R}^+$,

$$Cl\mathcal{N}_\psi\left(\mathcal{I}^+\right) \subset In\mathbb{N}\left(t;t_0;\mathcal{I}^+\right),\ \forall (t,t_0) \in \mathfrak{T}_0 \times \mathfrak{T}_i, \tag{13.37}$$

which determines the ψ-neighborhood $\mathcal{N}_\psi\left(\mathcal{I}^+\right)$ of the set $\mathcal{I}^+$ such that the system motion $\mathcal{S}(.;t_0;\mathbf{S}_0)$ starting from

$$\mathbb{N}\left(t_0;t_0;\mathcal{I}^+\right) \backslash Cl\mathcal{N}_\psi\left(\mathcal{I}^+\right)$$

never enters the neighborhood $\mathcal{N}_\psi\left(\mathcal{I}^+\right)$ of the set $\mathcal{I}^+$:

$$\mathbf{S}(t;t_0;\mathbf{S}_0) \notin \mathcal{N}_\psi\left(\mathcal{I}^+\right),\ \forall (t,t_0,\mathbf{S}_0) \in \mathfrak{T}_0 \times \mathfrak{T}_i \times \mathbb{N}\left(t_0;t_0;\mathcal{I}^+\right) \backslash Cl\mathcal{N}_\psi\left(\mathcal{I}^+\right). \tag{13.38}$$

Positive definiteness of the function $W(.)$ relative to the set $\mathcal{I}^+$ on $\mathfrak{T}_0 \times \mathfrak{T}_i \times \mathbb{N}(t;t_0;\mathcal{I}^+)$ and $\mathbb{N}(t;t_0;\mathcal{I}^+) \supseteq \mathcal{N}(t_0;\mathcal{I}^+)$ guarantee the following in view of iv) of Definition 266:

$$\forall \theta \in]0,\psi],\ \exists \Delta \in \mathfrak{R}^+,\ \Delta = \Delta(\theta) \Longrightarrow$$
$$\zeta_{InfW}(\theta) = inf\left\{W(t;\mathbf{S}) : (t,t_0,\mathbf{S}) \in \mathfrak{T}_0 \times \mathfrak{T}_i \times \left[\mathbb{N}\left(t_0;t_0;\mathcal{I}^+\right) \backslash Cl\mathcal{N}_\theta\left(\mathcal{I}^+\right)\right]\right\} \in \mathfrak{R}^+,$$
$$\zeta_{InfW}(\theta) \geq \Delta. \tag{13.39}$$

This, (13.38), and Equation (12.11) imply for $\theta = \psi$:

$$\frac{dV\left[t,\mathbf{S}(t;t_0;\mathbf{S}_0)\right]}{dt} \leq -\zeta_{InfW}(\psi),\ \forall (t,t_0,\mathbf{S}_0) \in \mathfrak{T}_0 \times \mathfrak{T}_i \times \mathbb{N}\left(t_0;t_0;\mathcal{I}^+\right) \backslash Cl\mathcal{N}_\psi\left(\mathcal{I}^+\right)$$

i.e.,

$$V\left[t,\mathbf{S}(t;t_0;\mathbf{S}_0)\right] \leq V(t_0,\mathbf{S}_0) - \zeta_{InfW}(\psi) \leq V(t_0,\mathbf{S}_0) - \Delta(t-t_0).$$

In the limit as $t \longrightarrow \infty$, due to $\Delta \in \mathfrak{R}^+$,

$$lim\left\{V\left[t,\mathbf{S}(t;t_0;\mathbf{S}_0)\right] : t \longrightarrow \infty\right\} = -\infty,$$

which contradicts the positive definiteness of the function $V(.)$ relative to the set $\mathcal{I}^+$ on $\mathfrak{T}_0 \times \mathfrak{T}_i \times \mathcal{N}(t;t_0;\mathcal{I}^+)$ in view of

$$\mathcal{N}\left(t;t_0;\mathcal{I}^+\right) \subseteq \mathbb{N}\left(t_0;t_0;\mathcal{I}^+\right).$$

The assumed ψ-neighborhood $\mathcal{N}_\psi\left(\mathcal{I}^+\right)$ does not exist that, together with the strict monotonous decreasing value of the function $V(.)$ along system motions, (12.11) and (12.12), implies

$$lim\left\{V(t;t_0) : t \longrightarrow \infty\right\} = 0,\ \forall t_0 \in \mathfrak{T}_i.$$

and proves the attraction of the set $\mathcal{I}^+$ on $\mathfrak{T}_i$ (Definition 225) due to the positive definiteness of the function $V(.)$ relative to the set $\mathcal{I}^+$. The set $\mathcal{I}^+$ has $\mathcal{D}_A(t_0;\mathcal{I}^+)$ on $\mathfrak{T}_i$ (Definition 226). Stability and attraction of the set $\mathcal{I}^+$ verify its asymptotic stability, Definition 234, and the existence of its asymptotic stability domain $\mathcal{D}_{AS}(t_0;\mathcal{I}^+)$ on $\mathfrak{T}_i$, Definition 235, so that $\mathcal{D}_A(t_0;\mathcal{I}^+) \equiv \mathcal{D}_{AS}(t_0;\mathcal{I}^+)$ and

$$\mathcal{D}_A(\mathfrak{T}_i;\mathcal{I}^+) = \mathcal{D}_{AS}(\mathfrak{T}_i;\mathcal{I}^+)$$

(Lemma B, Section 12.5,).

In order to show that both $\mathcal{N}(t_0;\mathcal{I}^+) \equiv \mathcal{D}_{AS}(t_0;\mathcal{I}^+)$ and

$$\mathcal{N}(\mathfrak{T}_i;\mathcal{I}^+) \equiv \mathcal{D}_{UA}(\mathfrak{T}_i;\mathcal{I}^+)$$

we proceed as follows. The condition (ii) of Strong smoothness property **295** guarantees $\mathcal{D}_{AS}(t_0;\mathcal{I}^+) \subseteq \mathcal{N}(t_0;\mathcal{I}^+)$, $\forall t_0 \in \mathfrak{T}_i$. Let $t_0 \in \mathfrak{T}_i$ be arbitrary and fixed. If $\partial\mathcal{N}(t_0;\mathcal{I}^+) = \phi$, then $\mathcal{N}(t_0;\mathcal{I}^+) = \mathfrak{R}^n$. Hence, $\mathcal{D}_{AS}(t_0;\mathcal{I}^+) \subseteq \mathcal{N}(t_0;\mathcal{I}^+)$ is then only possible. If $\mathcal{D}_{AS}(t_0;\mathcal{I}^+) \subset \mathcal{N}(t_0;\mathcal{I}^+)$, then $\partial\mathcal{D}_{AS}(t_0;\mathcal{I}^+) \cap \mathcal{N}(t_0;\mathcal{I}^+) \neq \phi$, which implies $V(t_0,\mathbf{S}) \longrightarrow \infty$ [because the function $V(.)$ is the solution to (12.11), (12.12), as shown in the proof of the necessity], which contradicts the continuity of the function $V(.)$ on $\mathcal{N}(t_0;\mathcal{I}^+)$ ensured by the condition 4-a,i). This implies $\partial\mathcal{D}_{AS}(t_0;\mathcal{I}^+) \cap \mathcal{N}(t_0;\mathcal{I}^+) = \partial\mathcal{D}_{AS}(t_0;\mathcal{I}^+) \cap \mathfrak{R}^n = \phi$. Because $\mathcal{D}_{AS}(t_0;\mathcal{I}^+)$ is an open connected neighborhood of $\mathcal{I}^+$ and $\mathcal{I}^+$ is a compact connected set, $\mathcal{D}_{AS}(t_0;\mathcal{I}^+) = \mathfrak{R}^n$; i.e., $\mathcal{D}_{AS}(t_0;\mathcal{I}^+) = \mathcal{N}(t_0;\mathcal{I}^+)$. Let it be now supposed that $\partial\mathcal{N}(t_0;\mathcal{I}^+) \neq \phi$. i.e., $\mathcal{N}(t_0;\mathcal{I}^+) \subset \mathfrak{R}^n$. If we assume now $\partial\mathcal{D}_{AS}(t_0;\mathcal{I}^+) = \phi$, then $\mathcal{D}_{AS}(t_0;\mathcal{I}^+) = \mathfrak{R}^n$. This implies $\partial\mathcal{N}(t_0;\mathcal{I}^+) \cap \mathcal{D}_{AS}(t_0;\mathcal{I}^+) \neq \phi$. This and the condition 4-a,ii) show that there is a set $L \subseteq \partial\mathcal{N}(t_0;\mathcal{I}^+) \cap \mathcal{D}_{AS}(t_0;\mathcal{I}^+)$ such that $V(t_0,\mathbf{S}) \longrightarrow \infty$ as $\mathbf{S} \longrightarrow L \subseteq \partial\mathcal{N}(t_0;\mathcal{I}^+) \cap \mathcal{D}_{AS}(t_0;\mathcal{I}^+)$. This is impossible because the function $V(.)$ is the unique solution of (12.11), (12.12), which is continuous on $\mathfrak{T}_i \times \mathcal{D}_{AS}(t;\mathcal{I}^+)$, as shown in the details in the proof of the necessity. Assume $\partial\mathcal{D}_{AS}(t_0;\mathcal{I}^+) = \phi$ fails. Let $\partial\mathcal{D}_{AS}(t_0;\mathcal{I}^+) \neq \phi$ be considered. If $\partial\mathcal{D}_{AS}(t_0;\mathcal{I}^+) \cap \partial\mathcal{N}(t_0;\mathcal{I}^+) = \phi$, then either $\mathcal{D}_{AS}(t_0;\mathcal{I}^+) \subset \mathcal{N}(t_0;\mathcal{I}^+)$ or $\mathcal{D}_{AS}(t_0;\mathcal{I}^+) \supset \mathcal{N}(t_0;\mathcal{I}^+)$ because both are open connected neighborhoods of $\mathcal{I}^+$ and their boundaries are nonempty. The last two cases are impossible due to the positive definiteness of the function $V(.)$ on $\mathfrak{T}_i \times \mathcal{N}(t;t_0;\mathcal{I}^+)$ and its construction via the equations (12.11), (12.12) as shown above. If $\partial\mathcal{D}_{AS}(t_0;\mathcal{I}^+) \cap \partial\mathcal{N}(t_0;\mathcal{I}^+) \neq \phi$, then either $\partial\mathcal{D}_{AS}(t_0;\mathcal{I}^+) = \partial\mathcal{N}(t_0;\mathcal{I}^+)$, which implies $\mathcal{D}_{AS}(t_0;\mathcal{I}^+) = \mathcal{N}(t_0;\mathcal{I}^+)$, or $\partial\mathcal{D}_{AS}(t_0;\mathcal{I}^+) \cap \mathcal{N}(t_0;\mathcal{I}^+) \neq \phi$ and/or $\mathcal{D}_{AS}(t_0;\mathcal{I}^+) \cap \partial\mathcal{N}(t_0;\mathcal{I}^+) \neq \phi$. If $\partial\mathcal{D}_{AS}(t_0;\mathcal{I}^+) \cap \mathcal{N}(t_0;\mathcal{I}^+) \neq \phi$, then it means that the function $V(.)$ blows up (to ∞) on $\mathcal{N}(t_0;\mathcal{I}^+)$, which contradicts its continuity on $\mathcal{N}(t_0;\mathcal{I}^+)$ due to the condition 4-a,i. If $\mathcal{D}_{AS}(t_0;\mathcal{I}^+) \cap \partial\mathcal{N}(t_0;\mathcal{I}^+) \neq \phi$, then it means that $V(.)$ blows up on $\mathcal{D}_{AS}(t_0;\mathcal{I}^+)$; which is impossible due to (13.14) because $V(.)$ is generated by (12.11), (12.12). Hence, $\partial\mathcal{D}_{AS}(t_0;\mathcal{I}^+) = \partial\mathcal{N}(t_0;\mathcal{I}^+)$, which implies $\mathcal{D}_{AS}(t_0;\mathcal{I}^+) = \mathcal{N}(t_0;\mathcal{I}^+)$, which holds as the overall result. Now, $\mathcal{N}(\mathfrak{T}_i;\mathcal{I}^+) = \cap[\mathcal{N}(t_0;\mathcal{I}^+) : t_0 \in \mathfrak{T}_i]$/the condition 2) of the theorem statement/and the conditions of Definition 204, Definition 212, Definition 226, Definition 229, Definition 235 and Definition 238 imply $\mathcal{D}_{UAS}(\mathfrak{T}_i;\mathcal{I}^+) \equiv \mathcal{N}(\mathfrak{T}_i;\mathcal{I}^+)$. Positive definiteness of $W(.)$ on $\mathrm{N}(t;t_0;\mathcal{I}^+)$, $W(.) \in L(\mathfrak{T}_0,\mathfrak{T}_i,\mathrm{N},\mathcal{I}^+)$, the equation (12.11), $\mathcal{N}(t;\mathcal{I}^+) \subseteq \mathrm{N}(t;t;\mathcal{I}^+)$, $\forall t \in \mathfrak{T}_i$, the condition 4-a,i) and a) of Lemma A, Section 12.4, imply

$$V[t_0+\tau, \mathbf{S}(t_0+\tau;t_0;\mathbf{S}_0)] \leq V(t_0;\mathbf{S}_0) - \xi(\varsigma;W;V)\,\tau(t_0;\mathbf{S}_0;\varsigma;W),$$

where $\varsigma \in \mathfrak{R}^+$ is arbitrarily small,

$$\xi(\varsigma;W;V) =$$

$$= min\left\{V(t;\mathbf{S}) : (t;\mathbf{S}) \in \mathfrak{T}_i \times \left[\mathcal{N}(\mathfrak{T}_i;\mathcal{I}^+) \backslash \mathfrak{V}_\psi(\mathfrak{T}_i;\mathcal{I}^+)\right]\right\} \in \mathfrak{R}^+,$$

$$\psi = \varphi_1(\zeta),\ \varphi_1(.) \in \mathcal{K},$$

$$\mathfrak{V}_\psi(\mathfrak{T}_i;\mathcal{I}^+) = \cap\left[\mathfrak{V}_\psi(t;\mathcal{I}^+) : t \in \mathfrak{T}_i\right],$$

$$\varphi_1\left[d\left(\mathbf{S},\mathcal{I}^+\right)\right] \leq V\left(t;\mathbf{S}\right),\ \forall\left(t;\mathbf{S}\right) \in \mathfrak{T}_i \times \mathcal{N}\left(t;\mathcal{I}^+\right),$$

$$\varphi_2\left[d\left(\mathbf{S},\mathcal{I}^+\right)\right] \geq V\left(t;\mathbf{S}\right),\ \forall\left(t;\mathbf{S}\right) \in \mathfrak{T}_i \times \mathcal{N}\left(\mathfrak{T}_i;\mathcal{I}^+\right),\ \varphi_2\left(.\right) \in \mathcal{K},$$

so that

$$\tau\left(t_0;\mathbf{S}_0;\varsigma;W\right) \leq \left\{\varphi_2\left[d\left(\mathbf{S}_0,\mathcal{I}^+\right)\right] - \varphi_1\left(\zeta\right)\right\}\xi^{-1}\left(\varsigma;W;V\right),$$

$$\begin{gathered} sup\left[\tau_m\left(t_0;\mathbf{S}_0;\varsigma;W\right) : t_0 \in \mathfrak{T}_i\right] \\ \leq \left[\varphi_2\left\{d\left[\mathbf{S}_0,\mathcal{I}^+\right]\right\} - \varphi_1\left(\zeta\right)\right]\xi^{-1}\left(\varsigma;W;V\right) < \infty, \\ \forall \mathbf{S}_0 \in \mathcal{N}\left(\mathfrak{T}_i;\mathcal{I}^+\right), \end{gathered}$$

and, therefore, the conditions of Definition 228, Definition 229, Definition 237 and Definition 238 are satisfied. This completes the proof of the sufficiency of the conditions 1-4,a). The sufficiency of the conditions 1-3 and 4b) is implied by the sufficiency of 1-4a) and (12.26)-(12.28), which completes the proof. ∎

The direct determination of the system Lyapunov function $V(.)$ follows.

Property 307 ***Additional property of the function*** **f** $(.)$

Let the function $W(.)$ defined by (13.40),

$$W\left(t,\mathbf{S}\right) = \mathbf{f}^T\left(t,\mathbf{S}\right)\mathbf{f}\left(t,\mathbf{S}\right), \tag{13.40}$$

be such a decrescent positive definite function relative to $\mathcal{I}^+$ on

$$\mathfrak{T}_0 \times \mathfrak{T}_i \times \mathbb{N}\left(t;t_0;\mathcal{I}^+\right)$$

that $W(.) \in L\left(\mathfrak{T}_0;\mathfrak{T}_i,\mathbb{N},\mathcal{I}^+\right)$.

Theorem 308 ***Uniform asymptotic stability of the set $\mathcal{I}^+$ on $\mathfrak{T}_i$ of the system (5.22), (5.23) with Strong smoothness property and with Property 307***

*Let the system **(5.22), (5.23)** possess Strong smoothness property 295 together with Property 307.*

In order for the set $\mathcal{I}^+$ to be uniformly asymptotically stable on $\mathfrak{T}_i$, for a set $\mathcal{N}\left(t_0;\mathcal{I}^+\right)$ to be the asymptotic stability domain $\mathcal{D}_{AS}\left(t_0;\mathcal{I}^+\right)$ of $\mathcal{I}^+$ at $t_0 \in \mathfrak{T}_i$:

$$\mathcal{N}\left(t_0;\mathcal{I}^+\right) \equiv \mathcal{D}_{AS}\left(t_0;\mathcal{I}^+\right),\ t_0 \in \mathfrak{T}_i,$$

and for the set $\mathcal{N}\left(\mathfrak{T}_i;\mathcal{I}^+\right)$ to be the domain $\mathcal{D}_{UAS}\left(\mathfrak{T}_i;\mathcal{I}^+\right)$ of the uniform asymptotic stability of $\mathcal{I}^+$ on $\mathfrak{T}_i$:

$$\mathcal{N}\left(\mathfrak{T}_i;\mathcal{I}^+\right) = \mathcal{D}_{UAS}\left(\mathfrak{T}_i;\mathcal{I}^+\right)$$

it is both necessary and sufficient that:

1) the set $\mathcal{N}\left(t_0;\mathcal{I}^+\right)$ is an open, continuous, connected neighborhood of $\mathcal{I}^+$ and $\mathcal{N}\left(t_0;\mathcal{I}^+\right) \subseteq \mathbb{N}\left(t_0;t_0;\mathcal{I}^+\right)$ for every $t_0 \in \mathfrak{T}_i$,

2) the set $\mathcal{N}\left(\mathfrak{T}_i;\mathcal{I}^+\right) = \cap\left\{\mathcal{N}\left(t_0;\mathcal{I}^+\right) : t_0 \in \mathfrak{T}_i\right\}$ is a connected neighborhood of $\mathcal{I}^+$ on $\mathfrak{T}_i$ such that

$$\mathcal{N}\left(\mathfrak{T}_i;\mathcal{I}^+\right) \subseteq \mathbb{N}\left(\mathfrak{T}_i;\mathfrak{T}_i;\ \mathcal{I}^+\right), \tag{13.41}$$

3) $\mathbf{f}\left(t,\mathbf{S}\right) = \mathbf{0}_n$ for all $\left(t,t_0\right) \in \mathfrak{T}_0 \times \mathfrak{T}_i$, is possible only for $\mathbf{S} \notin \left\{\mathcal{N}\left(t;\mathcal{I}^+\right)\backslash\mathcal{I}^+\right\}$,

4) for the function $W(.)$ defined by (13.40):

a) the equations (12.11) and (12.12) have the unique solution function $V(.)$ defined along the system motions $\mathcal{S}\left(.;t_0;\mathbf{S}_0\right)$ by Equation (13.42):

$$V(t,\mathbf{S}) = V(t_0,\mathbf{S}_0) - \int_{\mathbf{S}_0}^{\mathbf{S}} \mathbf{f}^T\left(t,\mathbf{S}\right)d\mathbf{S},\ \mathbf{S} \equiv \mathbf{S}\left(t\right) \equiv \mathbf{S}\left(t;t_0;\mathbf{S}_0\right), \tag{13.42}$$

with the following properties:

(i) $V(.)$ *is a decrescent positive definite function relative to* $\mathcal{I}^+$ *on* $\mathfrak{T}_0 \times \mathfrak{T}_i \times \mathcal{N}(t; t_0; \mathcal{I}^+)$,

and

(ii) if the boundary $\partial\mathcal{N}(t_0; \mathcal{I}^+)$ *of* $\mathcal{N}(t_0; \mathcal{I}^+)$ *is nonempty, then* $\mathbf{S} \longrightarrow \partial\mathcal{N}(t; \mathcal{I}^+)$, $\mathbf{S} \in \mathcal{N}(t; \mathcal{I}^+)$, *implies* $V(t, \mathbf{S}) \longrightarrow \infty$, $\forall t \in \mathfrak{T}_i$.

Proof. The function $\mathbf{f}(.)$ is continuous on $\mathfrak{T} \times \mathfrak{R}^n$ due to **(4.142)**, Section 4.11. Theorem 308 is a special case of Theorem 306 due to Property 307. ∎

Theorem 308 shows the direct link between the system Lyapunov function $V(.)$ and the system dynamics **(4.142)**. It defines the system Lyapunov function in the integral form (13.42) and as the solution of the vector differential equation (12.11) with the boundary condition (12.12).

13.1.2 System with Weak smoothness property

The systems **(5.22)**, **(5.23)** that possess Weak smoothness property 293 form another class of the systems, which is different from the class of the systems obeying Strong smoothness property **295**. The conditions of Theorem 306 and Theorem 308 change slightly for them, as follows.

Theorem 309 ***Uniform asymptotic stability of the set*** $\mathcal{I}^+$ ***on*** $\mathfrak{T}_0 \times \mathfrak{T}_i$ ***of the system (5.22), (5.23) with Weak smoothness property***

Let the system ***(5.22), (5.23)*** *possess Weak smoothness property 293 on* $\mathfrak{R}^n$.

In order for the set $\mathcal{I}^+$ *to be uniformly asymptotically stable on* $\mathfrak{T}_i$, *for a set* $\mathcal{N}(t_0; \mathcal{I}^+)$ *to be its asymptotic stability domain* $\mathcal{D}_{AS}(t_0; \mathcal{I}^+)$ *at* $t_0 \in \mathfrak{T}_i$:

$$\mathcal{N}(t_0; \mathcal{I}^+) = \mathcal{D}_{AS}(t_0; \mathcal{I}^+), \; t_0 \in \mathfrak{T}_i,$$

and for the set $\mathcal{N}(\mathfrak{T}_i; \mathcal{I}^+)$ *to be the domain* $\mathcal{D}_{UAS}(\mathfrak{T}_i; \mathcal{I}^+)$ *of its uniform asymptotic stability on* $\mathfrak{T}_i$:

$$\mathcal{N}(\mathfrak{T}_i; \mathcal{I}^+) \equiv \mathcal{D}_{UAS}(\mathfrak{T}_i; \mathcal{I}^+)$$

it is both necessary and sufficient that:

1) the set $\mathcal{N}(t_0; \mathcal{I}^+)$ *is a continuous open connected neighborhood of the set* $\mathcal{I}^+$,

2) the set $\mathcal{N}(\mathfrak{T}_i; \mathcal{I}^+)$ *is a connected neighborhood of* $\mathcal{I}^+$,

3) $\mathcal{I}^+$ *is the unique positive invariant set in* $\mathcal{N}(t_0; \mathcal{I}^+)$ *relative to system motions on* $\mathfrak{T}_i$,

4) for any decrescent positive definite function $W(.)$ *relative to* $\mathcal{I}^+$ *on* $\mathfrak{T}_0 \times \mathfrak{T}_i \times \mathfrak{R}^n$, *which obeys:*

a) $W(.) \in L(\mathfrak{T}_0, \mathfrak{T}_i, \mathfrak{R}^n, \mathcal{I}^+)$, *the equations (12.11) and (12.12) have the unique solution function* $V(.)$ *with the following properties:*

(i) $V(.)$ *is a decrescent positive definite function relative to* $\mathcal{I}^+$ *on* $\mathfrak{T}_0 \times \mathfrak{T}_i \times \mathfrak{R}^n$,

and

(ii) if the boundary $\partial\mathcal{N}(t_0; \mathcal{I}^+)$ *of* $\mathcal{N}[t_0; \mathcal{I}^+(t_0; t_0; \mathcal{I}^+)]$ *is nonempty, then* $\mathbf{S} \rightarrow \partial\mathcal{N}[t; \mathcal{I}^+(t; t; \mathcal{I}^+)]$, $\mathbf{S} \in \mathcal{N}[t; \mathcal{I}^+(t; t; \mathcal{I}^+)]$, *implies* $V(t, \mathbf{S}) \rightarrow \infty$, $\forall t \in \mathfrak{T}_i$,

or obeys:

b) $W(.) \in E(\mathfrak{T}_0, \mathfrak{T}_i, \mathfrak{R}^n, \mathcal{I}^+)$, *the equations (12.19) and (12.20) have the unique solution function* $U(.)$ *with the following properties:*

(i) $U(.)$ *is a decrescent positive definite function relative to* $\mathcal{I}^+$ *on* $\mathfrak{T}_0 \times \mathfrak{T}_i \times \mathfrak{R}^n$,

and

(ii) if the boundary $\partial\mathcal{N}(t_0;\mathcal{I}^+)$ *of* $\mathcal{N}(t_0;\mathcal{I}^+)$ *is nonempty, then*

$$\mathbf{S} \to \partial\mathcal{N}\left[t;\mathcal{I}^+\left(t;t;\mathcal{I}^+\right)\right], \mathbf{S} \in \mathcal{N}\left[t;\mathcal{I}^+\left(t;t;\mathcal{I}^+\right)\right],$$

implies $U(t,\mathbf{S}) \to 1$, $\forall t \in \mathfrak{T}_i$.

Proof. This proof relies largely on the proof 13.1.1 of Theorem 306. Let the system **(5.22), (5.23)** possess Weak smoothness property 293.

Necessity. Let the set $\mathcal{I}^+$ be uniformly asymptotically stable on $\mathfrak{T}_i$. Definition 234, Definition 235, Definition 237 and Definition 238 hold. They imply that the set $\mathcal{I}^+$ has the domain $\mathcal{D}_{AS}(t_0;\mathcal{I}^+)$ of the asymptotic stability at every $t_0 \in \mathfrak{T}_i$ and the domain $\mathcal{D}_{UAS}(\mathfrak{T}_i;\mathcal{I}^+)$ of the uniform asymptotic stability on $\mathfrak{T}_i$. The preceding definitions, Definition 201, Definition 204, Definition 211, Definition 212, Definition 225, Definition 226, Definition 228 and Definition 229, show that the system also has the stability domain $\mathcal{D}_S(t_0;\mathcal{I}^+)$ and the attraction domain $\mathcal{D}_A(t_0;\mathcal{I}^+)$ at every $t_0 \in \mathfrak{T}_i$, the uniform stability domain $\mathcal{D}_{US}(\mathfrak{T}_i;\mathcal{I}^+)$, and the uniform attraction domain $\mathcal{D}_{UA}(\mathfrak{T}_i;\mathcal{I}^+)$ on $\mathfrak{T}_i$ so that (13.2) holds. Let

$$\mathcal{N}\left(t_0;\mathcal{I}^+\right) \equiv \mathcal{D}_{AS}\left(t_0;\mathcal{I}^+\right) \Longrightarrow \mathcal{N}\left(\mathfrak{T}_i;\mathcal{I}^+\right) \equiv \mathcal{D}_{UAS}\left(\mathfrak{T}_i;\mathcal{I}^+\right). \tag{13.43}$$

The neighborhood $\mathcal{N}(t_0;\mathcal{I}^+)$ is a continuous connected neighborhood of the set $\mathcal{I}^+$ on $\mathfrak{T}_i$ (Lemma 304) and $\mathcal{N}(\mathfrak{T}_i;\mathcal{I}^+)$ is a connected neighborhood of $\mathcal{I}^+$ (Definition 238). Let $W(.) \in L(\mathfrak{T}_0,\mathfrak{T}_i,\mathfrak{R}^n,\mathcal{I}^+)$, or $W(.) \in E(\mathfrak{T}_0,\mathfrak{T}_i,\mathfrak{R}^n,\mathcal{I}^+)$, Definition 298, be an arbitrarily selected positive definite decrescent function on $\mathfrak{T}_i \times \mathfrak{R}^n$. From now on we should repeat the necessity part of the proof 13.1.1 with $\mathbf{N}(\mathfrak{T}_i;\mathcal{I}^+)$ replaced by $\mathfrak{R}^n$.

Sufficiency. Let the conditions 1) through 4-a) hold. Then the conditions 1) through 4) of Theorem 306 are satisfied for $\mathbf{N}(\mathfrak{T}_i;\mathcal{I}^+)$ replaced by $\mathfrak{R}^n$. ■

Comment 310 ***The comparison of Theorem 309 and Theorem 306***

Weak smoothness property 293 imposes a more relaxed condition on the system properties than Strong smoothness property ***295*** *does. However, the conditions 1) through 4) of Theorem 309 are slightly sharper than the corresponding conditions of Theorem 306. In the former case, the properties of the generating function* $W(.)$*, and the properties of, by it induced, solution functions* $V(.)$ *and* $U(.)$ *should hold on the whole space* $\mathfrak{R}^n$*, whereas in the latter case it should hold on the set* $\mathcal{N}(t;t_0;\mathcal{I}^+)$*, which is a subset of* $\mathfrak{R}^n$*,* $\mathcal{N}(t;t_0;\mathcal{I}^+) \subseteq \mathfrak{R}^n$.

13.2 Applications

13.2.1 Power-energy application

Theorem 309 permits a direct power-energy application to a dynamical system that possesses either Strong or Weak smoothness property 12.2.

Theorem 311 *Let the system possess Weak smoothness property 293, let its power* $P(.)$ *assume the role of the function* $[-W(.)]$*, and let its energy* $E(.)$ *play the role of the function* $V(.)$ *in Equations (2a), (2b), Definition 298. Let the system satisfy all conditions of Theorem 309. If the system power* $P(.)$ *is negative definite, then for the set* $\mathcal{I}^+$ *of the system to be uniformly asymptotically stable on* $\mathfrak{T}_i$ *it is necessary and sufficient that its energy* $E(.)$ *is positive definite.*

This theorem follows directly from Theorem 309.

In order to apply this theorem to a physical dynamical time-varying system, the following should be known:

1. mathematical model of the system,

2. system energy, and

3. system power.

Their determination demands separate research and or engineering work that is out of the scope of this and theoretical (Section 0.5).

Let the set $\mathfrak{S}_{P_0}$ be the largest subset of $\mathfrak{R}^n = \mathcal{I}^+$ such that it contains all system states $\mathbf{S}$ for which the system power $P(\mathbf{S})$ equals zero,

$$\mathfrak{S}_{P_0} = \{\mathbf{S} \in \mathfrak{R}^n / \mathcal{I}^+ : P(\mathbf{S}) = 0\}. \tag{13.44}$$

Corollary 312 *If the set $\mathfrak{S}_{P_0}$, Equation 13.44, is not a positive invariant set of the system, then Theorem 312 is still valid.*

13.2.2 Application to linear systems

Let us show how the well-known Lyapunov matrix theorem 248 (Section 10.4) results from Theorem 308:

Theorem 313 ***Lyapunov matrix theorem for the ISO time-invariant linear systems***

In order for the unique equilibrium state $\mathbf{S} = \mathbf{0}_\rho$ of the ISO time-invariant linear system

$$\frac{d\mathbf{S}}{dt} = A\mathbf{S},\ A \in \mathfrak{R}^{n\times n},\ detA \neq 0, \tag{13.45}$$

to be asymptotically stable (in the whole), it is necessary and sufficient that for any positive definite matrix $G = G^T \in \mathfrak{R}^{n\times n}$ the matrix solution H of the Lyapunov matrix equation (13.46),

$$A^T H + H^T A = -G \tag{13.46}$$

is also a positive definite symmetric matrix H, $H = H^T \in \mathfrak{R}^{n\times n}$.

Proof. In this case, $\mathbf{f}(t, \mathbf{S}) \equiv A\mathbf{S}$. Equation (13.42) simplifies to

$$V(t, \mathbf{S}) = -\int_{\mathbf{0}_n}^{\mathbf{S}} \mathbf{f}^T(t, \mathbf{S}; \mathbf{I})\, d\mathbf{S}, = -\int_{\mathbf{0}_n}^{\mathbf{S}} \mathbf{S}^T A^T d\mathbf{S} = -\int_{\mathbf{0}_n}^{\infty} \mathbf{S}^T A^T A\mathbf{S} dt = V(\mathbf{S})$$

along the system solution $\mathbf{S}(t; \mathbf{S}_0) = e^{At}\mathbf{S}_0$,

$$V(\mathbf{S}) = -\int_0^\infty \mathbf{S}_0^T e^{A^T t} A^T A e^{At} \mathbf{S}_0 dt,$$

or

$$V(\mathbf{S}) = -\frac{1}{2}\mathbf{S}_0^T \left\{ \int_0^\infty \left[A^T \left(e^{At}\right)^T e^{At} A + A^T \left(e^{At}\right)^T e^{At} A \right] dt \right\} \mathbf{S}_0,\ \forall \mathbf{S}_0 \in \mathfrak{R}^n.$$

Necessity. Let the unique equilibrium state $\mathbf{S} = \mathbf{0}_n$ of the *ISO* *time*-invariant linear system be asymptotically stable in the whole. Then $\mathbf{S}(\infty; \mathbf{S}_0) = e^{A\infty}\mathbf{S}_0 = \mathbf{0}_n$, which implies

$$V(\mathbf{S}) = -\frac{1}{2}\mathbf{S}_0^T \left\{ \underbrace{\left[A^T \int_0^\infty \left(e^{At}\right)^T e^{At} dt A \right]}_{-H^T=-H} + \underbrace{\int_0^\infty A^T \left(e^{At}\right)^T e^{At} dt A}_{-H^T=-H} \right\} \mathbf{S}_0 =$$
$$= \mathbf{S}_0^T H \mathbf{S}_0.$$

and negative definiteness of

$$A^T \left[\int_0^\infty \left(e^{At}\right)^T e^{At} dt \right] A = -H^T = -H$$

and positive definiteness of the symmetric matrix H,

$$H = -A^T \left[\int_0^\infty \left(e^{At}\right)^T e^{At} dt \right] A = H^T.$$

This proves the necessity part of the theorem.

Sufficiency. Let, for any positive definite symmetric matrix $G = G^T$, the solution matrix $H = H^T$ of Equation 13.46 be positive definite. Let $V(\mathbf{S}) = \mathbf{S}^T H \mathbf{S}$. Its total time derivative along the motions of the system (13.45) reads for $\mathbf{S}(t) = \mathbf{S}(t; \mathbf{S}_0) = e^{At}\mathbf{S}_0 \neq \mathbf{0}_n$ for every $t \in \mathfrak{T}$:

$$\frac{d}{dt}V([\mathbf{S}(t)] = \left[\frac{d}{dt}\mathbf{S}(t)\right]^T H\mathbf{S} + \mathbf{S}^T(t) H \left[\frac{d}{dt}\mathbf{S}(t)\right] =$$
$$= [A\mathbf{S}(t)]^T H\mathbf{S} + \mathbf{S}^T(t) H [A\mathbf{S}(t)] = [\mathbf{S}(t)]^T \left(A^T H + HA\right) \mathbf{S}(t) =$$
$$= -[\mathbf{S}(t; \mathbf{S}_0)]^T G\mathbf{S}(t; \mathbf{S}_0), \ \forall (t, \mathbf{S}_0) \in \mathfrak{T} \times \mathfrak{R}^n \Longrightarrow$$
$$\frac{d}{dt}V(\mathbf{S}) = -\mathbf{S}^T G\mathbf{S} = -W(\mathbf{S}) < 0, \ \forall (\mathbf{S} \neq \mathbf{0}_n) \in \mathfrak{R}^n, \ W(\mathbf{0}_n) = 0.$$

This verifies that the *time*-invariant positive definite, hence decrescent, functions $V(.)$ and $W(.)$, $V(\mathbf{S}) = \mathbf{S}^T H\mathbf{S}$ and $W(\mathbf{S}) = \mathbf{S}^T G\mathbf{S}$ obey all the conditions of Theorem 308 that implies the global asymptotic stability of the set $\mathfrak{O} = \{\mathbf{0}_n\}$, i.e., of the zero equilibrium state. ■

13.3 Future research

This theory can be further developed for the following:
- Absolute stability domains of time-invariant sets.
- Exponential stability domains of time-invariant sets.
- Asymptotic stability domains of time-varying sets.
- Exponential stability domains of time-varying sets.
- Absolute stability domains of time-varying sets.

In order to apply this theory to an engineering dynamical system, it is necessary to determine the system energy and power. If power is not negative definite, control should be incorporated to assure the power negative definitness. Then the energy should be positive definite without or with added control. This approach avoids the need for solving the partial differential equation, which should not be a problem for the artificial intelligence application.

Note that this book is an excellent basis for new development of control theories on trackability and tracking developed in [173], [232], [243], [244].

Part VII

APPENDICES

Appendix A

Notation

The meaning of the notation is explained in the text at its first use. What follows is the complete list of the notation.

The notations "$\mathfrak{T}_\tau$", "$\forall t \in \mathfrak{T}_\tau$", and "on $\mathfrak{T}_\tau$" are to be omitted if and only if $\mathfrak{T}_\tau = \mathfrak{T}$. The instant τ can be equal to the initial instant t_0: $\tau = t_0$ is permitted, but not required.

A.1 Abbreviations

We present the abbreviations used most frequently in the book.

A	*attraction*
AS	*asymptotic stability*
$\mathcal{F}$	*the family of all set functions* $\Upsilon(.) : \mathfrak{T} \longrightarrow 2^{\mathfrak{R}^n}$
$\mathcal{K}_{[0,\alpha[}$	*the class of functions* $\varphi(.) : \mathfrak{R}_+ \longrightarrow \mathfrak{R}_+$ *defined, continuous, and strictly increasing on* $[0, \alpha[$ *such that vanish at the origin,* $\varphi(0) = 0, 0 < \alpha \leq \infty$
$\mathcal{K} = \mathcal{K}_{[0,\infty[}$	
S	stability
$\mathcal{S}$	*a dynamical system*
$\mathcal{S}(.; t_0, \mathbf{S}_0; \mathbf{I})$	a *motion (solution)* that describes the state behavior of the system (5.22), (5.23) under the action of an unknown input vector $\mathbf{I}$, $$\mathcal{S}(.) : \mathfrak{T}_0 \times \mathfrak{T}_i \times \mathfrak{R}^n \times \mathcal{J} \to \mathfrak{R}^n, \ \mathbf{S}(t_0; t_0; \mathbf{S}_0; \mathbf{I}) \equiv \mathbf{S}_0$$
$\mathcal{S}(.; t_0, \mathbf{S}_0)$	a *motion (solution)* that describes the state behavior of the system (5.22), (5.23) under the action of a known and fixed nominal input vector $\mathbf{I}_N$, $$\mathcal{S}(.) : \mathfrak{T}_0 \times \mathfrak{T}_i \times \mathfrak{R}^n \to \mathfrak{R}^n, \ \mathbf{S}(t_0; t_0; \mathbf{S}_0) \equiv \mathbf{S}_0$$
$\mathfrak{S}_{E0}$	the *zero* equilibrium singleton set: $$\mathfrak{S}_{E0} = \{\mathbf{0}_n\}$$
UA	*uniform attraction*
UAS	*uniform asymptotic stability*
US	*uniform stability*
$\mathcal{Y}(.)$	a mapping called *output response* that describes the output system behavior, $$\mathcal{Y}(.) : \mathfrak{T}_0 \times \mathfrak{T}_i \times \mathfrak{Y} \times \mathcal{J} \to \mathfrak{Y}, \ \mathbf{Y}(t_0; t_0; \mathbf{Y}_0; \mathbf{I}) \equiv \mathbf{Y}_0$$
w.r.t.	*with respect to*

DOI: 10.1201/9781003744528-A

BLACKBOARD LETTER

$\mathbb{I}_A(\mathfrak{T}_i;\mathcal{I}^+)$	*the integral space attraction domain of the set* $\mathcal{I}^+$ *on* $\mathfrak{T}_i$, $\mathbb{I}_A(\mathfrak{T}_i;\mathcal{I}^+) \subseteq \mathfrak{T} \times \mathfrak{R}^n$
$\mathbb{I}_{AS}(\mathfrak{T}_i;\mathcal{I}^+)$	*the integral space asymptotic stability domain of the set* $\mathcal{I}^+$ *on* $\mathfrak{T}_i$, $\mathbb{I}_{AS}(\mathfrak{T}_i;\mathcal{I}^+) \subseteq \mathfrak{T} \times \mathfrak{R}^n$
$\mathbb{I}_S(\mathfrak{T}_i;\mathcal{I}^+)$	*the integral space stability domain of the set* $\mathcal{I}^+$ *on* $\mathfrak{T}_i$, $\mathbb{I}_S(\mathfrak{T}_i;\mathcal{I}^+) \subseteq \mathfrak{T} \times \mathfrak{R}^n$
$\mathbb{N}$	*neighborhood.*

A.1.1 $\mathcal{CALIGRAPHIC}$ letters

$\mathcal{C}$	*the set of all complex numbers,*
$\mathcal{I}^+$	*the constant positive invariant set,* $\mathcal{I}^+ \subset \mathfrak{R}^n$
$\mathcal{J}$	the family of the input vector functions $\mathbf{I}(.)$
$\mathcal{N}$	*neighborhood.*

A.1.2 $\mathfrak{Fraktur}$ letters

Capital Fraktur letters are used for spaces or for sets.

$\mathfrak{B} \subseteq \mathfrak{R}^n$	*a nonempty subset of* $\mathfrak{R}^n$
$\mathfrak{B}_\xi(\mathbf{a})$	an *open hyperball with the radius* ξ *centered at the point* $\mathbf{a} \in \mathfrak{R}^n$ *in the space* $\mathfrak{R}^n$, $\mathfrak{B}_\xi(\mathbf{a}) = \{\mathbf{w} : \|\mathbf{w} - \mathbf{a}\| < \xi\}$
$\mathfrak{B}_\xi$	*an open hyperball with the radius* ξ *centered at the origin of the corresponding space,* $\mathfrak{B}_\xi = \mathfrak{B}_\xi(\mathbf{0}_n)$
$\mathfrak{C}$	*the family of all continuous functions on* $\mathfrak{T}$
$\mathfrak{C}^{ki}(\mathfrak{S})$	*the family of all functions defined, continuous and k-times continuously differentiable on the set* $\mathfrak{S} \subseteq \mathfrak{R}^i$, $\mathfrak{C}^{ki}(\mathfrak{R}^i) = \mathfrak{C}^k(\mathfrak{R}^i) = \mathfrak{C}^{ki}$
$\mathfrak{C}^k(\mathfrak{T}_0)$	*the family of all functions defined, continuous and* k*-times continuously differentiable on* $\mathfrak{T}_0$, $\mathfrak{C}^k = \mathfrak{C}^k(\mathfrak{T})$ and $\mathfrak{C}^0 = \mathfrak{C}$
$\mathfrak{C}^0(\mathfrak{S}) = \mathfrak{C}(\mathfrak{S})$	*the family of all functions defined and continuous on the set* $\mathfrak{S}$, $\mathfrak{C}^0(\mathfrak{R}^i) = \mathfrak{C}^{0,i}$
$\mathfrak{C}^{k-}(\mathfrak{R}^i)$	*the family of all functions defined and continuous everywhere on* $\mathfrak{R}^i$ and k*-times continuously differentiable on* $\mathfrak{R}^i \backslash \{\mathbf{0}_i\}$, which have defined and continuous derivatives at the origin $\mathbf{0}_i$ of $\mathfrak{R}^i$ up to the order $(k-1)$, which have defined both the left and the right k-th order derivative at the origin $\mathbf{0}_i$
$\mathfrak{d}_m = \mathfrak{d}_m(t;\mathfrak{S}) \in \mathfrak{R}_+ \cup \{\infty\}$	*the inner diameter of a time-varying set* $\mathfrak{S}(t)$ *at a moment* $t \in \mathfrak{T}$
$\mathfrak{d}_m(\mathfrak{T}_\tau;\mathfrak{S})$	*the inner diameter of a time-varying set* $\mathfrak{S}(t)$ *on* $\mathfrak{T}_\tau$, $$\mathfrak{d}_m(\mathfrak{T}_\tau;\mathfrak{S}) = inf\left[\mathfrak{d}_m(t;\mathfrak{S}) : t \in \mathfrak{T}_\tau\right]$$
$\mathfrak{d}_M = \mathfrak{d}_M(t;\mathfrak{S}) \in \mathfrak{R}_+ \cup \{\infty\}$	*the outer diameter of a time-varying set* $\mathfrak{S}(t)$ *at a moment* $t \in \mathfrak{T}$
$\mathfrak{d}_M = \mathfrak{d}_M(\mathfrak{T}_\tau;\mathfrak{S})$	*the outer diameter of a time-varying set* $\mathfrak{S}(t)$ *on* $\mathfrak{T}_\tau$, $$\mathfrak{d}_M(\mathfrak{T}_\tau;\mathfrak{S}) = sup\left[\mathfrak{d}_M(t;\mathfrak{S}) : t \in \mathfrak{T}_\tau\right]$$
$\mathfrak{D}_{..}(.;\mathcal{I}^+)$	*domain of a stability property of a constant positive invariant set* $\mathcal{I}^+$, $\mathfrak{D}_{..}(.;\mathcal{I}^+) \subseteq \mathfrak{R}^n$

$\mathfrak{D}_A(t_0;\mathcal{I}^+)$ — *the domain of attraction of the set* $\mathcal{I}^+$ at $t_0 \in \mathfrak{T}_i$

$\mathfrak{D}_S(t_0;\mathcal{I}^+)$ — *the domain of stability of the set* $\mathcal{I}^+$ at $t_0 \in \mathfrak{T}_i$

$\mathfrak{D}_S(t;t_0;\mathcal{I}^+)$ — *the time development of the stability domain* $\mathfrak{D}_S(t_0;\mathcal{I}^+)$ *of the set* $\mathcal{I}^+$ *along system motions* $\mathcal{S}(.;t_0;\mathbf{S}_0)$ *at any moment* $t \in \mathfrak{T}_i$

$\mathfrak{D}_{SS}(t_0;\mathcal{I}^+)$ — *the domain of strict stability of the set* $\mathcal{I}^+$ at $t_0 \in \mathfrak{T}_i$

$\mathfrak{D}_{SS}(t;t_0;\mathcal{I}^+)$ — *the time development of the strict stability domain* $\mathfrak{D}_S(t_0;\mathcal{I}^+)$ *of the set* $\mathcal{I}^+$ *along system motions* $\mathcal{S}(.;t_0;\mathbf{S}_0)$ *at any moment* $t \in \mathfrak{T}_0$

$\mathfrak{D}_{SS}(t;t_0;\mathcal{I}^+)$ — *the time development of the strict stability domain* $\mathfrak{D}_S(t_0;\mathcal{I}^+)$ *of the set* $\mathcal{I}^+$ *along system motions* $\mathcal{S}(.;t_0;\mathbf{S}_0)$ *at any moment* $t \in \mathfrak{T}_i$

$\mathfrak{D}_{US}[\mathfrak{T}_i;\mathcal{I}^+]$ — *the domain of stability of the set* $\mathcal{I}^+$ on $\mathfrak{T}_i$

$\mathfrak{D}_{USS}[\mathfrak{T}_i;\mathcal{I}^+]$ — *the domain of strict stability of the set* $\mathcal{I}^+$ on $\mathfrak{T}_i$

$\mathfrak{L}$ — *a constant set*

$\mathfrak{L}[\mathbf{S}_0,\mathbf{I}(.)]$ — *the largest temporal set of the initial moments* t_0 *for which the given system has a solution* for $\mathbf{I}(.) \in \mathcal{J}$, $\mathfrak{L}[\mathbf{S}_0,\mathbf{I}(.)] \subseteq \mathfrak{T}$,

$\mathfrak{L}^c$ — *an arbitrary nonempty connected compact subinterval of* $\mathfrak{L}$

$\mathfrak{N}_{(.)} \subseteq \mathfrak{R}^n$ — *a neighborhood of the origin* $\mathbf{s} = \mathbf{0}_n$

$\mathfrak{N}_\varepsilon(\mathfrak{S})$ — *the* ε*-neighborhood of the set* $\mathfrak{S}$,

$$\mathfrak{N}_\varepsilon(\mathfrak{S}) = \{\mathbf{s} : \mathbf{s} \in \mathfrak{R}^n,\ d(\mathbf{s},\mathfrak{S}) < \varepsilon\}$$

$\mathfrak{N}(\varepsilon;\mathfrak{S})$ — *a connected neighborhood of the set* $\mathfrak{S}$, which is determined by ε and the outer radius of which cannot be greater than ε, such that

$$0 < \varepsilon_1 < \varepsilon_2 \Longrightarrow \mathfrak{N}(\varepsilon_1;\mathfrak{S}) \subseteq \mathfrak{N}(\varepsilon_2;\mathfrak{S}),$$
$$\varepsilon \longrightarrow 0^+ \Longrightarrow \mathfrak{N}(\varepsilon;\mathfrak{S}) \longrightarrow \mathfrak{S},\ \mathfrak{N}(\varepsilon;\mathfrak{S}) \subseteq \mathfrak{N}_\varepsilon(\mathfrak{S})$$

$\mathfrak{N}(\varepsilon;t;\mathfrak{S})$ — *the connected neighborhood of a time-varying bounded set* $\mathfrak{S}(t)$ determined by ε at $t \in \mathfrak{T}_\tau$

$\mathfrak{N}_\varepsilon[\mathfrak{S}(t)]$ — *the* ε*-neighborhood of a time-varying bounded set* $\mathfrak{S}(t)$ at $t \in \mathfrak{T}_\tau$,

$$\mathfrak{N}_\varepsilon[\mathfrak{S}(t)] = \{\mathbf{s} : \mathbf{s} \in \mathfrak{R}^n,\ d[\mathbf{s},\mathfrak{S}(t)] < \varepsilon\} \supset Cl\mathfrak{S}(t), \forall t \in \mathfrak{T}_\tau$$

$\mathfrak{N}_\varepsilon[\mathbf{S}_d(t)]$ — *the* ε*- neighborhood of* $\mathbf{S}_d(t)$ *at a moment* $t \in \mathfrak{T}$, $\mathfrak{N}_\varepsilon[\mathbf{S}_d(t)] \subseteq \mathfrak{R}^n$

$\mathbb{N}(t;t_0;\mathcal{I}^+)$ — *an open, continuous, connected time-varying neighborhood of the constant positive invariant set* $\mathcal{I}^+$

$\mathbb{N}(\mathfrak{T}_0;\mathfrak{T}_i;\mathcal{I}^+)$ — *an open connected constant neighborhood of the set* $\mathcal{I}^+$ *on* $\mathfrak{T}_0 \times \mathfrak{T}_i$, which is induced by $\mathbb{N}(t;t_0;\mathcal{I}^+)$, $\mathbb{N}(t;t_0;\mathcal{I}^+) \supseteq \mathfrak{N}(t;t_0;\mathcal{I}^+)$

$$\mathbb{N}(\mathfrak{T}_0;\mathfrak{T}_i;\mathcal{I}^+) = \mathbb{N}_m(\mathfrak{T}_0;\mathfrak{T}_i;\mathcal{I}^+) = \cap\{\mathbb{N}(t;t_0;\mathcal{I}^+) : (t,t_0) \in \mathfrak{T}_0 \times \mathfrak{T}_i\},$$
$$\mathbb{N}(t;t_0;\mathcal{I}^+) \supseteq \mathfrak{N}(t;t_0;\mathcal{I}^+),\ \forall (t,t_0) \in \mathfrak{T}_0 \times \mathfrak{T}_i$$

$\mathbb{N}(\mathfrak{T}_0;\mathfrak{T}_i;\mathcal{I}^+)$ — *an open connected neighborhood of the set* $\mathcal{I}^+$, which contains $\mathfrak{N}(\mathfrak{T}_0;\mathfrak{T}_i;\mathcal{I}^+)$,

$$Cl\mathcal{I}^+ \subset In\mathbb{N}(\mathfrak{T}_0;\mathfrak{T}_i;\mathcal{I}^+),\ \mathbb{N}(\mathfrak{T}_0;\mathfrak{T}_i;\mathcal{I}^+) \supseteq \mathfrak{N}(\mathfrak{T}_0;\mathfrak{T}_i;\mathcal{I}^+)$$

$\mathfrak{N}(t;t_0;\mathcal{I}^+)$ — *an open, continuous, connected time-varying neighborhood of the constant positive invariant set* $\mathcal{I}^+$

$\mathfrak{N}(\mathfrak{T}_0;\mathfrak{T}_i;\mathcal{I}^+)$ — *an open connected constant neighborhood of the set* $\mathcal{I}^+$ on $\mathfrak{T}_0 \times \mathfrak{T}_i$, which is induced by $\mathfrak{N}(t;t_0;\mathcal{I}^+)$, $\mathfrak{N}(t;t_0;\mathcal{I}^+) \subseteq \mathbb{N}(t;t_0;\mathcal{I}^+)$

$$\mathfrak{N}\left(\mathfrak{T}_0;\mathfrak{T}_i;\mathcal{I}^+\right) = \mathfrak{N}_m\left(\mathfrak{T}_0;\mathfrak{T}_i;\mathcal{I}^+\right) = \cap\left\{\mathfrak{N}\left(t;t_0;\mathcal{I}^+\right) : (t,t_0) \in \mathfrak{T}_0 \times \mathfrak{T}_i\right\},$$
$$\mathfrak{N}\left(t;t_0;\mathcal{I}^+\right) \subseteq \mathbb{N}\left(t;t_0;\mathcal{I}^+\right),\ \forall (t,t_0) \in \mathfrak{T}_0 \times \mathfrak{T}_i$$

$\mathfrak{N}(\mathfrak{T}_0;\mathfrak{T}_i;\mathcal{I}^+)$ — an open connected neighborhood of the set $\mathcal{I}^+$, which is a subset of $\mathbb{N}(\mathfrak{T}_0;\mathfrak{T}_i;\mathcal{I}^+)$,

$$Cl\mathcal{I}^+ \subset In\mathfrak{N}\left(\mathfrak{T}_0;\mathfrak{T}_i;\mathcal{I}^+\right),\ \mathfrak{N}\left(\mathfrak{T}_0;\mathfrak{T}_i;\mathcal{I}^+\right) \subseteq \mathbb{N}\left(\mathfrak{T}_0;\mathfrak{T}_i;\mathcal{I}^+\right)$$

$\mathfrak{N}[t;t_0;\mathcal{I}^+]$ — an open, contin*uous, connected time-varying neighborhood of the constant set* $\mathcal{I}^+$

$\mathfrak{O} \subset \mathfrak{R}^n$ — *the singleton containing only the origin* $\mathbf{0}_n$ of $\mathfrak{R}^n$, $\mathfrak{O} = \{\mathbf{0}_n\}$

$\mathfrak{P}(\mathfrak{S})$ — *the power set of the set* $\mathfrak{S}$ denoted by $\mathfrak{P}(\mathfrak{S})$ or by $2^{\mathfrak{S}}$,

$$\mathfrak{P}(\mathfrak{S}) = 2^{\mathfrak{S}} \Longrightarrow \mathfrak{P}(\mathfrak{R}^n) = 2^{\mathfrak{R}^n}$$

$\mathfrak{R}$ — *the set of all real numbers*

$\mathfrak{R}^+$ — *the set of all positive real numbers*

$\mathfrak{R}_+$ — *the set of all non-negative real numbers*

$\mathfrak{R}^-$ — *the set of all negative real numbers*

$\mathfrak{R}^n$ — *the* $n-$*dimensional real vector space,*

$\mathfrak{R}^{M\times N}$ — *the* $M \times N$ *matrix space, the zero matrix in which is* $O_{M\times N}$. *If* $M = N$ *then for short* $O_{N\times N} = O_N$

$\mathfrak{S}$ — *the constant set or a symbolic notation for the time-varying connected set* $\mathfrak{S}(t) \subseteq \mathfrak{R}^n$ or $\mathfrak{S}(t;t_0;\mathfrak{S}_0)$,

$\mathfrak{S}(t)$ — *time-varying set,* $\mathfrak{S}(t) \subseteq \mathfrak{R}^n$ at a moment $t \in \mathfrak{T}$ if the initial moment t_0 was at $inf\mathfrak{T} = -\infty$, or at the moment $t \in \mathfrak{T}_0$ if the initial moment $t_0 \in \mathfrak{T}_i$ is fixed

$\mathfrak{S}(t;t_0;\mathfrak{S}_0)$ — *time-varying connected set,* $\mathfrak{S}(t;t_0;\mathfrak{S}_0) \subseteq \mathfrak{R}^n, \forall t \in \mathfrak{T}_0, \forall t_0 \in \mathfrak{T}_i$. It is defined on $\mathfrak{T}_0$, $\forall t_0 \in \mathfrak{T}_i$. Its initial set value $\mathfrak{S}_0$ fulfills: $\mathfrak{S}(t_0;t_0;\mathfrak{S}_0) = \mathfrak{S}(t_0) = \mathfrak{S}_0$. If $\mathfrak{T}_i = \mathfrak{T}$, then for short $\mathfrak{S}(t;t_0;\mathfrak{S}_0) \equiv \mathfrak{S}(t)$

$\mathfrak{S}(\mathfrak{L})$ — *the intersection of* $\mathfrak{S}(t)$ *over* $\mathfrak{L}$,

$$\mathfrak{S}(\mathfrak{L}) = \cap[\mathfrak{S}(t) : t \in \mathfrak{L}] \subseteq \mathfrak{R}^n,\ \mathfrak{S}(\mathfrak{L}) \neq \phi$$

$\mathfrak{S}_E = \{\mathbf{S}_E\}$ — *the equilibrium singleton set* that contains the single equilibrium state $\mathbf{S}_E$

$\mathfrak{S}_{E0}$ — if and only if $\mathbf{S}_E = \mathbf{0}_n$, then the equilibrium singleton set is denoted by $\mathfrak{S}_{E0} = \{\mathbf{0}_n\}$

$\mathfrak{S}_m(t;\mathfrak{T}_i)$ — *the lower set limit of the set* $\mathfrak{S}(t;t_0;\mathfrak{S}_0)$ *on* $\mathfrak{T}_i$ at a moment $t \in \mathfrak{T}_0$,

$$\mathfrak{S}_m(t;\mathfrak{T}_i) = \cap[\mathfrak{S}(t;t_0;\mathfrak{S}_0) : t_0 \in \mathfrak{T}_i],\ t \in \mathfrak{T}_0$$

$\mathfrak{S}_m(\mathfrak{T}_0; t_0; \mathfrak{S}_0)$ — *the lower set limit of the set* $\mathfrak{S}(t; t_0; \mathfrak{S}_0)$ *on* $\mathfrak{T}_0$ for $t_0 \in \mathfrak{T}_i$,

$$\mathfrak{S}_m(\mathfrak{T}_0; t_0) = \cap [\mathfrak{S}(t; t_0; \mathfrak{S}_0) : t \in \mathfrak{T}_0],\ t_0 \in \mathfrak{T}_i$$

$\mathfrak{S}_m(\mathfrak{T}_0; \mathfrak{T}_i)$ — *the lower set limit of the set* $\mathfrak{S}(t; t_0; \mathfrak{S}_0)$ *on* $\mathfrak{T}_0 \times \mathfrak{T}_i$,

$$\mathfrak{S}_m(\mathfrak{T}_0; \mathfrak{T}_i) = \cap [\mathfrak{S}(t; t_0; \mathfrak{S}_0) : (t; t_0) \in \mathfrak{T}_0 \times \mathfrak{T}_i]$$

$\mathfrak{S}_{0m}(\mathfrak{T}_i)$ — *the lower set limit of the set* $\mathfrak{S}(t; t_0; \mathfrak{S}_0)$ *on* $\mathfrak{T}_i \times \mathfrak{T}_i$,

$$\mathfrak{S}_{0m}(\mathfrak{T}_i) = \mathfrak{S}_{0m}(\mathfrak{T}_i; \mathfrak{T}_i) = \cap [\mathfrak{S}(t_0; t_0; \mathfrak{S}_0) : t_0 \in \mathfrak{T}_i] \\ = \cap [\mathfrak{S}_0 : t_0 \in \mathfrak{T}_i]$$

$\mathfrak{S}_M(t; \mathfrak{T}_i)$ — *the upper set limit of the set* $\mathfrak{S}(t; t_0; \mathfrak{S}_0)$ *on* $\mathfrak{T}_i$,

$$\mathfrak{S}_M(t; \mathfrak{T}_i) = \cup [\mathfrak{S}(t; t_0; \mathfrak{S}_0) : t_0 \in \mathfrak{T}_i],\ t \in \mathfrak{T}_0$$

$\mathfrak{S}_M(\mathfrak{T}_0; t_0; \mathfrak{S}_0)$ — *the upper set limit of the set* $\mathfrak{S}(t; t_0; \mathfrak{S}_0)$ *on* $\mathfrak{T}_0$ for $t_0 \in \mathfrak{T}_i$,

$$\mathfrak{S}_M(\mathfrak{T}_0; t_0; \mathfrak{S}_0) = \cup [\mathfrak{S}(t; t_0; \mathfrak{S}_0) : t \in \mathfrak{T}_0],\ t_0 \in \mathfrak{T}_i$$

$\mathfrak{S}_M(\mathfrak{T}_0; \mathfrak{T}_i)$ — *the upper set limit of the set* $\mathfrak{S}(t; t_0; \mathfrak{S}_0)$ *on* $\mathfrak{T}_0$,

$$\mathfrak{S}_M(\mathfrak{T}_0; \mathfrak{T}_i) = \cup [\mathfrak{S}(t; t_0; \mathfrak{S}_0) : (t; t_0) \in \mathfrak{T}_0 \times \mathfrak{T}_i]$$

$\mathfrak{S}_{0M}(\mathfrak{T}_i)$ — *the upper set limit of the set* $\mathfrak{S}(t; t_0; \mathfrak{S}_0)$ *on* $\mathfrak{T}_i \times \mathfrak{T}_i$,

$$\mathfrak{S}_{0M} = \mathfrak{S}_{0M}(\mathfrak{T}_i; \mathfrak{T}_i) = \cup [\mathfrak{S}(t_0; t_0; \mathfrak{S}_0) : t_0 \in \mathfrak{T}_i] \\ = \cup [\mathfrak{S}_0 : t_0 \in \mathfrak{T}_i]$$

$\mathfrak{T}$ — *the accepted reference time set,* the arbitrary element of which is an arbitrary moment t and the time unit of which is seconds, s, $1_t = s$, $t\langle s\rangle$,

$$\mathfrak{T} = \{t : t[T]\langle s\rangle, num\ t \in \mathfrak{R},\ dt > 0\},\ inf\ \mathfrak{T} = -\infty,\ sup\mathfrak{T} = \infty$$

$\mathfrak{T}_i$ — *the set of the initial instants* t_0, which can, but need not, be unbounded subset of $\mathfrak{T}$,

$$\mathfrak{T}_i = \{t : t[T]\langle s\rangle,\ t \in \mathfrak{T}\} \subseteq \mathfrak{T},\ inf\mathfrak{T}_i \geq -\infty,\ sup\mathfrak{T}_i \leq \infty,$$

throughout the book

$$sup\mathfrak{T}_i = \infty$$

if it is not otherwise specially accepted

$\mathfrak{T}_0$ — *the largest connected temporal set on which the system has a unique solution at every* $t \in \mathfrak{T}_0$ *passing through* $\mathbf{S}_0$ *at* $t = t_0$,

$$\mathfrak{T}_0 = \{t :\ t \in \mathfrak{T},\ t \geq t_0\},\ t_0 \in \mathfrak{T}_i,\ \mathfrak{T}_0 \subseteq \mathfrak{T}_i$$

$\mathfrak{T}_0^0$ — *the largest subset of* $\mathfrak{T}_0$ *with the zero measure*

$\mathfrak{T}_\tau^-$ — *a nonempty connected subset of* $\mathfrak{T}$, $\mathfrak{T}_\tau^- = \{t :\ t \in \mathfrak{T},\ -\infty < t \leq \tau\ < \infty\}$

$\mathfrak{T}_\tau$ *a nonempty connected subset of* $\mathfrak{T}$, $\mathfrak{T}_\tau = \{t: \ t \in \mathfrak{T}, \ \tau \leq t \ < \infty\}$

$\mathfrak{T}_{t^*}^-$ *a set composed of a sequence* t_k *converging to* t^* *from the left as* $k \longrightarrow \infty$ if $t^* > -\infty$, ($t^* = \infty$ is permitted)

$$\mathfrak{T}_{t^*}^- = \{t_k : t_k \in \mathfrak{T}, \ t_k < t^*, \ k = 0, 1, 2, \ \cdots \ , \ k \longrightarrow \infty \Longrightarrow \ t_k \longrightarrow t^*\}, \ t^* \in]-\infty, \infty]$$

$\mathfrak{T}_{t^*}^+$ *a set composed of a sequence* t_k *converging to* t^* *from the right as* $k \longrightarrow \infty$ if $t^* < \infty$, ($t^* = -\infty$ is permitted),

$$\mathfrak{T}_{t^*}^+ = \{t_k : t_k \in \mathfrak{T}, \ t_k > t^*, \ k = 0, 1, 2, \ \cdots \ , \ k \longrightarrow \infty \Longrightarrow \ t_k \longrightarrow t^*\}, \ t^* \in [-\infty, \infty[$$

$\mathfrak{T}_{t^*}$ *a set composed of a sequence* t_k *converging to* t^* *as the integer* $k \longrightarrow \infty$,

$$\mathfrak{T}_{t^*} = \{t_k : t_k \in \mathfrak{T}, \ k = 0, 1, 2, \ \cdots \ , \ k \longrightarrow \infty \Longrightarrow \ t_k \longrightarrow t^*\}, \ t^* \in [-\infty, \infty]$$

$\partial\mathfrak{V}_\xi(t;\tau)$ *the boundary of the set* $\mathfrak{V}_\xi(t;\tau)$

$\mathfrak{V}_\xi(t;\tau) \subseteq \mathfrak{R}^n$ *the set associated with the function* $V(.) : \mathfrak{T}_\tau \times \mathfrak{R}^n \longrightarrow \mathfrak{R}$ *and with its value* $\xi \in \mathfrak{R}^+$, or $\xi = \infty$, at $t \in \mathfrak{T}_\tau$ so that it is the largest open connected neighborhood $\mathfrak{V}_\xi(t;\tau)$ of the origin $\mathbf{0}_n$ on $\mathfrak{T}_\tau$ such that:

1. If $\xi \in \mathfrak{R}^+$, then the boundary $\partial\mathfrak{V}_\xi(t;\tau)$ of $\mathfrak{V}_\xi(t;\tau)$ is the hypersurface $V(t, \mathbf{s}) = \xi$:

$$\xi \in \mathfrak{R}^+ \Longrightarrow \partial\mathfrak{V}_\xi(t;\tau) = \{\mathbf{s} : V(t, \mathbf{s}) = \xi\}, \ \forall t \in \mathfrak{T}_\tau,$$
$$V(t, \mathbf{s}) = \xi \Longleftrightarrow \mathbf{s} \in \partial\mathfrak{V}_\xi(t;\tau), \ \forall t \in \mathfrak{T}_\tau,$$

2. The intersection $\mathfrak{V}_\xi(\mathfrak{T}_\tau)$ of $\mathfrak{V}_\xi(t;\tau)$ over $\mathfrak{T}_\tau$,

$$\mathfrak{V}_\xi(\mathfrak{T}_\tau) = \cap[\mathfrak{V}_\xi(t;\tau) : t \in \mathfrak{T}_\tau],$$

is also a neighborhood of the origin $\mathbf{0}_n$:

$$\exists\zeta \in \mathfrak{R}^+ \Longrightarrow \mathfrak{B}_\zeta \subset \mathfrak{V}_\xi(\mathfrak{T}_\tau),$$

and

3. $V(t, \mathbf{s})$ is strictly less than ξ on $\mathfrak{V}_\xi(t;\tau)$ at every instant $t \in \mathfrak{T}_\tau$:

$$V(t, \mathbf{s}) < \xi, \ \forall(\mathbf{s}, \ t) \in \mathfrak{V}_\xi(t;\tau) \times \mathfrak{T}_\tau,$$

where the expression "*on* $\mathfrak{T}_\tau$" and the notation "$;\tau$" are to be omitted if and only if$\mathfrak{T}_\tau = \mathfrak{T}$

$\mathfrak{Y} \subseteq \mathfrak{R}^N$ *the total output vector space of any system*

$\mathfrak{Y}_d^k$ *a given, or to be determined, family of all bounded and $(k+1)$-times continuously differentiable realizable total desired output vector functions $\mathbf{Y}_d(.) \in \mathfrak{C}^{k+1}$* such that they and their first $k+1$ derivatives obey *TCUP*,

$$\mathfrak{Y}_d^k \subset \mathfrak{C}^{k+1}(\mathfrak{R}^N),$$

i.e., it is a family of all bounded continuously differentiable realizable total desired extended output vector functions $\mathbf{Y}_d^k(.) \in \mathfrak{C}^1$ such that they and their derivatives obey *TCUP*

$\mathfrak{Y}_{d0}^k$ *the set of the initial vector values $\mathbf{Y}_{d0}^k$ of $\mathbf{Y}_d^k(t)$ of every $\mathbf{Y}_d(.) \in \mathfrak{Y}_d^k$*. If, and only if, $\mathbf{Y}_d(.) \in \mathfrak{Y}_d^k$, then $\mathbf{Y}_{d0}^k = \mathcal{Y}_d^k\left(t_0; t_0; \mathbf{Y}_{d0}^k\right) \in \mathfrak{Y}_{d0}^k$

$\mathfrak{Y}_d = \mathfrak{Y}_d^0$ *the family of all bounded continuous and continuously differentiable realizable desired total output vector functions $\mathbf{Y}_d(.)$,* such that they and their first derivatives obey *TCUP*,

$$\mathfrak{Y}_d \subset \mathfrak{C}^1(\mathfrak{R}^N).$$

A.1.3 Greek letters

Lower-case Greek letters denote scalars.

α	*a natural number, the order of the () system*
δ_{ij}	*the Kronecker delta,* $\delta_{ij} = 1$ for $i = j$, and $\delta_{ij} = 0$ for $i \neq j$
θ	*a nonnegative integer*
η	*a nonnegative integer*
μ	*a nonnegative integer*
τ	*a subsidiary notation for time t*
ρ	*a natural number, the dimension of the vector* $\mathbf{R}$
ϕ	*the empty set*
$\varphi(.) : \mathfrak{R}_+ \longrightarrow \mathfrak{R}_+$	*a comparison function*

A.1.4 Roman letters

Lower-case Roman letters designate scalars, but if they are bold, then they represent vectors. Capital Roman letters are used for matrices, but if they are bold, then they represent vectors.

$E(.) : \mathfrak{T} \longrightarrow \mathfrak{R}$ *the (dimensionless) instantaneous accumulated energy of the system at a moment $t \in \mathfrak{T}$*

$gradV(t, \mathbf{S}) \in \mathfrak{R}^n$ *the gradient of the function $V(.)$,*

$$gradV(t, \mathbf{S}) = \left[\frac{\partial V(t, \mathbf{S})}{\partial S_1} .. \frac{\partial V(t, \mathbf{S})}{\partial S_2} \quad \cdots \quad \frac{\partial V(t, \mathbf{S})}{\partial S_n} \right]^T$$

$\mathbf{i} \in \mathfrak{R}^M$	*the input deviation vector,* $\mathbf{i} = \mathbf{I} - \mathbf{I}_N$,
$I_i \in R$	*the total value of the i-th input variable of the system*
$\mathbf{I} \in R^M$	*the total real input vector of the system,*
$\mathbf{I}_N \in \mathfrak{R}^M$	*the total nominal input vector of the system*
$\mathbf{I}^\mu \in \mathfrak{R}^{(\mu+1)M}$	*the extended total input vector*

$$\mathbf{I}^\mu = \left[\mathbf{I}^T \vdots \mathbf{I}^{(1)T} \vdots \ldots \vdots \mathbf{I}^{(\mu)T} \right]^T \in \mathfrak{R}^{(\mu+1)M}$$

N	*the dimension of the output vector Y*
$\mathbf{N}$	*neighborhood*
S_i	*the i-th state variable, the value of which is measured in the total scale,*
$\mathbf{S}$	*the state vector,* $$\mathbf{S} = [S_1 \ S_2 \ \cdots \ S_n]^T \in \mathfrak{R}^n$$
$t \in \mathfrak{T}$	*time, an arbitrary instant, an arbitrary moment*
$t_0 \in \mathfrak{T}$	*an accepted initial moment*
t^-	*the left-hand side of the moment t, $t^- = \lim(t - \varepsilon : \varepsilon \longrightarrow 0, \ \varepsilon > 0)$*
t^+	*the right-hand side of the moment t, $t^+ = \lim(t + \varepsilon : \varepsilon \longrightarrow 0, \ \varepsilon > 0)$*
t_{zero}	*the relative zero moment,*
$\overline{t} \in \mathfrak{R}$	*the nondimensional mathematical temporal variable (the mathematical time),* $$\overline{t} = \frac{t}{1_t}[-], \ num \ \overline{t} = num \ t$$
T_t	*a chosen time scale with the corresponding time unit 1_t*
T	*the temporal (the time) dimension*
$v_t(v_\tau)$	*the speed of the evolution (of the flow) of the time value and of its numerical value*
$V(.) : \mathfrak{R}^n \to \mathfrak{R}$	a *time-invariant semidefinite or definite function*
$W(.) : \mathfrak{R}^n \to \mathfrak{R}$	a *time-invariant semidefinite or definite function*
Y_j	*the total value of the j-th output variable of the plant*
$\mathbf{Y} \in \mathfrak{R}^N$	*the real total output vector of the system,* $$\mathbf{Y} = [Y_1 \ Y_2 \ \cdots \ Y_N]^T \in \mathfrak{R}^N$$
$\mathbf{Y}_d \in \mathfrak{R}^N$	*the desired total output vector of the system.*

A.2 Names

Almost at every $t \in \mathfrak{T}_0$ means: "*at every* $t \in \left(\mathfrak{T}_0 \backslash \mathfrak{T}_0^0\right)$", where $\mathfrak{T}_0^0$ is the largest subset of $\mathfrak{T}_0$ with the zero measure

Lyapunov's method and methodology *the original method and methodologies established by Lyapunov on time-invariant sets*

Lyapunov method *Lyapunov's method and its extensions to both time-invariant sets and time-varying sets*

Uniform attraction means ***temporal uniform attraction***.

A.3 Symbols

∞	*infinite value,* $\infty 0 = 0$
$\{.\}$	*large brackets,* or *a set*
$(.)$	*an arbitrary variable,* or *an index*
(a, b)	*an interval* that can be $]a, b[$ or $]a, b]$ or $[a, b[$ or $[a, b]$: $$(a, b) \in \{]a, b[, \quad]a, b], \quad [a, b[, \quad [a, b]\}$$
$].[$	*an open interval*
$].]$	*a semiopen semiclosed interval* that is open on the left and closed on the right

$[.[$ *a semiclosed semiopen interval* that is closed on the left and open on the right

$[.]$ *a closed interval*, or *the notation for the physical dimension*

$\wedge$ *and*, $\alpha \wedge \beta$ means "α *and* β"

$\exists$ *there exist(s); there is (there are),*

$\nexists$ *there does not exist (do not exist); there is not (there are not),*

$\exists!$ *there exists exactly one,*

$\langle 1_t \rangle$ the notation for *the units*; e.g., if the *time* unit is second s then $t\langle 1_t \rangle = t\langle s \rangle$,

$||.|| : \mathfrak{R}^n \to \mathfrak{R}_+$ *the norm on the vector space* $\mathfrak{R}^n$, it can be any norm on $\mathfrak{R}^n$. It can be *the sum norm* $||\mathbf{s}||_1$ *on* $\mathfrak{R}^n$,

$$||\mathbf{s}||_1 = \sum_{k=1}^{k=n} |s_k|,$$

or *the Euclidean norm* $||.||_2$ *on* $\mathfrak{R}^n$,

$$||\mathbf{s}||_2 = \left(\mathbf{s}^T\mathbf{s}\right)^{1/2} = \sqrt{\sum_{k=1}^{k=n} |s_k^2|} = \left(s_1^2 + s_2^2 + \cdots + s_n^2\right)^{1/2},$$

or *the max norm* $||\mathbf{s}||_\infty$ *on* $\mathfrak{R}^n$,

$$||\mathbf{s}||_\infty = max\left\{s_k : k = 1, 2, ..., n\right\}$$

$exp(\alpha t)$ $exp(\alpha t) \equiv e^{\alpha t}$

$Ev_{\mathfrak{T}_i}\mathfrak{S}(t_0)$ the *time evolution* $Ev_{\mathfrak{T}_i}\mathfrak{S}(t_0)$ of a set $\mathfrak{S}(t)$ over $\mathfrak{T}_i$,

$$Ev_{\mathfrak{T}_i}\mathfrak{S}(t_0) = \{(t_0, \mathbf{s}) : \ (t_0, \mathbf{s}) \in \mathfrak{T}_i \times \mathfrak{S}(t_0)\} \subseteq \mathfrak{T}_i \times \mathfrak{R}^n$$

$x\,[.]$ the notation for *the physical dimension of the variable* x, if $x = t$, then $x\,[.] = t\,[\mathrm{T}]$, $x\,[\text{-}]$ means that the variable x is nondimensional (dimensionless)

$det(.)$ *the determinant of the square matrix* $(.)$

inf *infimum,*

$min\,(n, r)$ *the smaller number between* n *and* r,

$$min\,(n, r) = \left\{ \begin{array}{c} n,\ n \leq r, \\ r,\ r \leq n \end{array} \right\}$$

$num\ x$ *the numerical value of the variable* $x \in \mathfrak{R}$

$sign(.) : \mathfrak{R} \to \{-1, 0, 1\}$ *the scalar signum function,*

$$sign\ x = |x|^{-1} x\ if\ x \neq 0,\ and\ sign\ 0 = 0$$

sup *supremum*

A.4 Matrix and vector notation

$\mathbf{0}_k \in \mathfrak{R}^k$ *the zero vector in* R^k, $\mathbf{0}_k = [0\ \ 0\ \ \cdots\ \ 0]^T \in \mathfrak{R}^k$

$O_{M\times N}$ *the zero matrix in the* $M \times N$ *dimensional matrix space* $R^{M\times N}$

O_N *the zero matrix in* $\mathfrak{R}^{N\times N}$, $O_N = O_{N\times N}$

$\mathbf{1}_k \in \mathfrak{R}^k$ *the unit vector in* R^k, $\mathbf{1}_k = [1\ \ 1\ \ \cdots\ \ 1]^T \in \mathfrak{R}^k$

$\mathbf{S} = [S_1\ \ S_2\ \ \cdots\ \ S_n]^T \in \mathfrak{R}^n$, $S_i \in \mathfrak{R}$, $\forall i = 1, 2, \cdots, n.$

A.5 Sets

$\mathfrak{A}\backslash\mathfrak{B}$ — *the set difference* between the set $\mathfrak{A}$ and the set $\mathfrak{B}$ is the set of all vectors $\mathbf{S}$ in $\mathfrak{A}$ that do not belong to $\mathfrak{B}$,

$$\mathfrak{A}\backslash\mathfrak{B} = \{\mathbf{S}: \ \mathbf{S} \in \mathfrak{A}, \ \mathbf{S} \notin \mathfrak{B}\}$$

$\mathfrak{A}\Delta\mathfrak{B}$ — *the symmetric set difference* between the set $\mathfrak{A}$ and the set $\mathfrak{B}$ is the set of all vectors $\mathbf{S}$ in $\mathfrak{A}$, which do not belong to $\mathfrak{B}$, and of all vectors that are in $\mathfrak{B}$ but not in $\mathfrak{A}$,

$$\mathfrak{A}\Delta\mathfrak{B} = (\mathfrak{A}\backslash\mathfrak{B}) \cup (\mathfrak{B}\backslash\mathfrak{A})$$

$\mathfrak{B} \subseteq \mathfrak{R}^n$ — *a nonempty subset of* $\mathfrak{R}^n$

$\mathfrak{B}_\xi(z)$ — an *open hyperball with the radius* ξ *centered at the point* z *in the corresponding space,*

$$\mathfrak{B}_\xi(\mathbf{s}) = \{\mathbf{w}: \ \|\mathbf{w} - \mathbf{s}\| < \xi\}$$

$\mathfrak{B}_\xi$ — *an open hyperball with the radius* ξ *centered at the origin of the corresponding space,*

$$\mathfrak{B}_\xi = \mathfrak{B}_\xi(\mathbf{0})$$

$Cl\mathfrak{S}$ — *the closure of the set* $\mathfrak{S}$

$Ev_{\mathfrak{T}_i}\mathfrak{S}(t)$ — *the time evolution of* $\mathfrak{D}(t;.)$ in t over the *time* set $\mathfrak{T}_i$,

$$Ev_{\mathfrak{T}_i}\mathfrak{S}(t) = \{(t, \mathbf{S}): \ (t, \mathbf{S}) \in \mathfrak{T}_i \times \mathfrak{S}(t)\} \subseteq \mathfrak{T}_i \times \mathfrak{R}^n.$$

$\mathfrak{S}_{E0}$ — *the zero equilibrium singleton set,* $\mathfrak{S}_{E0} = \{\mathbf{0}_n\}$

$\partial\mathfrak{S}$ — *the boundary of the set* $\mathfrak{S}$

$In\mathfrak{S}$ — *the interior of the set* $\mathfrak{S}$

$\mathfrak{N}(\mathfrak{S})$ — *a neighborhood of the set* $\mathfrak{S}$, $\mathfrak{S} \subset \mathfrak{N}(\mathfrak{S})$

$\mathfrak{N}_\varepsilon(\mathfrak{S})$ — *the* ε*-neighborhood of the set* $\mathfrak{S}$,

$$\mathfrak{N}_\varepsilon(\mathfrak{S}) = \{\mathbf{S} : d(\mathbf{S}, \mathfrak{S}) < \varepsilon\} \supset \mathfrak{S}$$

$\mathfrak{N}_\eta[\mathfrak{S}(t)]$ — *the* η*−neighborhood of the set* $\mathfrak{S}(t)$ *at* $t \in \mathfrak{T}$,

$$\mathfrak{N}_\eta[\mathfrak{S}(t)] = \{\mathbf{S} : \mathbf{S} \in \mathfrak{R}^n, \ d[\mathbf{S}, \mathfrak{S}(t)] < \eta\}$$

$\mathfrak{P}(\mathfrak{S})$ — *the power set of the set* $\mathfrak{S}$, which is denoted also by $2^{\mathfrak{S}}$, is the set of all subsets of a set $\mathfrak{S}$

$\mathfrak{S}(.): \ \mathfrak{T} \to \mathfrak{P}(\mathfrak{S})$ — a *set-valued function*

$\mathfrak{S}(t)$ — *the set value of the set-valued function* $\mathfrak{S}(.)$ at a moment $t \in \mathfrak{T}$, which is a *time-varying set* (at the moment $t \in \mathfrak{T}$)

$d[\mathbf{S}, \mathfrak{S}(t)]$ — *the distance* of $\mathbf{S}$ from $\mathfrak{S}(t)$ at the moment $t \in \mathfrak{T}$,

$$d[\mathbf{S}, \mathfrak{S}(t)] = inf\,[\|\mathbf{S} - \mathbf{W}\| : \mathbf{W} \in \mathfrak{S}(t)]$$

$d_m[\mathfrak{S}_1(t), \mathfrak{S}_2(t)]$ — *the inner set distance between the sets* $\mathfrak{S}_1(t)$ *and* $\mathfrak{S}_2(t)$ *at the moment* $t \in \mathfrak{T}$,

$$\begin{aligned} & d_m[\mathfrak{S}_1(t), \mathfrak{S}_2(t)] \\ &= inf\,\{d(\mathbf{X}, \mathbf{Y}) : \mathbf{X} \in \mathfrak{S}_1(t), \ \mathbf{Y} \in \mathfrak{S}_2(t)\} = \\ &= min \left\{ \begin{array}{c} inf\,\langle d[\mathbf{S}, \mathfrak{S}_1(t)] : \mathbf{S} \in \mathfrak{S}_2(t) \rangle, \\ inf\,\langle d[\mathbf{S}, \mathfrak{S}_2(t)] : \mathbf{S} \in \mathfrak{S}_1(t) \rangle \end{array} \right\} \end{aligned}$$

$d_M\left[\mathfrak{S}_1(t), \mathfrak{S}_2(t)\right]$ *the outer set distance between the sets* $\mathfrak{S}_1(t)$ *and* $\mathfrak{S}_2(t)$ *at the moment* $t \in \mathfrak{T}$,

$$d_M\left[\mathfrak{S}_1(t), \mathfrak{S}_2(t)\right] =$$
$$= sup\left\{d\left(\mathbf{X}, \mathbf{Y}\right) : \mathbf{X} \in \mathfrak{S}_1(t),\ \mathbf{Y} \in \mathfrak{S}_2(t)\right\} =$$
$$= max\left\{\begin{array}{l} sup\left\langle d\left[\mathbf{S}, \mathfrak{S}_1(t)\right] : \mathbf{S} \in \mathfrak{S}_2(t)\right\rangle, \\ sup\left\langle d\left[\mathbf{S}, \mathfrak{S}_2(t)\right] : \mathbf{S} \in \mathfrak{S}_1(t)\right\rangle \end{array}\right\}$$

$\partial\mathfrak{V}_\xi\left(t;\tau\right)$ *the boundary of the set* $\mathfrak{V}_\xi\left(t;\tau\right)$

$\mathfrak{V}_\xi\left(t;\tau\right) \subseteq \mathfrak{R}^n$ *the set associated with the function* $V(.) : \mathfrak{T}_\tau \times \mathfrak{R}^n \longrightarrow \mathfrak{R}$ *and with its set value determined by* $\xi \in \mathfrak{R}^+$, or $\xi = \infty$, at $t \in \mathfrak{T}_\tau$ so that it is the largest open connected neighborhood $\mathfrak{V}_\xi\left(t;\tau\right)$ of the origin $\mathbf{0}_n$ on $\mathfrak{T}_\tau$ such that

1. If $\xi \in \mathfrak{R}^+$, then the boundary $\partial\mathfrak{V}_\xi\left(t;\tau\right)$ of $\mathfrak{V}_\xi\left(t;\tau\right)$ is the hypersurface $V\left(t, \mathbf{S}\right) = \xi$,

$$\xi \in \mathfrak{R}^+ \Longrightarrow \partial\mathfrak{V}_\xi\left(t;\tau\right) = \left\{\mathbf{S} : V\left(t, \mathbf{S}\right) = \xi\right\},\ \forall t \in \mathfrak{T}_\tau,$$
$$V\left(t, \mathbf{S}\right) = \xi \Longleftrightarrow \mathbf{S} \in \partial\mathfrak{V}_\xi\left(t;\tau\right)$$

2. The intersection $\mathfrak{V}_\xi\left(\mathfrak{T}_\tau\right)$ of $\mathfrak{V}_\xi\left(t;\tau\right)$ on $\mathfrak{T}_\tau$,

$$\mathfrak{V}_\xi\left(\mathfrak{T}_\tau\right) = \cap\left[\mathfrak{V}_\xi\left(t;\tau\right) : t \in \mathfrak{T}_\tau\right],$$

is also a neighborhood of the origin $\mathbf{0}_n$,

$$\exists\zeta \in \mathfrak{R}^+ \Longrightarrow \mathfrak{B}_\zeta \subset \mathfrak{V}_\xi\left(\mathfrak{T}_\tau\right)$$

and

3. $V\left(t, \mathbf{S}\right)$ is strictly less than ξ on $\mathfrak{V}_\xi\left(t;\tau\right)$ at every instant $t \in \mathfrak{T}_\tau$,

$$V\left(t, \mathbf{S}\right) < \xi,\ \forall\left(\mathbf{S},\ t\right) \in \mathfrak{V}_\xi\left(t;\tau\right) \times \mathfrak{T}_\tau$$

where the expression "on $\mathfrak{T}_\tau$" and the notation "; τ" are to be omitted if and only if $\mathfrak{T}_\tau = \mathfrak{T}$

A.6 Units

1_x *the unit of a physical variable* x

1_t *the time unit of the reference time axis* T, if second s is the *time* unit then $1_t = s$

Appendix B

Dini derivatives

B.1 Definitions of derivatives of functions

The term *function* determines a real-valued function $f(.) : \mathfrak{T} \times \mathfrak{R}^n \rightarrow \mathfrak{R}$.

Note 314 *The notation t^- designates mathematically the left-hand side of the moment t,*

$$t^- = lim(t - \varepsilon : \varepsilon \longrightarrow 0, \varepsilon > 0). \tag{B.1}$$

The notation t^+ denotes mathematically the right-hand side of the moment t,

$$t^+ = lim(t + \varepsilon : \varepsilon \longrightarrow 0, \varepsilon > 0). \tag{B.2}$$

Note 315 *From the physical point of view, the left-hand side of the moment t is the beginning of the (infinitesimal) duration of the moment t. The right-hand side of the moment t represents the end of the (infinitesimal) duration of the moment t.*

Let $t^* \in \mathfrak{R} \cup \{-\infty\} \cup \{\infty\}$. Let $\mathfrak{T}^-_{\longrightarrow t^*}$ be a set composed of a sequence t_k converging to t^* from the left as $k \longrightarrow \infty$ if $t^* > -\infty$, ($t^* = \infty$ is permitted),

$$\mathfrak{T}^-_{\longrightarrow t^*} = \left\{t_k : t_k \in \mathfrak{T},\ t_k < t^*,\ k = 0, 1, 2, \cdots,\ k \longrightarrow \infty \Longrightarrow\ t_k \longrightarrow t^{*-}\right\},$$
$$t^* \in]-\infty, \infty],$$

and, analogously, let $\mathfrak{T}^+_{\longrightarrow t^*}$ be a set composed of a sequence t_k converging to t^* from the right as $k \longrightarrow \infty$ if $t^* < \infty$, ($t^* = -\infty$ is permitted),

$$\mathfrak{T}^+_{\longrightarrow t^*} = \left\{t_k : t_k \in \mathfrak{T},\ t_k > t^*,\ k = 0, 1, 2, \cdots,\ k \longrightarrow \infty \Longrightarrow\ t_k \longrightarrow t^{*+}\right\},$$
$$t^* \in [-\infty, \infty[.$$

We accept the following definition from [247, Definition 3.5, p. 54].

Definition 316 *a) A number $\alpha \in \mathfrak{R}$ is **the left, [right], partial limit** of a function $f(.)$, $f(.) : \mathfrak{R} \longrightarrow \mathfrak{R}$, over a sequence set $\mathfrak{T}^-_{\longrightarrow t^*}$, $\left[\mathfrak{T}^+_{\longrightarrow t^*}\right]$, if, and only if, for every $\varepsilon \in \mathfrak{R}^+$ there is an integer N such that $k > N$ implies $|f(t_k) - \alpha| < \varepsilon$.*

b) The symbol $\alpha = -\infty$, $[\alpha = \infty]$, is the left [right] partial limit of the function $f(.)$ over a sequence set $\mathfrak{T}^-_{\longrightarrow t^*}$, $\left(\mathfrak{T}^+_{\longrightarrow t^*}\right)$, if, and only if, for every $\varepsilon \in \mathfrak{R}^+$ there is an integer N such that, respectively, $k > N$ implies $f(t_k) < -\varepsilon^{-1}$, $\left[f(t_k) > \varepsilon^{-1}\right]$.

DOI: 10.1201/9781003744528-B

c) The greatest [the smallest] partial limit of the function $f(.)$ over all the sequence sets $\mathfrak{T}^{-}_{\longrightarrow t^*}$ is its left upper [left lower] limit at $t = t^*$, which is denoted, respectively, by

$$lim\ sup\ \left[f(t) : t \longrightarrow t^{*-}\right] = \overline{lim}\ \left[f(t) : t \longrightarrow t^{*-}\right],$$

$$lim\ inf\ \left[f(t) : t \longrightarrow t^{*-}\right] = \underline{lim}\ \left[f(t) : t \longrightarrow t^{*-}\right].$$

d) The greatest [the smallest] partial limit of the function $f(.)$ over all the sequence sets $\mathfrak{T}^{+}_{\longrightarrow t^*}$ is its right upper [right lower] limit at $t = t^*$, which is denoted, respectively, by

$$lim\ sup\ \left[f(t) : t \longrightarrow t^{*+}\right] = \overline{lim}\ \left[f(t) : t \longrightarrow t^{*+}\right],$$

$$lim\ inf\ \left[f(t) : t \longrightarrow t^{*+}\right] = \underline{lim}\ \left[f(t) : t \longrightarrow t^{*+}\right].$$

Example 317 *Let $f(0) = 0$ and $f(t) = 5sign\left[7sin\left(t^{-4}\right)\right]$ for $t \neq 0$. Then:*

$$\overline{lim}\ \left[f(t) : t \longrightarrow 0^-\right] = \overline{lim}\ \left[f(t) : t \longrightarrow 0^+\right] = 5,$$
$$\underline{lim}\ \left[f(t) : t \longrightarrow 0^-\right] = \underline{lim}\ \left[f(t) : t \longrightarrow 0^+\right] = -5.$$

The definition of Dini derivatives follows by referring to [247], [369], [534].

Definition 318 ***Forward-time derivatives***

Let $V(.)$ be a continuous scalar function of $(t, \mathbf{z}) \in \mathfrak{T} \times \mathfrak{R}^K$, $V(t, \mathbf{z}) \in \mathfrak{C}\left(\mathfrak{T} \times \mathfrak{R}^K\right)$. Let $\mathcal{Z}\left(.; t_0; \mathbf{z}; \mathbf{i}\right)$ be a motion of a dynamical system through $\mathbf{z}$ at $t = t_0$, $\mathcal{Z}\left(t; t_0; \mathbf{z}; \mathbf{i}\right) \equiv \mathbf{z}(t)$, $\mathcal{Z}\left(t_0; t_0; \mathbf{z}; \mathbf{i}\right) \equiv \mathbf{z}$. Then:

- The forward-time upper left Dini derivative $D_l^+ V(t, \mathbf{z})$ of $V(.)$ along $\mathcal{Z}\left(.; t_0; \mathbf{z}; \mathbf{i}\right)$ at $(t, \mathbf{z})$ is

$$D_l^+ V(t, \mathbf{z}) = lim\ sup\left\{\frac{V\left[t + \theta, \mathcal{Z}\left(t + \theta; t; \mathbf{z}; \mathbf{i}\right)\right] - V\left(t, \mathbf{z}\right)}{\theta} : \theta \longrightarrow 0^-\right\}.$$

- The forward-time lower left Dini derivative $D_l^- V(t, \mathbf{z})$ of $V(.)$ along $\mathcal{Z}\left(.; t_0; \mathbf{z}; \mathbf{i}\right)$ at $(t, \mathbf{z})$ is

$$D_l^- V(t, \mathbf{z}) = lim\ inf\left\{\frac{V\left[t + \theta, \mathcal{Z}\left(t + \theta; t; \mathbf{z}; \mathbf{i}\right)\right] - V\left(t, \mathbf{z}\right)}{\theta} : \theta \longrightarrow 0^-\right\}.$$

- The forward-time upper right Dini derivative $D_r^+ V(t, \mathbf{z})$ of $V(.)$ along $\mathcal{Z}\left(.; t_0; \mathbf{z}; \mathbf{i}\right)$ at $(t, \mathbf{z})$ is

$$D_r^+ V(t, \mathbf{z}) = lim\ sup\left\{\frac{V\left[t + \theta, \mathcal{Z}\left(t + \theta; t; \mathbf{z}; \mathbf{i}\right)\right] - V\left(t, \mathbf{z}\right)}{\theta} : \theta \longrightarrow 0^+\right\}.$$

It is denoted throughout the book simpler by $D^+ V(t, \mathbf{z})$:

$$D^+ V(t, \mathbf{z}) = D_r^+ V(t, \mathbf{z}) = lim\ sup\left\{\frac{V\left[t + \theta, \mathcal{Z}\left(t + \theta; t; \mathbf{z}; \mathbf{i}\right)\right] - V\left(t, \mathbf{z}\right)}{\theta} : \theta \longrightarrow 0^+\right\}.$$

- The forward-time lower right Dini derivative $D_r^- V(t, \mathbf{z})$ of $V(.)$ along $\mathcal{Z}\left(.; t_0; \mathbf{z}; \mathbf{i}\right)$ at $(t, \mathbf{z})$ is

$$D_r^- V(t, \mathbf{z}) = lim\ inf\left\{\frac{V\left[t + \theta, \mathcal{Z}\left(t + \theta; t; \mathbf{z}; \mathbf{i}\right)\right] - V\left(t, \mathbf{z}\right)}{\theta} : \theta \longrightarrow 0^+\right\}.$$

- *The function* $V(.)$ *has the forward-time left Dini derivative along the motion* $\mathcal{Z}(.;t_0;\mathbf{z};\mathbf{i})$ *at* $(t,\mathbf{z})$, *which is denoted by* $D_lV(t,\mathbf{z})$ *if, and only if,* $D_l^+V(t,\mathbf{z}) = D_l^-V(t,\mathbf{z})$ *and then*

$$D_lV(t,\mathbf{z}) = D_l^+V(t,\mathbf{z}) = D_l^-V(t,\mathbf{z}).$$

- *The function* $V(.)$ *has the forward-time right Dini derivative along the motion* $\mathcal{Z}(.;t_0;\mathbf{z};\mathbf{i})$ *at* $(t,\mathbf{z})$, *which is denoted by* $D_rV(t,\mathbf{z})$ *if, and only if,* $D_r^+V(t,\mathbf{z}) = D_r^-V(t,\mathbf{z})$ *and then*

$$D_rV(t,\mathbf{z}) = D_r^+V(t,\mathbf{z}) = D_r^-V(t,\mathbf{z}).$$

- *The function* $V(.)$ *has the forward-time upper Dini derivative along the motion* $\mathcal{Z}(.;t_0;\mathbf{z};\mathbf{i})$ *at* $(t,\mathbf{z})$, *which is denoted by* $D^+V(t,\mathbf{z})$ *if, and only if,* $D_r^+V(t,\mathbf{z}) = D_r^-V(t,\mathbf{z})$ *and then*

$$D_rV(t,\mathbf{z}) = D_r^+V(t,\mathbf{z}) = D_r^-V(t,\mathbf{z}).$$

- *The function* $V(.)$ *has the forward-time Eulerian derivative along the motion* $\mathcal{Z}(.;t_0;\mathbf{z};\mathbf{i})$ *at* $(t,\mathbf{z})$, *which is denoted by* $d_fV(t,\mathbf{z})/dt$ *or by* $V_f^{(1)}(t,\mathbf{z})$ *or simpler by* $dV(t,\mathbf{z})/dt$ *or by* $V^{(1)}(t,\mathbf{z})$ *if, and only if, it has both* $D_lV(t,\mathbf{z})$ *and* $D_rV(t,\mathbf{z})$ *such that* $D_lV(t,\mathbf{z}) = D_rV(t,\mathbf{z})$ *and then*

$$\frac{dV(t,\mathbf{z})}{dt} = V^{(1)}(t,\mathbf{z}) = \frac{d_fV(t,\mathbf{z})}{dt} = V_f^{(1)}(t,\mathbf{z}) = D_lV(t,\mathbf{z}) = D_rV(t,\mathbf{z}).$$

$D^*V(t,\mathbf{z})$ *means that either* $D^-V(t,\mathbf{z})$ *or* $D^+V(t,\mathbf{z})$ *can be equally used. Analogously,* $D_*V(t,\mathbf{z})$ *signifies that either* $D_-V(t,\mathbf{z})$ *or* $D_+V(t,\mathbf{z})$ *can be equally used.*

Note 319 *From the physical point of view, only the forward-time derivatives have the sense [237], [236] so that* $V_f^{(1)}(t,\mathbf{z})$ *is simply denoted by* $V^{(1)}(t,\mathbf{z})$,

$$\frac{dv(t,\mathbf{z})}{dt} = V^{(1)}(t,\mathbf{z}) = \frac{d_fV(t,\mathbf{z})}{dt} = V_f^{(1)}(t,\mathbf{z}) = D_lV(t,\mathbf{z}) = D_rV(t,\mathbf{z}). \quad \text{(B.3)}$$

From the mathematical point of view we can also introduce the backward-time derivatives as follows.

Definition 320 ***Backward-time derivatives***

Let $V(.)$ *be a continuous scalar function of* $(t,\mathbf{z}) \in \mathfrak{T} \times \mathfrak{R}^K$, $V(t,\mathbf{z}) \in \mathfrak{C}(\mathfrak{T} \times \mathfrak{R}^K)$. *Let* $\mathcal{Z}(.;t_0;\mathbf{z};\mathbf{i})$ *be a motion of a dynamical system through* $\mathbf{z}$ *at* $t = t_0, \mathcal{Z}(t;t_0;\mathbf{z};\mathbf{i}) \equiv \mathbf{z}(t)$, $\mathcal{Z}(t_0;t_0;\mathbf{z};\mathbf{i}) \equiv \mathbf{z}$. *Then:*

- *The backward-time upper left Dini derivative* ${}^-D_lv(t,\mathbf{z})$ *of* $V(.)$ *along* $\mathcal{Z}(.;t_0;\mathbf{z};\mathbf{i})$ *at* $(t,\mathbf{z})$ *is*

$${}^+D_lv(t,\mathbf{z}) = lim\ sup\left\{\frac{V[t-\theta,\mathcal{Z}(t-\theta;t_0;\mathbf{z};\mathbf{i})] - V(t,\mathbf{z})}{\theta} : \theta \longrightarrow 0^-\right\}.$$

- *The backward-time lower left Dini derivative* ${}_-D_lv(t,\mathbf{z})$ *of* $V(.)$ *along* $\mathcal{Z}(.;t_0;\mathbf{z};\mathbf{i})$ *at* $(t,\mathbf{z})$ *is*

$${}_-D_lv(t,\mathbf{z}) = lim\ inf\left\{\frac{V[t-\theta,\mathcal{Z}(t-\theta;t_0;\mathbf{z};\mathbf{i})] - V(t,\mathbf{z})}{\theta} : \theta \longrightarrow 0^-\right\}.$$

- *The backward-time upper right Dini derivative* ${}^+D_rv(t,\mathbf{z})$ *of* $V(.)$ *along* $\mathcal{Z}(.;t_0;\mathbf{z};\mathbf{i})$ *at* $(t,\mathbf{z})$ *is*

$${}^+D_rv(t,\mathbf{z}) = lim\ sup\left\{\frac{V[t-\theta,\mathcal{Z}(t-\theta;t_0;\mathbf{z};\mathbf{i})] - V(t,\mathbf{z})}{\theta} : \theta \longrightarrow 0^+\right\}.$$

- The backward-time lower right Dini derivative $_{-}D_r v(t,\mathbf{z})$ *of* $V(.)$ *along the motion* $\mathcal{Z}(.;t_0;\mathbf{z};\mathbf{i})$ *at* $(t,\mathbf{z})$ *is*

$$_{-}D_r v(t,\mathbf{z}) = lim\ inf\left\{\frac{V\left[t-\theta,\mathcal{Z}\left(t-\theta;t_0;\mathbf{z};\mathbf{i}\right)\right]-V\left(t,\mathbf{z}\right)}{\theta}:\theta\longrightarrow 0^+\right\}.$$

- The function $V(.)$ *has the backward-time left Dini derivative along* $\mathcal{Z}(.;t_0;\mathbf{z};\mathbf{i})$ *at* $(t,\mathbf{z})$*, which is denoted by* $_bD_lV(t,\mathbf{z})$ *if, and only if,* $^{-}D_lV(t,\mathbf{z}) = {}_{-}D_lV(t,\mathbf{z})$ *and then*

$$_bD_lV(t,\mathbf{z}) = {}^{+}D_lV(t,\mathbf{z}) = {}_{-}D_lV(t,\mathbf{z}).$$

- The function $V(.)$ *has the backward-time right Dini derivative along* $\mathcal{Z}(.;t_0;\mathbf{z};\mathbf{i})$ *at* $(t,\mathbf{z})$*, which is denoted by* $_bD_rV(t,\mathbf{z})$ *if, and only if,* $^{+}D_rV(t,\mathbf{z}) = {}_{-}D_rV(t,\mathbf{z})$ *and then*

$$_bD_rV(t,\mathbf{z}) = {}^{+}D_rV(t,\mathbf{z}) = {}_{-}D_rV(t,\mathbf{z}).$$

- The function $V(.)$ *has the backward-time Eulerian derivative along* $\mathcal{Z}(.;t_0;\mathbf{z};\mathbf{i})$ *at* $(t,\mathbf{z})$*, which is denoted by* $d_bV(t,\mathbf{z})/dt$ *or by* $V_b^{(1)}(t,\mathbf{z})$ *if, and only if, it has both* $_lDV(t,\mathbf{z})$ *and* $_rDV(t,\mathbf{z})$ *such that* $_bD_lV(t,\mathbf{z}) = {}_bD_rV(t,\mathbf{z})$ *and then*

$$\frac{d_bV(t,\mathbf{z})}{dt} = V_b^{(1)}(t,\mathbf{z}) = {}_bD_lV(t,\mathbf{z}) = {}_bD_rV(t,\mathbf{z}).$$

$^{*}Dv(t,\mathbf{z})$ *means that either* $^{-}Dv(t,\mathbf{z})$ *or* $^{+}Dv(t,\mathbf{z})$ *can be equally used. Analogously,* $_{*}DV(t,\mathbf{z})$ *signifies that either* $_{-}DV(t,\mathbf{z})$ *or* $_{+}DV(t,\mathbf{z})$ *can be equally used.*

B.2 Properties: Dini derivatives of functions

Definition 318, Definition 320 and the following relationships:

$$\sup f(.) = -\inf\left[-f(.)\right],\ \inf f(.) = -\sup\left[-f(.)\right]$$

yield:

$$\begin{gathered}
^{-}DV(t,\mathbf{z}) = -D_{+}V(t,\mathbf{z}),\ {}_{-}DV(t,\mathbf{z}) = -D^{+}V(t,\mathbf{z}),\\
^{+}DV(t,\mathbf{z}) = -D_{-}V(t,\mathbf{z}),\ {}_{+}DV(t,\mathbf{z}) = -D^{-}V(t,\mathbf{z}),\\
_{l}DV(t,\mathbf{z}) = -D_rV(t,\mathbf{z}),\ {}_{r}DV(t,\mathbf{z}) = -D_lV(t,\mathbf{z}),\\
V_b^{(1)}(t,\mathbf{z}) = -V_f^{(1)}(t,\mathbf{z}).
\end{gathered}$$

Let $\delta_{ij} = 1$ if $i = j$, $\delta_{ij} = 0$ if $i \neq j$, $\Delta_j\mathbf{z} = \Delta s\left[\delta_{1j}\ \ \delta_{2j}....\delta_{kj}\right]^T$, and, by referring to [170, p. 84],

$$D_t^{+}V(t,\mathbf{z}) = \lim\ \sup\left\{\frac{V\left(t+\theta,\mathbf{z}\right)-V\left(t,\mathbf{z}\right)}{\theta}:\theta\longrightarrow 0^+\right\},$$

$$D_s^{+}V(t,\mathbf{z}) = \sum_{j=1}^{j=k}\lim\ \sup\left\{\frac{V\left(t,\mathbf{z}+\Delta_j\mathbf{z}\right)-V\left(t,\mathbf{z}\right)}{\Delta s}\frac{\Delta s}{\theta}:\theta\longrightarrow 0^+,\ \Delta s\longrightarrow 0\right\}.$$

These definitions and the fact that

$$\sup\left[\ f_1(t)+f_2(t)\right] \leq \sup\ f_1(t)+\sup\ f_2(t)$$

imply, [170, p. 84],

$$D_t^+ V(t, \mathbf{z}) \leq D_t^+ V(t, \mathbf{z}) + D_s^+ V(t, \mathbf{z}). \tag{B.4}$$

Let

$\mathfrak{I}^c$ any connected nonempty compact subset of the output integral space $\mathfrak{I} \subseteq \mathfrak{T} \times \mathfrak{R}^K$

$\mathfrak{L}^c$ any connected nonempty compact subset of $\mathfrak{T}$

$\mathfrak{T}^c$ any connected nonempty compact subset of $\mathfrak{T}$

$\mathfrak{C}_Z(\mathfrak{T}^c \times \mathfrak{L}^c \times \mathfrak{I}^c)$ be the family of the motions $\mathcal{Z}(t; t_0; \mathbf{z}; \mathbf{i})$ continuous in $\mathbf{z} \in \mathfrak{R}^K, \forall (t; t_0) \in \mathfrak{T}^c \times \mathfrak{L}^c$

$\mathfrak{C}_t(\mathfrak{T}^c \times \mathfrak{L}^c \times \mathfrak{I}^c)$ be the family of the motions $\mathcal{Z}(t; t_0; \mathbf{z}; \mathbf{i})$ continuous in time $t \in \mathfrak{T}^c$, $\forall (t_0; \mathbf{z}_0) \in \mathfrak{L}^c \times \mathfrak{I}^c$

$In\mathfrak{I}$ be the interior of $\mathfrak{I}$.

The following result is presented in [247, Lemma 4.15, p. 114] by noting that there is a typographical error in the proof. In the fourth line of the proof in [247, Lemma 4.15, p. 114] there should have been written Lemma 4.14 instead of Lemma 4.15. The complete proof is in [322, 6.2 Lemma, p. 30].

Lemma 321 ***The fundamental lemma on Dini derivatives***

Let $\mathcal{Z}(.; t_0; \mathbf{z}_0; \mathbf{i})$ be a motion of a dynamical system through $\mathbf{z}_0$ at $t = t_0$, $\mathcal{Z}(t; t_0; \mathbf{z}_0; \mathbf{i}) \equiv \mathbf{z}(t)$, $\mathcal{Z}(t_0; t_0; \mathbf{z}_0; \mathbf{i}) \equiv \mathbf{z}(t_0) \equiv \mathbf{z}_0$. Let $\mathcal{Z}(t; t_0; \mathbf{z}_0; \mathbf{i})$ be continuous in time $t \in \mathfrak{T}^c$, $\forall (t_0; \mathbf{z}_0) \in \mathfrak{L}^c \times \mathfrak{I}^c$, i.e., $\mathcal{Z}(t; t_0; \mathbf{z}_0; \mathbf{i}) \in \mathfrak{C}_t(\mathfrak{T}^c \times \mathfrak{L}^c \times \mathfrak{I}^c \times \mathcal{J}^i)$, and be continuous in $\mathbf{z}_0 \in \mathfrak{I}^c, \forall (t; t_0, \mathbf{i}) \in \mathfrak{T}^c \times \mathfrak{L}^c \times \mathcal{J}^i$; i.e., $\mathcal{Z}(t; t_0; \mathbf{z}_0; \mathbf{i}) \in \mathfrak{C}_s(\mathfrak{T}^c \times \mathfrak{L}^c \times \mathfrak{I}^c \times \mathcal{J}^i)$. Let $V(t, \mathbf{z}) \in \mathfrak{C}(\mathfrak{I})$. If i) or ii) holds,

i) $D_+ V(t, \mathbf{z}) \leq 0, \forall (t; \mathbf{z}) \in \mathfrak{T}^c \times In\mathfrak{I}$,

ii) $D^+ V(t, \mathbf{z}) \leq 0, \forall (t; \mathbf{z}) \in \mathfrak{T}^c \times In\mathfrak{I}$,

then

a) $V(.)$ is non-increasing on $[t_0, t_f[$, where t_f is the first moment when $\mathcal{Z}(t; t_0; \mathbf{z}_0; \mathbf{i})$ is not in $In\mathfrak{I}$,

$$\mathcal{Z}(t; t_0; \mathbf{z}_0; \mathbf{i}) \in In\mathfrak{I}, \ \forall t \in [t_0, t_f[, \text{ and } \mathcal{Z}(t_f; t_0; \mathbf{z}_0) \notin In\mathfrak{I},$$

b) $V(.)$ is differentiable almost everywhere on $[t_0, t_f[$,

and

c) either

$$1. \quad \int_{t_0}^{t} D_+ V[\sigma, \mathcal{Z}(\sigma; t_0; \mathbf{z}_0)] \geq V[t, \mathcal{Z}(t; t_0; \mathbf{z}_0; \mathbf{i})] - V(t_0; \mathbf{z}_0), \ \forall t \in [t_0, t_f[,$$

$$2. \quad \int_{t_0}^{t} D^+ V[\sigma, \mathcal{Z}(\sigma; t_0; \mathbf{z}_0)] \geq V[t, \mathcal{Z}(t; t_0; \mathbf{z}_0; \mathbf{i})] - V(t_0; \mathbf{z}_0), \ \forall t \in [t_0, t_f[,$$

respectively.

B.3 Dini derivatives of set-valued functions

The mapping $\mathfrak{S}(.) : \mathfrak{T} \times 2^{\mathfrak{R}^n} \to 2^{\mathfrak{R}^n}$ determines a set-valued function $\mathfrak{S}(.)$, the set value of which is a *time*-varying set $\mathfrak{S}(t)$ at a moment $t \in \mathfrak{T}$.

We accept the following definition by referring to Definition 316:

Definition 322 *a) A set $\mathfrak{A} \in 2^{\mathfrak{R}^n}$ is **the left, [right], partial limit set** of a set-valued function $\mathfrak{S}(.)$, $\mathfrak{S}(.) : \mathfrak{T} \times 2^{\mathfrak{R}^n} \to 2^{\mathfrak{R}^n}$, over a sequence set $\mathfrak{T}^-_{\to t^*}$, $\left[\mathfrak{T}^+_{\to t^*}\right]$, if, and only if, for every $\mathfrak{E} \in 2^{\mathfrak{R}^n}$ there is an integer N such that $k > N$ implies $\mathfrak{S}(t_k) \setminus \mathfrak{A} \subset \mathfrak{E}$.*

b) A set $\mathfrak{A} \in 2^{\mathfrak{R}^n}$ *is* ***the left, [right], partial limit set*** *of a set-valued function* $\mathfrak{S}(.)$, $\mathfrak{S}(.) : \mathfrak{T} \times 2^{\mathfrak{R}^n} \rightarrow 2^{\mathfrak{R}^n}$, *over a sequence set* $\mathfrak{T}^-_{\longrightarrow t^*}$, $\left(\mathfrak{T}^+_{\longrightarrow t^*}\right)$, *if, and only if, for every* $\mathfrak{E} \in 2^{\mathfrak{R}^n}$ *there is an integer* N *such that, respectively,* $k > N$ *implies* $\mathfrak{S}(t_k) \subset \mathfrak{E}$, $[\mathfrak{S}(t_k) \supset \mathfrak{E}]$.

c) The greatest [the smallest] partial limit set $\overline{\overline{\mathfrak{A}}}_l$ $(\underline{\underline{\mathfrak{A}}}_l)$ *of a set-valued function* $\mathfrak{S}(.)$ *over all the sequence sets* $\mathfrak{T}^-_{\longrightarrow t^*}$ *is its left upper [left lower] limit set at* $t = t^*$, *which is denoted, respectively, by*

$$lim\ sup\ \left[\mathfrak{S}(t) : t \longrightarrow t^{*-}\right] = \overline{lim}\ \left[\mathfrak{S}(t) : t \longrightarrow t^{*-}\right] = \overline{\overline{\mathfrak{A}}}_l,$$

$$lim\ inf\ \left[\mathfrak{S}(t) : t \longrightarrow t^{*-}\right] = \underline{lim}\ \left[\mathfrak{S}(t) : t \longrightarrow t^{*-}\right] = \underline{\underline{\mathfrak{A}}}_l.$$

d) The greatest [the smallest] partial limit set $\overline{\overline{\mathfrak{A}}}_r$ $(\underline{\underline{\mathfrak{A}}}_r)$ *of a set-valued* $\mathfrak{S}(.)$ *over all the sequence sets* $\mathfrak{T}^+_{\longrightarrow t^*}$ *is its right upper [right lower] limit at* $t = t^*$, *which is denoted, respectively, by*

$$lim\ sup\ \left[\mathfrak{S}(t) : t \longrightarrow t^{*+}\right] = \overline{lim}\ \left[\mathfrak{S}(t) : t \longrightarrow t^{*+}\right] = \overline{\overline{\mathfrak{A}}}_r,$$

$$lim\ inf\ \left[\mathfrak{S}\,(t) : t \longrightarrow t^{*+}\right] = \underline{lim}\ \left[f(t) : t \longrightarrow t^{*+}\right] = \underline{\underline{\mathfrak{A}}}_r.$$

The definition of Dini derivatives of a set-valued function follows by referring to Definition 318 (Section B.1):

Definition 323 ***Forward-time derivatives of a set-valued function***

Let $\mathfrak{S}(.)$ *be a continuous set-valued function of* $(t, \mathfrak{B}) \in \mathfrak{T} \times 2^{\mathfrak{R}^n}$, $\mathfrak{S}(t, \mathfrak{B}) \in \mathfrak{C}\left(\mathfrak{T} \times 2^{\mathfrak{R}^n}\right)$. *Let* $\mathfrak{S}\,(.; t_0; \mathfrak{B})$ *be a time variation of* $\mathfrak{S}(t, \mathfrak{B})$ *passing through set* $\mathfrak{B}$ *at* $t = t_0$, $\mathfrak{S}\,(t_0; t_0; \mathfrak{B}) \equiv \mathfrak{B}$. *Then:*

- The forward-time upper left Dini derivative $D_l^+\mathfrak{S}(t; \mathfrak{B})$ *of* $\mathfrak{S}(.)$ *along* $\mathfrak{S}\,(.; t_0; \mathfrak{B})$ *at* $(t, \mathfrak{B})$ *is*

$$D_l^+\mathfrak{S}(t; \mathfrak{B}) = lim\ sup \left\{ \frac{\mathfrak{S}\left[t + \theta, \mathfrak{S}\left(t + \theta; t; \mathfrak{B}\right)\right] - \mathfrak{S}\left(t, \mathfrak{B}\right)}{\theta} : \theta \longrightarrow 0^- \right\}.$$

- The forward-time lower left Dini derivative $D_l^-\mathfrak{S}(t, \mathfrak{B}$ *of* $\mathfrak{S}.)$ *along* $\mathfrak{S}\,(.; t_0; \mathfrak{B})$ *at* $(t, \mathfrak{B})$ *is*

$$D_l^-\mathfrak{S}(t, \mathfrak{B}) = lim\ inf \left\{ \frac{\mathfrak{S}\left[t + \theta, \mathfrak{S}\left(t + 0; t; \mathfrak{B}\right)\right] - \mathfrak{S}\left(t, \mathfrak{B}\right)}{\theta} : \theta \longrightarrow 0 \right\}.$$

- The forward-time upper right Dini derivative $D_r^+\mathfrak{S}(t, \mathfrak{B}$ *of* $\mathfrak{S}(.)$ *along* $\mathfrak{S}\,(.; t_0; \mathfrak{B})$ *at* $(t, \mathfrak{B})$ *is*

$$D_r^+\mathfrak{S}(t, \mathfrak{B}) = lim\ sup \left\{ \frac{\mathfrak{S}\left[t + \theta, \mathfrak{S}\left(t + \theta; t; \mathfrak{B}\right)\right] - \mathfrak{S}\left(t, \mathfrak{B}\right)}{\theta} : \theta \longrightarrow 0^+ \right\}.$$

It is denoted throughout the book simpler by $D^+\mathfrak{S}(t, \mathfrak{B})$:

$$D^+\mathfrak{S}(t, \mathfrak{B}) = D_r^+\mathfrak{S}(t, \mathfrak{B}) = lim\ sup \left\{ \frac{\mathfrak{S}\left[t + \theta, \mathfrak{S}\left(t + \theta; t; \mathfrak{B}\right)\right] - \mathfrak{S}\left(t, \mathfrak{B}\right)}{\theta} : \theta \longrightarrow 0^+ \right\}.$$

- The forward-time lower right Dini derivative $D_+^-\mathfrak{S}(t, \mathfrak{B}$ *of* $\mathfrak{S}(.)$ *along* $\mathfrak{S}\,(.; t_0; \mathfrak{B})$ *at* $(t, \mathfrak{B})$ *is*

$$D_+^-\mathfrak{S}(t, \mathfrak{B}) = lim\ inf \left\{ \frac{\mathfrak{S}\left[t + \theta, \mathfrak{S}\left(t + \theta; t; \mathfrak{B}\right)\right] - \mathfrak{S}\left(t, \mathfrak{B}\right)}{\theta} : \theta \longrightarrow 0^+ \right\}.$$

- *The set-valued function* $\mathfrak{S}(.)$ *has the forward-time upper Dini derivative along the motion* $\mathfrak{S}(.;t_0;\mathfrak{B})$ *at* $(t,\mathfrak{B})$, *which is denoted by* $D^+\mathfrak{S}(t,\mathfrak{B})$ *if, and only if,* $D_l^+\mathfrak{S}(t,\mathfrak{B}) = D_r^+\mathfrak{S}(t,\mathfrak{B})$ *and then*

$$D^+\mathfrak{S}(t,\mathfrak{B}) = D_l^+\mathfrak{S}(t,\mathfrak{B}) = D_r^+\mathfrak{S}(t,\mathfrak{B}).$$

- *The set-valued function* $\mathfrak{S}(.)$ *has the forward-time left Dini derivative along the motion* $\mathfrak{S}(.;t_0;\mathfrak{B})$ *at* $(t,\mathfrak{B})$, *which is denoted by* $D_l\mathfrak{S}(t,\mathfrak{B})$ *if, and only if,* $D_l^+\mathfrak{S}(t,\mathfrak{B}) = D_l^-\mathfrak{S}(t,\mathfrak{B})$ *and then*

$$D_l\mathfrak{S}(t,\mathfrak{B}) = D_l^+\mathfrak{S}(t,\mathfrak{B}) = D_l^-\mathfrak{S}(t,\mathfrak{B}).$$

- *The set-valued function* $\mathfrak{S}(.)$ *has the forward-time right Dini derivative along the motion* $\mathfrak{S}(.;t_0;\mathfrak{B})$ *at* $(t,\mathfrak{B})$, *which is denoted by* $D_r\mathfrak{S}(t,\mathfrak{B})$ *if, and only if,* $D_r^+\mathfrak{S}(t,\mathfrak{B}) = D_r^-\mathfrak{S}(t,\mathfrak{B})$ *and then*

$$D_r\mathfrak{S}(t,\mathfrak{B}) = D_r^+\mathfrak{S}(t,\mathfrak{B}) = D_r^-\mathfrak{S}(t,\mathfrak{B}).$$

- *The set-valued function* $\mathfrak{S}(.)$ *has the forward-time Eulerian derivative, for short: Eulerian derivative, along the motion* $\mathfrak{S}(.;t_0;\mathfrak{B})$ *at* $(t,\mathfrak{B})$, *which is denoted by* $d_f\mathfrak{S}(t,\mathfrak{B})/dt$ *or by* $\mathfrak{S}^{(1)}(t,\mathfrak{B})$ *or simpler by* $d\mathfrak{S}(t,\mathfrak{B})/dt$ *if, and only if, it has both* $D_l\mathfrak{S}(t,\mathfrak{B})$ *and* $D_r\mathfrak{S}(t,\mathfrak{B})$ *such that* $D_l\mathfrak{S}(t,\mathfrak{B}) = D_r\mathfrak{S}(t,\mathfrak{B})$ *and then*

$$\frac{d_f\mathfrak{S}(t,\mathfrak{B})}{dt} = \mathfrak{S}_f^{(1)}(t,\mathfrak{B}) = D_l\mathfrak{S}(t,\mathfrak{B}) = D_r\mathfrak{S}(t,\mathfrak{B}) = \mathfrak{S}^{(1)}(t,\mathfrak{B}) = \frac{d\mathfrak{S}(t,\mathfrak{B})}{dt}. \quad \text{(B.5)}$$

$D^*\mathfrak{S}(t,\mathfrak{B})$ *means that either* $D^-\mathfrak{S}(t,\mathfrak{B})$ *or* $D^+\mathfrak{S}(t,\mathfrak{B})$ *can be equally used. Analogously,* $D_*\mathfrak{S}(t,\mathfrak{B})$ *signifies that either* $D_-\mathfrak{S}(t,\mathfrak{B})$ *or* $D_+\mathfrak{S}(t,\mathfrak{B})$ *can be equally used.*

Note 324 *From the physical point of view, only the forward-time derivatives have the sense [237], [236].*

From the mathematical point of view we can also introduce the backward-time derivatives as follows.

Definition 325 ***Backward-time derivatives of a set-valued function along motions***

Let $\mathfrak{S}(.)$ *be a continuous set-valued function of* $(t,\mathfrak{B}) \in \mathfrak{T} \times 2^{\mathfrak{R}^n}$, $\mathfrak{S}(t,\mathfrak{B}) \in \mathfrak{C}\left(\mathfrak{T} \times 2^{\mathfrak{R}^n}\right)$. *Let* $\mathfrak{S}(.;t_0;\mathfrak{B})$ *be a time variation of* $\mathfrak{S}(t,\mathfrak{B})$ *passing through set* $\mathfrak{B}$ *at* $t = t_0$, $\mathfrak{S}(t_0;t_0;\mathfrak{B}) \equiv \mathfrak{B}$. *Then:*

- *The backward-time upper left Dini derivative* ${}^+D_l\mathfrak{S}(t,\mathbf{z})$ *of* $\mathfrak{S}(.)$ *along* $\mathfrak{S}(.;t_0;\mathfrak{B})$ *at* $(t,\mathfrak{B})$ *is*

$${}^+D_l\mathfrak{S}(t,\mathbf{z}) = lim\ sup\left\{\frac{\mathfrak{S}\left[t-\theta,\mathfrak{S}\left(t-\theta;t;\mathfrak{B}\right)\right] - \mathfrak{S}\left(t,\mathfrak{B}\right)}{\theta} : \theta \longrightarrow 0^-\right\}.$$

- *The backward-time lower left Dini derivative* ${}_-D_l\mathfrak{S}(t,\mathbf{z})$ *of* $\mathfrak{S}(.)$ *along* $\mathfrak{S}(.;t_0;\mathfrak{B})$ *at* $(t,\mathfrak{B})$ *is*

$${}_-D_l\mathfrak{S}(t,\mathbf{z}) = lim\ inf\left\{\frac{\mathfrak{S}\left[t-\theta,\mathfrak{S}\left(t-\theta;t;\mathfrak{B}\right)\right] - \mathfrak{S}\left(t,\mathfrak{B}\right)}{\theta} : \theta \longrightarrow 0^-\right\}.$$

- *The backward-time upper right Dini derivative* ${}^+D_r\mathfrak{S}(t,\mathbf{z})$ *of* $\mathfrak{S}(.)$ *along* $\mathfrak{S}(.;t_0;\mathfrak{B})$ *at* $(t,\mathfrak{B})$ *is*

$${}^+D_r\mathfrak{S}(t,\mathbf{z}) = lim\ sup\left\{\frac{\mathfrak{S}\left[t-\theta,\mathfrak{S}\left(t-\theta;t;\mathfrak{B}\right)\right] - \mathfrak{S}\left(t,\mathfrak{B}\right)}{\theta} : \theta \longrightarrow 0^+\right\}.$$

- *The backward-time lower right Dini derivative* $_{-}D_r\mathfrak{S}(t,\mathbf{z})$ *of* $\mathfrak{S}(.)$ *along* $\mathfrak{S}(.;t_0;\mathfrak{B})$ *at* $(t,\mathfrak{B})$ *is*

$$_{-}D_r\mathfrak{S}(t,\mathbf{z}) = lim\ inf\left\{\frac{\mathfrak{S}\left[t-\theta,\mathfrak{S}\left(t-\theta;t;\mathfrak{B}\right)\right]-\mathfrak{S}\left(t,\mathfrak{B}\right)}{\theta} : \theta \longrightarrow 0^+\right\}.$$

- *The function* $\mathfrak{S}(.)$ *has the backward-time left Dini derivative along* $\mathfrak{S}(.;t_0;\mathfrak{B})$ *at* $(t,\mathfrak{B})$, *which is denoted by* $_bD_l\mathfrak{S}(t,\mathbf{z})$ *if, and only if,* $_{-}D_l\mathfrak{S}(t,\mathbf{z}) = {}^{+}D_l\mathfrak{S}(t,\mathbf{z})$ *and then*

$$_bD_l\mathfrak{S}(t,\mathbf{z}) = \ _{-}D_l\mathfrak{S}(t,\mathbf{z}) = \ ^{+}D_l\mathfrak{S}t,\mathbf{z}).$$

- *The function* $\mathfrak{S}(.)$ *has the backward-time right Dini derivative along* $\mathfrak{S}(.;t_0;\mathfrak{B})$ *at* $(t,\mathfrak{B})$, *which is denoted by* $_bD_r\mathfrak{S}(t,\mathbf{z})$ *if, and only if,* ${}^{-}D_r\mathfrak{S}(t,\mathbf{z}) = {}^{+}D_r\mathfrak{S}(t,\mathbf{z})$ *and then*

$$_bD_r\mathfrak{S}(t,\mathbf{z}) = \ _{-}D_r\mathfrak{S}(t,\mathbf{z}) = \ ^{+}D_r\mathfrak{S}(t,\mathbf{z}).$$

- *The function* $\mathfrak{S}(.)$ *has the backward-time Eulerian derivative along* $\mathcal{Z}(.;t_0;\mathbf{z};\mathbf{i})$ *at* $(t,\mathbf{z})$, *which is denoted by* $_bD\mathfrak{S}t,\mathbf{z})$ *or by* $\mathfrak{S}_b^{(1)}(t,\mathbf{z})$ *if, and only if, it has both* $_bD_l\mathfrak{S}(t,\mathbf{z})$ *and* $_bD_r\mathfrak{S}(t,\mathbf{z})$ *such that* $_bD_l\mathfrak{S}(t,\mathbf{z}) =_b D_r\mathfrak{S}(t,\mathbf{z})$ *and then*

$$_bD\mathfrak{S}t,\mathbf{z}) = \ _bD_l\mathfrak{S}(t,\mathbf{z}) = \ _bD_r\mathfrak{S}(t,\mathbf{z}) = \frac{d_b\mathfrak{S}(t,\mathbf{z})}{dt} = \mathfrak{S}_b^{(1)}(t,\mathbf{z}).$$

${}^{*}D\mathfrak{S}(t,\mathbf{z})$ *means that either* $_{-}D\mathfrak{S}(t,\mathbf{z})$ *or* ${}^{+}D\mathfrak{S}(t,\mathbf{z})$ *can be equally used. Analogously,* $_{*}D\mathfrak{S}(t,\mathbf{z})$ *signifies that either* $_{-}D\mathfrak{S}(t,\mathbf{z})$ *or* $_{+}D\mathfrak{S}(t,\mathbf{z})$ *can be equally used.*

Part VIII

BIBLIOGRAPHY

Bibliography

[1] M. A. Aizerman and F. R. Gantmacher, *Absolute Stability of Controls Systems* (in Russian), Moscow: A. N. SSSR, 1963.

[2] B. D. O. Anderson, "Stability of control systems with multiple nonlinearities," *J. Franklin Institute*, Vol. 282, No. 3, pp. 155-160, 1966.

[3] B. D. O. Anderson and J. B. Moore, "Construction of Liapunov functions for non-stationary systems containing non-inertial nonlinearities," (in Russian), *Avotmatika i Telemehanika*, No. 5, pp. 14-21, 1972.

[4] A. Astolfi and L. Marconi, Editors, *Analysis and Design of Nonlinear Control Systems*, In Honor of Alberto Isidori, Berlin: Springer, 2008.

[5] H. A. Antosiewcz, "A survey of Liapunov's second method," in *Contributions to the Theory of Nonlinear Oscillations*, Ed. S. Lesfchetz, Vol. IV, Princeton N. J: Princeton University Press, 1958.

[6] H. A. Antosiewcz, "Recent contributions to Lyapunov's second method," *Colloqies Internationaux sur les Vibrations Forces dans les Systemes Non-lineaires*, pp. 29-37, September 1964.

[7] P. J. Antsaklis and A. N. Michel, *Linear Systems*, New York: McGraw Hill, 1997.

[8] M. Aoki, "Control of large-scale dynamic systems by aggregation", *IEEE Trans. Aut. Control*, Vol. AC-13, No. 3, pp. 246-253, 1968.

[9] M. Araki, "Stability of large-scale non linear systems - quadratic - order theory of composite - systems method using M-matrices", *IEEE Transactions on Automatic Control*, Vol. AC-23, No. 2, pp. 129-142, 1978.

[10] M. Araki and B. Kondo, "Stability and transient behaviour of composite systems", *IEEE Transactions on Automatic Control*, Vol. AC-17, No. 4, 531-547, 1972.

[11] G. Arienti, C. Sutti, and G. P. Szegö, "On the numerical construction of Lyapunov function", *Preprints of the IV IFAC Congress*, Warszawa, Poland, pp. 3-20, 1969.

[12] M. A. Athans and P. L. Falb, *Optimal Control*, New York: McGraw–Hill, 1966, 1994, 2007.

[13] J.-P. Aubin and H Frankowska, *Set-Valued Analysis*, Springer Science & Business Media, Berlin, 1990, 2008.

[14] J. Awrejcewicz, *Applied Non-Linear Dynamical Systems*, Berlin: Springer, 2016.

[15] F. N. Bailey, "The application of Lyapunov's second method to interconnected systems", *J. SIAM Control*, Ser. A, Vol. 3, No. 3, pp. 443-462, 1966.

[16] F. N. Bailey, "The concept of aggregation in systems stability analysis", *2nd Asilomar Conf. on Circ. and Syst.*, pp. 570-576.

[17] Ye. A. Barbashin, *Introduction to the Theory of Stability* (in Russian), Moscow: Nauka, 1967.

[18] E. A. Barbashin, "On the construction of Lyapunov functions for nonlinear systems", *Proc. 1st International Congress: Theory of Continuous Systems-Special Mathematical Problems*, Moscow: Acad. Sci. U.S.S.R., pp. 742-750, 1961.

[19] Ye. A. Barbashin, *The Liapunov functions* (in Russian), Moscow: Nauka, 1970.

[20] Ye. A. Barbashin and N. N. Krasovskii, "On the stability of motion in the large" (in Russian), *Dokl. Akad. Nauk SSSR*, Vol. 86, No. 3, pp. 453-456, 1952.

[21] Ye. A. Barbashin and N. N. Krasovskii, "On the existence of Liapunov functions in the case of asymptotic stability in the whole" (in Russian), *Prikl. Mat. Meh.*, Vol. XVIII, pp. 345-350, 1954.

[22] S. Barnett, "Some topics in algebraic systems theory: a survey", in *Recent Mathematical Developments in Control*, Ed. D. J. Bell, New York: Academic Press, pp. 323-344, 1973.

[23] S. Barnett and C. Storey, *Matrix Methods in Stability Theory*, London: Nelson, 1970.

[24] G. Bartolini and A. Ferrara, "Multi-input sliding mode control of a class of uncertain nonlinear systems," *IEEE Trans. Automatic Control*, Vol. 41, No. 11, pp. 1662-1666, November 1996.

[25] G. Bartolini, A. Ferrara, V. I. Utkin and T. Zolezzi, "A control vector simplex approach to variable structure control of nonlinear systems," *Int. J. Robust and Nonlinear Control*, Vol. 7, pp. 321-335, 1997.

[26] R. Bellman, *Stability Theory of Differential Equations*, New York: McGraw Hill, 1953.

[27] R. Bellman, "Vector Lyapunov functions", *J.S.I.A.M. Control*, Ser. A, Vol. 1, No.1, pp. 32-34, 1962.

[28] A. M. Ben Lashier and C. Storey, "Final-stability with some applications", *J. Inst. Math. Applics.*, Vol. 9, pp. 397-410, 1972.

[29] M. Benrejeb, P. Borne and F. Laurent, "Sur une application de la représentation en flèche à l'analyse des processus", *R.A.I.R.O. Aut. SAC*, Vol. 16, No. 2, pp. 133-146, 1982.

[30] L. D. Berkovitz, *Optimal Control Theory*, New York: Springer Verlag, 2010.

[31] G. Bertoni, C. Bonivento and E. Sarti, "A graphical method for investigating the absolute stability of time-varying systems", *Ricerche di Automatica*, Vol. 1, No. 1, pp. 102-11, 1970.

[32] N. P. Bhatia, "Attraction and nonsaddle sets in dynamical systems", *J. Differential Equations*, Vol. 8, No. 2, 1970.

[33] N. P. Bhatia, "On asymptotic stability in dynamical systems", *Math. Systems Theory*, Vol. 1, No. 2, pp. 113-127, 1967.

[34] N. P. Bhatia and A. C. Lazer, "On global weak attractors in dynamical systems", in: J. K. Hale, J. P. LaSalle, eds., *Differential Equations and Dynamical Systems*, New York: Academic Press, pp. 321-325, 1967.

[35] N. P. Bhatia and G. P. Szegö, *Dynamical Systems: Stability Theory and Applications*, Berlin: Springer-Verlag, 1967.

[36] G. Bitsoris, *Principe de comparaison et stabilité de systèmes complexes*, Thèse d'Etat, Uniersity Paul Sabatier, Toulouse, France, No. 818, 1978.

[37] J. D. Blight and N. H. McClamroch, "Graphical stability criteria for large-scale nonlinear multiloop systems", *Preprints of the 6th IFAC World Congress*, Part 44.5, pp. 1-6.

[38] N. N. Bogoljubov and Yu. A. Mitropolsky, *Asymptotic methods in the Theory of Non-Linear Oscillations* (in Russian), Moscow: Nauka, 1974.

[39] V. Boltyanski, H. Martini and V. Soltan, *Geometric Methods and Optimization Problems*, New York: Springer, 2014.

[40] P. Bondi, P. Fergola and L. Gambardella, "Partial stability of large-scale systems", *IEEE Transactions on Automatic Control*, Vol. AC-24, No. 1, pp. 94-96, 1979.

[41] P. Borne and M. Benrejeb, "On the stability of a class of interconnected systems. Application to the forced working conditions", *IVth IFAC Symp. on Multivariable Technolog. Syst.*, Fredericton, pp. 261-265, 1977.

[42] P. Borne and M. Benrejeb, "Stability of non-linear composite systems", *1st World Conf. Math. Service of Man, Barcelona*, Techn. Session 4/6, pp. 1-12, 1977.

[43] P. Borne, M. Benrejeb and J. L. Coquerelle, "On the synthesis of a class of interconnected systems with structural variations: application to an electrical power plant", *MECO 78*, Acta Press, pp. 1-4, 1978.

[44] P. Borne, M. Dambrine, W. Perruquetti and J. P. Richard, "Vector Lyapunov functions: nonlinear, time-varying, ordinary and functional differential equations", *Stability Theory at the End of XXth Century*, Ed. A. A. Martynyuk, London: Taylor & Francis, Chap. 13, pp. 89-112, 2002.

[45] P. Borne, G. Dauphin-Tanguy and J.-P. Richard, "Estimation of attractors and invariant domains for perturbed complex and large scale systems", in *Stochastic Large Scale Engineering Systems*, Ed. S. Tzafestas and K. Watanabe, Marcel Dekker Inc., Chap. 10, pp. 307-316, 1992.

[46] P. Borne, G. Dauphin-Tanguy, J.-P. Richard, F. Rotella and I. Zambettakis, *Commande et Optimisation des Processus*, Paris: Éditions TECHNIP, 1990.

[47] P. Borne and J.-P. Richard, "On attractivity domains estimation for continuous systems", *IMACS Annals on Computing and Applied Mathematics*, Ed. P. Borne, S. Tzafestas, P. Breedveld and G. Dauphin-Tanguy, J. C. Baltzer AG Sc. Pub. Comp., Vol. 4, Sect. III, pp. 179-181, 1989.990.

[48] P. Borne, J.-P. Richard and N. E. Radhy, "Stability, stabilisation, regulation using vector norms", in *Nonlinear Systems*, Ed. A. J. Fossard and D. Normand-Cyrot, Chapman and Hall, Vol II, Chap. 2, 47 pp., 1996.

[49] R. K. Brayton and C.H. Tong, "Constructive stability and asymptotic stability of dynamical systems", *IEEE Transactions on Circuits and Systems*, Vol. CAS-21, No. 1, pp. 1121-1130, 1980.

[50] R. K. Brayton and C.H. Tong, "Stability of dynamical systems: a constructive approach", *IEEE Trans. Circuits Syst.*, Vol. 26, pp. 224-234, 1979.

[51] R. W. Brockett, "The status of stability theory of deterministic systems: a constructive approach", *IEEE Trans. Autom. Contr.*, Vol. AC-11, pp. 596-606, June 1966.

[52] W. L. Brogan, *Modern Control Theory*, New York: Quantum, 1974.

[53] A. E. Bryson, and Y. Ho, *Applied Optimal Control: Optimization, Estimation and Control* (Revised Printing), New York: Wiley, 1975.

[54] A. E. Bryson, and Y. C. Ho, *Applied Optimal Control.* Washington, DC: Hemisphere, 1975.

[55] Z. M. Buchevats and Ly. T. Gruyitch, *Linear Discrete-time Systems*, Boca Raton, FL: CRC Press, 2018.

[56] R. Bulirsch, A. Miele, J. Stoer, K. H. Well, Eds., *Optimal Control*, Basel: Springer Basel AG, 1993.

[57] C. Burgat , J. Bernussou, Lj. T. Grujić, J. C. Gentina et P. Borne, "Les perturbations struturelles arbitraires et périodiques", *R.A.I.R.O Automatique/Systems Analysis and Control*, Vol. 12, No. 3, pp. 245-267, 1978.

[58] C. Burgat et Lj. T. Grujić, "Estimations E_i du domaine de stabilité pour un système interconnecté de comparaison du type Lotka-Volterra", *C. R. Acad. Sc. Paris*, Vol. 288, Serie A 745-Serie A-747, 23 avril 1979.

[59] G. Burnand and G. Sarlos, "Determination of the domain of stability", *J. of Math. Analysis and Appl.*, Vol. 23, pp. 714-722, 1968.

[60] T. A. Burton, "On the construction of Lyapunov functions", *SIAM J. Appl. Math.*, Vol. 17, No. 6, pp. 1078-1085, 1969.

[61] Bushaw D., "Stabilities of Lyapunov and poisson types", *SIAM Review*, Vol. 11, No. 2, pp. 214-225, 1969.

[62] A. Cavallo and G. De Maria, "A sliding manifold approach to the feedback control of rigid robots," *Int. J. Robust and Nonlinear Control*, Vol. 6, pp. 501-516, 1996.

[63] L. Cesari, *Asymptotic Behavior and Stability Problems in Ordinary Differential Equations*, Berlin: Springer, 1959.

[64] F. L. Chernous'ko, I. M. Ananievski and S. A. Reshmin, *Control of Nonlinear Dynamical Systems*, Berlin: Springer, 2008.

[65] C.-T. Chen, *Linear System Theory and Design*, New York: Holt, Rinehart and Winston, Inc., 1984.

[66] Z. Chen and J. Huang, *Stabilization and Regulation of Nonlinear Systems*, Berlin: Springer, 2015.

[67] D. Cheng, X. Hu and T. Shen, *Analysis and Design of Nonlinear Control Systems*, Berlin: Springer, 2010.

[68] C. S. Chen and E. Kinnen, "Construction of Lyapunov functions", *J. Franklin Inst.*, Vol. 289, No. 2, pp. 133-146, 1970.

[69] N. G. Chetaev, "On stable trajectories of dynamics"(in Russian), Kazan State University, Kazan, USSR, *Mathematics*, 1, Vol. 91, no. 4, pp. 3–8, 1931.

[70] N. G. Chetaev, *Stability of Motion* (in Russian), Moscow: OGIZ, 1946. English translation, Oxford: Pergamon Press, 1961.

[71] H.-D. Chiang and L. F. C. Alberto, *Stability Regions of Nonlinear Dynamical Systems: Theory, Estimation, and Applications*, Cambridge: Cambridge University Press, 2015.

[72] H. D. Chiang and J. S. Thorp, "Stability regions of nonlinear dynamical systems: a constructive methodology", *IEEE Trans. on Aut. Contr.*, Vol. 34, No. 12, pp. 1229-1241, December 1989.

[73] P. S. M. Chin, "A general method to derive Lyapunov functions for non-linear systems", *Int. J. Control*, Vol. 44, No. 22, pp. 381-393, 1986.

[74] P. S. M. Chin, "Generalized integral method to derive Lyapunov functions for non-linear systems", *Int. J. Control*, Vol. 46, No. 22, pp. 933-943, 1987.

[75] E. A. Coddington and N. Levinson, *Theory of Ordinary Differential Equations*, New York: McGraw Hill, 1955 (Russian edition published in Moscow: Inostranoi Literaturi, 1958).

[76] W. A. Coppel, *Stability and Asymptotic Behaviour of Differential Equations*, Boston: D. C. Heatch and Co., 1965.

[77] J. L. Corne and N. Rouche, "Attractivity of closed sets proved by using a family of Liapunov functions", *J. Differential Equations*, Vol. 13, No. 2, pp. 231-246, 1973.

[78] G. Dauphin-Tanguy and Lj. T. Grujić, "Asymptotic stability via energy and power. Part II: bond-graph bridging for non-linear systems", *Proc. IFAC Conf. System, Structure and Control*, Nantes, France, pp. 96-101, 1995.

[79] G. Dauphin-Tanguy and Lj. T. Grujić, "Energetic stability and bond graphs. Part II: bond-graph approach to mechanical systems", *1996 IMACS-IEEE Conference on Computational Engineering in Systems Applications: Symposium on Modelling, Analysis and Simulation*, Lille, France, Vol. 2, pp.732-735, 1996.

[80] G. Dauphin-Tanguy and Lj. T. Gruyitch, "Strict monotonous asymptotic energetic stability. Part II: Bond graph resolutions", *Preprints 14 IFAC World Congress*, Vol. D, pp. 443-448, 5-9 July, 1999.

[81] E. J. Davison and E. Kurak, "A computational method for determining quadratic Lyapunov functions for non-linear systems", *Automatica*, Vol. 7, pp. 627-636, 1971.

[82] L. Debnath, *Integral Transformations and Their Applications*, Boca Raton: CRC Press, 1995.

[83] B. P. Demidovich, *Lectures on Mathematical Stability Theory* (in Russian), Moscow: Nauka, 1967.

[84] C. A. Desoer, "A generalization of Popov criterion", *IEEE Transactions on Automatic Control*, Vol. AC-10, No. 2, 182-85, 1965.

[85] C. A. Desoer and M. Vidyasagar, *Feedback Systems: Input-Output Properties*, New York: Academic Press, 1975.

[86] M. Djordjević, "Stability analysis of interconnected systems with possibly unstable subsystems", *Systems & Contr. Lett.*, Vol. 3, No. 3, pp. 165-169, 1983.

[87] M. Djordjević, "Stability analysis of large-scale systems whose subsystems may be unstable", *Large Scale Syst.*, Vol. 5, No. 3, pp. 255-262, 1983.

[88] M. Djordjević, "Stability analysis of nonlinear systems by the matrix Lyapunov method", *Proc. IMACS-IFAC Symp.*, IDN, Villeneuve d'Ascq, France, pp. 209-212, June 3-6, 1986

[89] P. Dorato, "Short-time stability", *IRE Trans. on Automatic Control*, p. 86, February 1967.

[90] P. Dorato, "Short time stability in linear time-varying systems", *IRE Itern. Conv. Rec.*, Pt. 4, pp. 83-87, 1961.

[91] A. G. Dewey and E. I. Jury, "A stability inequality for a class of nonlinear feedback system", *IEEE Transactions on Automatic Control*, Vol. AC-11, No. 1, 54 -62, 1966.

[92] F. Fallside and M. R. Patel, "Control engineering applications of V. I. Zubov's construction procedure for Lyapunov functions", *IEEE Transactions on Automatic Control*, Vol. AC-10, No. 2, 220 -222, 1965.

[93] M. Fiedler and V. Ptak, "On matrices with non-positive off-diagonal elements and positive principal minors", *Czech. Nat. J.*, Vol. 12, No. 87, pp. 382-400, 1962.

[94] G. M. Fikhtengol'c, *Course on Differential and Integral Calculus* (in Russian), Moscow: Fizmatgiz, 1958.

[95] W. R. Foster and M. L. Davies, "Estimating the domain of attraction for systems with multiple non-linearities", *Int. J. Contr.*, Vol. 15, No. 9, pp. 1001-1003, 1972.

[96] A. R. Gaiduk, "Absolute stability of control systems with multiple nonlinearities" (in Russian), *Avtomatika i Telemehanika*, No. 6, pp. 5-11, 1976.

[97] F. R. Gantmacher, *The Theory of Matrices*, Vol. 1, New York: Chelsea Publ. Co., 1960, 1974.

[98] F. R. Gantmacher, *The Theory of Matrices*, Vol. 2, New York: Chelsea Publ. Co., 1960, 1974.

[99] F. R. Gantmacher and V. A. Yakubovich, "Absolute stability of nonlinear control system" (in Russian), published in *Analytical Mechanics, Stability of Motions and Space Balistics*, Moscow: Nauka, 1965.

[100] D. P. Garg and M.J. Rabins, "Stability bounds for nonlinear systems designed via frequency domain stability criteria", *Transactions of ASMRE Journal of Dynamic Systems, Measurement and Control*, pp. 262-265, 1972.

[101] W. L. Garrard, "Finite time stability in control systems synthesis", *Proc. 1st IFAC World Congress*, London: Butterworth, pp. 21-29, 1961.

[102] A, H. Gelig, "Investigation of stability of non-linear discontinuous automatic regulating systems with non-unique equilibrium states" (in Russian), *Avtomatika i Telemehanika*, Vol. XXV, No. 2, pp. 153-160, 1964.

[103] R. Genesio and A. Vicino, "New techniques for constructing asymptotic stability regions for nonlinear systems", *IEEE Trans. Circ. Syst.*, Vol. Cas-31, No. 6, pp. 574-581, 1984.

[104] R. Genesio, M. Cartaglia and A. Vicino, "On the estimation of asymptotic stability regions: state of art and new proposals", *IEEE Trans. on Aut. Contr.*, Vol. AC-30, No. 8, pp. 747-755, August 1985.

[105] J. C. Gentnina, *Contribution à l'analyse et à la synthèse des systèmes continus non-linéaires de grande dimension*, Thèse de Doctorat es Sciences Physiques, University of Lille 1, France, 1976.

[106] J. C. Gentnina, P. Borne, C. Burgat, J. Benussou and Lj. T. Grujić, "Sur la stabilité des systèmes de grande dimension. Normes vectorielles", *RAIRO Automat./Syst. Anal. and Contr.*, Vol. 13, No. 1, pp. 57-75, 1979.

[107] J. C. Gentnina, P. Borne and F. Laurent, "On a study of stability and sensitivity of large continuous non-linear systems under uncertainty", *3rd IFAC Symp. on Sens. Adapt. and Optim.*, ISA, pp. 412-417, 1973.

[108] J. C. Gentnina, P. Borne and F. Laurent, "Stabilité des systèmes continus non linéaires de grande dimension", *RAIRO*, No. 3, pp. 69-77, August 1972.

[109] M. Gruber, "Path integrals and Lyapunov functionals", *IEEE Transactions on Automatic Control*, Vol. AC-14, No. 5, 465-478, 1969.

[110] Lj. T. Grujić, *Large Scale Systems Stability* (in Serbo-Croatian), D. Sci. Dissertation, Department of Mechanical Engineering, University of Belgrade, Belgrade, 1972.

[111] Lj. T. Grujić, "Absolute stability of non-stationary control systems of Lurie type", *Proc. ETAN 73*, Novi Sad, Serbia, Yugoslavia pp. 615-623, Dublin, Ireland, 1973.

[112] Lj. T. Grujić, "On multi-level absolute stability analysis of large-scale systems: I", *Automatika-Theoretical Supplement*, Zagreb, SFRY, pp. 67-72, 1974.

[113] Lj. T. Grujić, "On multi-level absolute stability analysis of large-scale systems: II", *Automatika-Theoretical Supplement*, Zagreb, SFRY, pp. 155-161, 1974.

[114] Lj. T. Grujić, "Non-Lyapunov stability analysis of large-scale systems on time-varying sets", *Int. J. Control*, Vol. 21, No. 3, pp. 401-405, 1975.

[115] Lj. T. Grujić, "Novel development of Lyapunov stability of motion", *Int. Jornal of Control*, Vol. 22, No. 4, pp. 525-549, 1975.

[116] Lj. T. Grujić, "General stability analysis of large-scale systems", *Proc. IFAC Symp. on Large Scale Systems*, Udine, Italy, pp. 203-213, 1976.

[117] Lj. T. Grujić, "Is the stability theory consistent and complete", *Presented at Ist World Conference on Mathematics at the Service of Man, Barcelona*, Spain, 1976: Proc. Ist World Conf. on Math. at the Service of Man, Univ. Politec. Barcelonne, Vol. 1, pp. 124-144, 1980.

[118] Lj. T. Grujić, "Necessary and sufficient Lyapunov like conditions for the absolute stability and Aizerman conjecture" (in Russian), *Mathematical Physics*, Ukraine, Vol. 28, pp. 7-20, 1980.

[119] Lj. T. Grujić, "Lyapunov-like solutions for stability problems of the most general Lurie-Postnikov systems", *Int. J. Systems Science*, Vol. 12, No. 7, pp. 813-833, 1981.

[120] Lj. T. Grujić, "On absolute stability and Aizerman conjecture", *Automatica*, Vol. 17, No. 2, pp. 335-349, 1981.

[121] Lj. T. Grujić, “Continuity, contractility and invariability of time-varying sets and systems theory”, Presented at *the International Symp. on Operator Theory of Networks and Systems*, Concordia University, Montreal, Canada, 1975.

[122] Lj. T. Grujić, “Absolute stability of non-stationary systems: resolutions and applications”, *Proceedings of 1978 JACC*, Philadelphia, pp. 327-337, 1978.

[123] Lj. T. Grujić, “A lemma and absolute stability of non-stationary systems”, Presented at Ist World Conference on Mathematics at the Service of Man, Barcelona, Spain, *1976: Proc. Ist World Conf. on Math. at the Service of Man*, Univ. Politec. Barcelonne, Vol. 1, pp. 162-170, 1980.

[124] Lj. T. Grujić, “Concepts of stability domains” (in Serbo-Croatian), Zagreb (Croatia): *Automatika*, Vol. 26, No. 1-2, pp. 5-10, 1985.

[125] Lj. T. Grujić, “Exact determination of a Lyapunov function and the asymptotic stability domain”, *Int. J. Systems Sc.*, Vol. 23, No. 11, pp. 1871-1888, 1992.

[126] Lj. T. Grujić (Ly. T. Gruyitch), *Continuous time control systems*, Lecture notes for the course "DNEL4CN2: Control Systems", Durban, RSA: Department of Electrical Engineering, University of Natal, South Africa, 1993.

[127] Lj. T. Grujić, "Exact both construction of Lyapunov function and asymptotic stability domain determination", *Proc. 1993 IEEE/SMC International Conference on System, Man and Cybernetics*, Le Touquet, France, Vol. I, pp. 331-336, October 17-20, 1993.

[128] Lj. T. Grujić, “Complete exact solution to the Lyapunov stability problem: Time-varying nonlinear systems with differentiable motions”, *Nonlinear Analysis, Theory, Methods & Applications*, Vol. 22, No. 8, pp. 971-981, 1994.

[129] Lj. T. Grujić, “Exact Solutions for asymptotic stability: Non-linear systems”, *Int. J. Non-Linear Mechanics*, Vol. 30, No. 1, 45-56, 1995.

[130] Lj. T. Grujić, “Comparative stability analysis: concepts and approaches”, *The Conference on Stability and Stabilization*, Université de Picardie Jules Verne, Amiens, France, November 20, 1996.

[131] Lj. T. Grujić, “Large scale systems stability”, *Applied Modelling and Simulation of Technological Systems*, Editors: P. Borne and S.G. Tzafestas, Elsevier Science Publishers, B.V. (North Holland), IMACS, pp. 233-247, 1987.

[132] Lj. T. Grujić, On large-scale systems stability”, *12th IMACS World Congress*, Paris, pp. 224-229, 1988.

[133] Lj. T. Grujić, "On large-scale systems stability", Computing and computers for control systems, editors: P. Borne et al., J. C. Baltzer AG, *Scientific Publishing Co., IMACS*, pp. 201-206, 1989.

[134] Lj. T. Grujić, “On determination of Lyapunov functions and the asymptotic stability domain”, *Proc. 3rd Conf. on Systems, Automatic Control and Measurement*, SAUM (in Serbian), Vrnjachka Banya, Serbia, Yugoslavia, pp. 12-20, 1989.

[135] Lj. T. Grujić, "New Lyapunov method based methodology for asymptotic stability", *Proc. IFAC Conference on System Structure and Control*, Nantes, France, pp. 536-541, July 5-7, 1995.

[136] Lj. T. Grujić, "New Lyapunov methodology and exact construction of a Lyapunov function: Exponential stability", *Problems of the Nonlinear Analysis in Engineering Systems*, Kazan, Russia, Vol. 1, pp. 9-16, 1995.

[137] Lj. T. Grujić, "New Lyapunov methodology: Asymptotic stability of hopfield neural networks", *1996 IEEE Symposium on Industrial Electronics*, Warsaw, Vol. 1, pp. 127-130, 1996.

[138] Lj. T. Grujić, "New stability methodology for neural networks", *1996 IMACS - IEEE Conference on Computational Engineering in Systems Applications: Symposium on Robotics and Cybernetics*, Lille, France, pp. 820-822, 9-12 July, 1996.

[139] Lj. T. Grujić, "New approach to asymptotic stability: Time-varying nonlinear systems", *International J. of Mathematics and Mathematical Sciences*, Vol. 20, No. 2, pp. 347-366, 1997.

[140] Lj. T. Grujić, "Consistent Lyapunov methodology for time-invariant nonlinear systems", *Avtomatika i Telemehanika* (in Russian), No. 12, December, pp. 35-73, 1997.

[141] Lj. T. Grujić, "New Lyapunov methodology: Absolute stability of Lurie-Postnykov systems", *IFAC–IFIP-IMACS Conference on Control of Industrial Systems*, Belfort, France, Vol. 1, pp. 533-538, May 20-22, 1997.

[142] Lj. T. Grujić, "Novel Lyapunov stability methodology for nonlinear systems: Complete solutions", *Nonlinear Analysis, Theory, Methods & Applications: Proc. 2nd Second World Congress of Nonlinear Analysts*, Elsevier Science Ltd., Vol. 30, No. 8, pp. 5315-5325, 1997.

[143] Lj. T. Grujić, "On new approaches to large-scale systems stability analysis", Proc. *13th IMACS World Congress on Computation and Applied Mathematics*, Vol. 3, pp. 1250-1251, 1991,

[144] Lj. T. Grujić, "On practical stability", *Proc. 5th ASILOMAR Conf.* California, pp. 174-178, 1971.

[145] Lj. T. Grujić, "On practical stability", *Intern. J. Control*, Vol. 17, No. 4, pp. 881-887, 1973.

[146] Lj. T. Grujić, "On solutions to Lyapunov stability problems", *Facta, Universitatis, Series: Mechanics, Automatic Control and Robotics*, University of Nish, Serbia, Yugoslavia, Vol. 1, No. 2, pp. 121-138, 1992.

[147] Lj. T. Grujić, "Practical stability analysis of large-scale systems", *Proc. JUREMA, Zagreb, Croatia*, Yugoslavia, pp. 31-39, 1973.

[148] Lj. T. Grujić, "Practical stability of automatic control systems", *Proc. Mašinstvo 1873-1973; (in Serbian), Faculty of Mechanical Engineering*, Belgrade, Yugoslavia, pp. B-21 to B-26, 1973.

[149] Lj. T. Grujić, "Practical stability of large-scale systems" (in Serbo-Croatian), *Proc. XXI Conf. ETAN*, Banja Luka, Bosnia and Herzegovina, Yugoslavia, pp. III.301-III.307, 1977.

[150] Lj. T. Grujić, "Practical stability with the settling time", *Automatika*, Zagreb-SFRY, Vol. 7, pp. 102-106, 1973.

[151] Lj. T. Grujić, "Practical stability with the settling time of composite systems", *Automatika*, Zagreb-SFRY, Vol. 16, No. 1-2, pp. 67-77, 1975.

[152] Lj. T. Grujić, "Sets and singularly perturbed systems", *Systems Science* (Wroclaw, Poland), Vol. 5, No. 4, pp. 327-338, 1979.

[153] Lj. T. Grujić, "Solutions for the Lurie-Postnikov and Aizerman problems", *Int, J. Systems Science*, Vol. 9, No. 12, pp. 1359-1372, 1978.

[154] Lj. T. Grujić, "Solutions to Lyapunov stability problems: nonlinear systems with continuous motions", *International Journal of Mathematics and Mathematical Sciences*, Orlando, USA, Vol. 17, No. 3, pp. 587-596, 1994.

[155] Lj. T. Grujić, "Solutions to Lyapunov stability problems: Nonlinear systems with differentiable motions", *Proc. 13th IMACS World Congress on Computation and Applied Mathematics*, Vol. 3, pp. 1228-1231, 1991, and in: *Computational and Applied Mathematics II: Differential Equations*, eds. W.F Ames and P.J. van der Houwen, Elsevier, Amsterdam, pp. 39-47, 1992.

[156] Lj. T. Grujić, "Solutions to Lyapunov stability problems: nonlinear systems with differentiable motions", *Int. J. Systems Science*, Vol. 23, No. 11, pp. 1874-1888, 1992.

[157] Lj. T. Grujić, "Solutions to Lyapunov stability problems: nonlinear systems with globally differentiable motions", in *The Lyapunov Functions Method and Applications*, ed. P. Borne and V. Matrosov, J.C. Baltzer AG, Scientific Publishing Co, IMACS, pp. 19-27, 1990.

[158] Lj. T. Grujić, "Solutions to Lyapunov stability problems of sets: nonlinear systems with differentiable motions", *Proc. 13th IMACS World Congress on Computation and Applied Mathematics*, Dublin, Vol. 3, pp. 1228-1231, and in *International Journal of Mathematics and Mathematical Sciences*, Orlando, USA, Vol. 17, No.1, pp.103-112, 1994.

[159] Lj. T. Grujić, "Solutions to Lyapunov stability problems: time-invariant systems", *Proc. 14th IMACS World Congress on Computation and Applied Mathematics*, Vol. 1, pp. 203-205, 1994.

[160] Lj. T. Grujić, "Solutions to Lyapunov stability problems via O-uniquely bounded sets", *Control-Theory and Advanced Technology*, Tokyo, Vol. 10, No. 4, Part 2, pp. 1069-1091, 1995.

[161] Lj. T. Grujić, "Some results on asymptotic stability of motion" *Automatika*, Zagreb, SFRY, pp. 281-288, 1972.

[162] Lj. T. Grujić, "Stability analysis of an orbital observatory" (in Serbo-Croatian), *Proc. 1st. Yugoslav Aero-space Congress*, Vol. I; *Automatika*, Zagreb, SFRY, Vol. 3-4, pp. 118-121, 1973.

[163] Lj. T. Grujić, "Stability analysis of large-scale systems with stable and unstable subsystems", *Int. J. Control*, Vol. 20, No. 3, pp. 453-463, 1974.

[164] Lj. T. Grujić, "Stability and instability of product sets", *Systems Science*, Vol. 3, No. 1, pp. 14-31, 1977.

[165] Lj. T. Grujić, "Stability and instability of product sets", *Inter. Symp. on Systems Science* III, Wroclaw, Poland, 1976.

[166] Lj. T. Grujić, "Stability domains concepts" (in Serbo-Croatian), *Automatika*, Zagreb, SFRY, Vol. 26, No. 1-2, pp. 5-10, 1985.

[167] Lj. T. Grujić, "Stability domains of general and large-scale systems", *Proc. of the IMACS/IFAC Symposium on Modelling and Simulation for Control and Distributed Parameter Systems*, eds. P. Borne and S. G. Tzafestas, Elsevier Science Publishers B. V. (North Holland), IMACS, pp. 317-327, 1987.

[168] Lj. T. Grujić, "Stability domains of general and large-scale time-invariant non-linear systems", *Proc. 2nd. Conf. on Systems, Automatic Control and Measurement (SAUM)*, Belgrade, Serbia, Yugoslavia: First Intern. Seminar Large - Scale Systems, pp. 55-71, 1986.

[169] Lj. T. Grujić, "Stability domains of general and large-scale stationary systems", in *Applied Modelling and Simulation of Technological Systems*, Institut Industriel du Nord (Lille, France), Vol. 1, pp. 267-272, 1986.

[170] Lj. T. Grujić, *Stability of Large-Scale Systems*, D. Sci. dissertation, Belgrade, Faculty of Mechanical Engineering, University of Belgrade, defended 1972, published 1974.

[171] Lj. T. Grujić, "Stability of product sets". *Proc. of 1975 Midwest Symp. on Circuits and Systems*, Concordia University, Montreal, Canada, pp. 254-258, 1975.

[172] Lj. T. Grujić, "Stability: physical and mathematical interpretation", *Proc. of The First International Conference on Neural, Parallel, and Scientific Computations, 1996 IMACS-IEEE Conference on Computational Engineering in Systems Applications: Symposium on Modelling, Analysis and Simulation*, Lille, France, Vol. 1, pp. 13-22, 1996.

[173] Lj. T. Grujić, "Stability versus tracking in automatic control systems"(in Serbo-Croatian), *Proc. JUREMA 29*, Part 1, Zagreb, pp. 1-4, 1984.

[174] Lj. T. Grujić, "State-Space domains of singularly perturbed systems", *Proc. 12th World Congress on Scientific Computation, IMACS*, Paris, pp. 83-91, July 18-22, 1988.

[175] Lj. T. Grujić, "The necessary and sufficient conditions for the exact construction of a Lyapunov function and the asymptotic stability domain", *Proc. 39th IEEE Conference on Decision and Control*, Brighton, Vol. 3, pp. 2885-2888, 1991.

[176] Lj. T. Grujić, "Time-varying continuous nonlinear systems: uniform asymptotic stability", *International Journal of Systems Science*, Vol. 26, No. 5, pp. 1103-1127, 1995; "Corrigendum", Ibid., Vol. 27, No. 7, p. 689, 1996.

[177] Lj. T. Grujić, "Time-varying sets, aggregation and stability of large-scale systems", *Proc. 1976 IEEE Intern. Symp. Circ. Syst.*, Münich, Germany, pp. 398-401, 1976.

[178] Lj. T. Grujić, "Uniform asymptotic stability of discrete large-scale systems", *IEEE Trans. SMC*, Vol. SMC-3, pp. 636-643, 1973.

[179] Lj. T. Grujić, "Uniform practical and finite-time stability of large-scale systems", *Int. J. Syst. Sci.*, Vol. 6, No. 2, 1975.

[180] Lj. T. Grujić, "Uniquely bounded sets", presented at *1st Conference on Mathematics at the Service of Man*, Barcelona, Spain, July 11-16, 1977; published at *Proc. of the 1st Conference on Mathematics at the Service of Man*, Univ. Politec., Barcelona, Vol I, pp. 145-161, 1980.

[181] Lj. T. Grujić, "Uniquely bounded sets and nonlinear systems", *Proc. 1978 IEEE Conference on Decision and Control*, San Diego, California, Vol. I, pp. 325-333, 1979.

[182] Lj. T. Grujić, "Un lemme matriciel réciproque; application à la stabilité absolue, *C. R. Acad. Sci*, Paris, Ser. A, Vol. 284, pp. 1409-1412, 1977.

[183] Lj. T. Grujić. P. Borne and J.-C. Gentina, "General aggregation of large-scale systems by vector Lyapunov functions and vector norms", *Int. J. Control*, Vol. 24, No. 4, pp. 529-550, 1976.

[184] Lj. T. Grujić. P. Borne and J.-C. Gentina, "Matrix approaches to the absolute stability of time-varying Lurie-Postnikov systems", *Int. J. Control.*, Vol. 30, pp. 967-980, 1979.

[185] Lj. T. Grujić, Z. Bucevac and Z. Ribar, "On necessary and sufficient conditions for absolute stability of discrete-time systems", *The Proceedings of 9th IFAC World Congress*, Vol. I: *Analysis and Synthesis of Control Systems*, Pergamon Press, pp. 201-205, 1985.

[186] Lj. T. Grujić and C. Burgat, "Lotka-Volterra-like approach to large-scale systems stability", *Int. J. Systems Sc.*, Vol. 11. No. 10, pp. 1131-1140, 1980.

[187] Lj. T. Grujić and C. Burgat, "Stability analysis of large-scale generalized Lotka-Volterra systems", *Ricerche di Automatica*, Italy, Vol. 9, No. 2, pp. 161-170, 1978.

[188] Lj. T. Grujić, M. Darwish and J. Fantin, "Coherence, vector Lyapunov functions and large-scale power systems", *Int. J. Syst. Sc.*, Vol. 10, No 3, pp. 351-362, 1979.

[189] Lj. T. Grujić and G. Dauphin-Tanguy, "Asymptotic stability via energy and Power. Part I: New Lyapunov methodology for nonlinear systems", *Proc. IFAC Conference on System Structure and Control*, Nantes, France, pp. 548-553, July 5-7, 1995.

[190] Lj. T. Grujić and G. Dauphin-Tanguy, "Energetic stability and bond graphs. Part I: Theoretical foundations", *1996 IMACS-IEEE Conference on Computational Engineering in Systems Applications: Symposium on Modelling, Analysis and Simulation*, Lille, France, Vol. 2, pp. 729-731, 9-12 July, 1996.

[191] Lj. T. Grujić, J.C. Gentina, P. Borne, "General aggregation of large-scale systems by vector Lyapunov functions and vector norms", *Int. J. Control*, Vol. 24, No. 4, pp. 529-550, 1976.

[192] Lj. T. Grujić, J.C. Gentina, P. Borne, C. Burgat et J. Bernussou, "Sur la stabilité des systèmes de grande dimension. Fonctions de Lyapunov vectorielles", *R.A.I.R.O. Automatique/Systems Analysis and Control*, Vol. 12, No. 4, pp. 319-348, 1978.

[193] Lj. T. Grujić and A. Kökösy, "Control synthesis for robot motion robust exponential stability. Part I: Modelling and synthesis", *IFAC-IFIP-IMACS Conference on Control of Industrial Systems*, Belfort, France, Vol. 3, pp. 704-709, May 20-22, 1997.

[194] Lj. T. Grujić and A. Kökösy, "Robust stabilizing robot control. Part I: Modelling and synthesis", *1996 IMACS-IEEE Conference on Computational Engineering in Systems Applications: Symposium on Modelling, Analysis and Simulation*, Lille, France, Vol. 2, pp. 1084-1088, 1996.

[195] Lj. T. Grujić, A. A. Martynyuk, and M. Ribbens-Pavella, *Large Scale Systems Stability under Structural and Singular Perturbations* (in Russian, Kiev: Naukova Dumka 1984), Berlin: Springer-Verlag, 1987.

[196] Lj. T. Grujić and A. N. Michel, "Discrete-time mathematical modeling of neural networks under pure structural variations", *Proc. 13th IMACS World Congress on Computation and Applied Mathematics*, Trinity College, Dublin, Ireland, Vol. 3, pp. 1256-1257, July 22-26, 1991.

[197] Lj. T. Grujić and A. N. Michel, "Exponential stability and trajectory bounds of neural networks under structural variations", *Proc. IEEE CDC*, Honolulu, Hawaii, pp. 1713-1718, December 1990.

[198] Lj. T. Grujić and A. N. Michel, "Qualitative analysis of neural networks under structural perturbations", *Proc. 1990 IEEE Intern. Symp. on Cyrc. and Syst.*, New Orleans, Vol. 1, pp. 391-394, May 1-3, 1990.

[199] Lj. T. Grujić and A. N. Michel, "Qualitative analysis of continuous-time neural networks under pure structural perturbations", *Proc. 13th IMACS World Congress on Computation and Applied Mathematics*, Trinity College, Dublin, Ireland, Vol. 3, pp. 1260-1261, July 22-26, 1991.

[200] Lj. T. Grujić and A. N. Michel, "Qualitative analysis of discrete-time neural networks under pure structural perturbations", *Proc. 13th IMACS World Congress on Computation and Applied Mathematics*, Trinity College, Dublin, Ireland, Vol. 3, pp. 1262-1263, July 22-26, 1991.

[201] Lj. T. Grujić and Dj. Petkovski, "On robustness of lurie systems with multiple non linearities", *Automatica*, Vol. 23, No. 3, pp. 327-334, 1987.

[202] Lj. T. Grujić and Dj. Petkovski, "On robustness of lurie systems with a single nonlinearity", *Control-Theory and Advanced Technology*, Vol. 2, No. 4, pp. 627-632, 1986.

[203] Lj. T. Grujić and Dj. Petkovski, "On stability robustness of Lurie systems", *12th IMACS World Congress*, Paris, Vol. 2, pp. 108-110, 1988.

[204] Lj. T. Grujić and Dj. Petkovski, "On stability robustness of lurie systems", *Computing and Computers for control systems*, ed. P. Borne et al., J.C. Baltzer AG, Scientific Publishing Co., IMACS, pp. 183-185, 1989.

[205] Lj. T. Grujić and Dj. Petkovski, "On various approaches to robustness of Lurie systems absolute stability", *1986 CDC*, Athens, Greece, pp. 2069-2070, 1986.

[206] Lj. T. Grujić and Dj. Petkovski, "Robust absolutely stable lurie systems", *Intern. Journal of Control*, Vol. 46, No. 1, pp. 357-368, 1987.

[207] Lj. T. Grujić and M. Ribbens-Pavella, "Asymptotic stability of large-scale systems: Part 1: domain estimations", *Elec. Power and Energy Systems*, Vol. 1, No. 3, pp. 151-157, 1979.

[208] Lj. T. Grujić and M. Ribbens-Pavella, "Asymptotic stability of large-scale power systems", *Automatika*, Zagreb, SFRY, No. 5-6, pp. 165-167, 1977.

[209] Lj. T. Grujić and M. Ribbens-Pavella, "Decomposition and aggregation of large-scale power systems", *Ibid*, pp. 225-228, 1977.

[210] Lj. T. Grujić and M. Ribbens-Pavella, "New approach to stability domain estimates of large-scale power systems", *Revue E*, Vol. VIII, No. 10, pp. 241-249, 1977.

[211] Lj. T. Grujić and M. Ribbens - Pavella, "Relaxed large-scale systems stability analysis applied to power systems", *IFAC VII World Congress*, pp. 27-34, 1978.

[212] Lj. T. Grujić and D. D. Šiljak, "Asymptotic stability and instability of large-scale systems", *IEEE Transactions on Automatic Control*, Vol. AC-18, No. 6, pp. 636-645, December 1973.

[213] Lj. T. Grujić and D. D. Šiljak, "Exponential stability of large-scale discrete systems", *Int. J. Control*, England, Vol. 9, No 3, pp. 481-491, 1974.

[214] Lj. T. Grujić and D. D. Šiljak, "On stability of discrete composite systems", *IEEE Transactions on Automatic Control*, Vol. AC-18, No. 5, pp. 522-524, October 1973.

[215] Lj. T. Grujić and D. D. Šiljak, "Stability of large-scale systems with stable and unstable subsystems", *Proc. 1972 Joint Automatic Control Conference*, Stanford, California, USA, pp. 550-555, 1972.

[216] Ly. T. Grouyitch et G. Dauphin-Tanguy, "Stabilité énergetique exponentielle", *Proceedings of the CIFA*, Lille, pp. 75-80, 5-7 July, 2000.

[217] Ly. T. Gruyitch, *Advances in the Linear Dynamic Systems Theory. Time-invariant continuous-time systems*, Tamarac, Florida: Llumina press, 2013.

[218] Ly. T. Gruyitch, "Aircraft natural control synthesis: Vector Lyapunov function approach", *Actual Problems of Airplane and Aerospace Systems: Processes, Models, Experiments*, Vol. 2, No. 6, Kazan, Russia, and Daytona Beach, FL, USA, pp. 1-9, 1998.

[219] Ly. T. Gruyitch, "A physical principle and consistent Lyapunov methodology: time-invariant nonlinear systems", *Proc. International Conference on Advances in Systems, Signals, Control and Computers*, 1, Durban, South Africa, pp. 42-50, 1998.

[220] Ly. T. Gruyitch, "Consistent Lyapunov methodology for exponential stability: PCUP approach", in *Advances in Stability Theory at the End of the 20th Century*, Ed. A. A. Martynyuk, London: Taylor and Francis, pp. 107-120, 2003.

[221] Ly. T. Gruyitch, "Consistent Lyapunov methodology for hopfield neural networks", *Neural, Parallel & Scientific Computations*, Vol. 7, pp. 195-210, 1999.

[222] Ly. T. Gruyitch, "Consistent Lyapunov methodology: global exponential stability of sets", *Proc. International Conference on Stability and Oscillations of Nonlinear Control Systems*, Moscow, pp 5-8, 1998.

[223] Ly. T. Gruyitch, "Consistent Lyapunov methodology: Non-differentiable non-linear systems", *Nonlinear Dynamics and Systems Theory*, Vol. 1, No. 1, pp. 1-22, 2001.

[224] Ly. T. Gruyitch, "Consistent Lyapunov methodology, time-varying nonlinear systems and sets", *Nonlinear Analysis*, Theory and Applications, Vol. 39, pp. 413-446, 2000.

[225] Ly. T. Gruyitch, *Einstein's Relativity Theory*. Correct, Paradoxical, and Wrong, ISBN 1-4122-0211-6, Trafford, Victoria, Canada, http://www.trafford.com/06-2239, 2006.

[226] Ly. T. Gruyitch, "Exponential quality of time-varying dynamical systems: Stability and tracking", in *Advances in Nonlinear Dynamics-Stability and Control: Theory, Methods and Applications*, Eds. S. Sivasundaram and A. A. Martynyuk, Amsterdam: Gordon and Breach Science Publishers, 5, pp. 51-61, 1997.

[227] Ly. T. Gruyitch, "Exponential stabilizing natural tracking control of robots: theory", *Proceedings of the Third ASCE Specialty Conference on Robotics for Challenging Environments*, held in Albuquerque, New Mexico, USA (Eds. Laura A. Demsetz, Raymond H. Bryne and John P. Wetzel), Reston, Virginia, USA: American Society of Civil Engineers (ASCE), pp. 286-292, April 26-30, 1998.

[228] Ly. T. Gruyitch, *Galilean-Newtonean Rebuttal to Einstein's Relativity Theory*, Cambridge: Cambridge International Science Publishing, 2015.

[229] Ly. T. Gruyitch,"Gaussian generalisations of the relativity theory fundaments with applications", *Proceedings of the VII International Conference: Physical Interpretations of Relativity Theory*, Ed. M. C. Duffy, *British Society for the Philosophy of Science*, London, pp. 125-136, September 15-18, 2000.

[230] Ly. T. Gruyitch, *Linear Continuous-time Systems*, CRC Press/Taylor & Francis Group, Boca Raton, Florida, USA, 2017.

[231] Ly. T. Gruyitch, "New development of vector Lyapunov functions and airplane control synthesis", Chapter 7 in *Advances in Dynamics and Control*, editor S. Sivasundaram, Boca Raton: Chapman & Hall/CRC, pp. 89-102, 2004.

[232] Ly. T. Gruyitch, *Nonlinear Systems Tracking*, Boca Raton: CRC press, Florida, USA, 2016.

[233] Ly. T. Gruyitch, *Observability and Controllability of General Linear Systems*, Boca Raton, FL: CRC Press, 2018.

[234] Ly. T. Gruyitch, "Physical continuity and uniqueness principle. Exponential natural tracking control", *Neural, Parallel & Scientific Computations*, 6, pp. 143-170, 1998.

[235] Ly. T. Gruyitch, *Time and Consistent Relativity. Physical and Mathematical Fundamentals*, Apple Academic Press, Waretown, NJ, USA and Oakville, Ontario, Canada, 2015.

[236] Ly. T. Gruyitch, *Time and Time Fields. Modeling, Relativity, and Systems Control,* Trafford, Victoria, Canada, 2007.

[237] Ly. T. Gruyitch, *Time. Fields, Relativity, and Systems*, Llumina, Coral Springs, Florida, USA, 2006.

[238] Ly. T. Gruyitch, "Time and uniform relativity theory fundaments", *Problems of Nonlinear Analysis in Engineering Systems*, 7, N° 2(14), Kazan, Russia, pp. 1-29, 2001.

[239] Ly. T. Gruyitch, "Time, relativity and physical principle: generalizations and applications", *Proc. V International Conference: Physical Interpretations of Relativity Theory* (Ed. M. C. Duffy), pp. 134-170, London, 11-14 September, 1998; (also in: Nelinijni Koluvannya, Vol. 2, No. 4, pp. 465-489, Kiev, Ukraine, 1999).

[240] Ly. T. Gruyitch, "Time, systems, and control: qualitative properties and methods", Chapter 2 in *Stability and Control of Dynamical Systems with Applications*, Editors D. Liu and P. J. Antsaklis, pp. 23-46, Boston: *Birkhâuser*, pp. 23-46, 2003.

[241] Ly. T. Gruyitch, "Time, Systems and Control", invited, submitted and accepted paper has had 50 pages, its abstract was published in *Abstracts of the papers of the VIII International seminar "Stability and oscillations of nonlinear control systems*, editor V. N. Thai, Moscow: IPU RAN, ISBN 5-201-14972-3, June 2-4, 2004.

[242] Ly. T. Gruyitch, TIME. TRUE RELATIVITY. Prepared to be submitted for publication.

[243] Ly. T. Gruyitch, *Trackability and Tracking of General Linear Systems*, Boca Raton, FL: CRC Press, 2018.

[244] Ly. T. Gruyitch, *Tracking Control of Linear Systems*, Boca Raton: CRC Press, Florida, USA, 2013.

[245] Ly. T. Gruyitch, TRUTH ON EINSTEIIN'S RELATIVITY THEORY, Prepared to be submited for publication.

[246] Ly. T. Gruyitch, "Vector Lyapunov function synthesis of aircraft control", *Proceedings of INPAA-98: Second International Conference on Nonlinear Problems in Aviation & Aerospace*, editor: Seenith Sivasundaram, ISBN: 0 9526643 1 3, Cambridge UK: European Conference Publications, Vol. 1, pp. 253-260, 1999.

[247] L. Gruyitch, J.-P. Richard, P. Borne and J.-C. Gentina, *Stability Domains*, Boca Raton: Chapman Hall/CRC, Florida: 2004.

[248] Ly. T. Gruyitch and G. Dauphin-Tanguy, "Strict monotonous asymptotic energetic stability. Part I: Theory", *Preprints 14 IFAC World Congress*, Vol. D, pp. 437-442, 5-9 July 1999.

[249] Ly. T. Gruyitch and A. Kökösy, "Conceptual development of vector Lyapunov functions and control synthesis for mechanical systems", *CD Proceedings of the 17th IMACS World Congress*, paper # T2-I-0572, Paris, pp. 1-8, July 11-15, 2005.

[250] Ly. T. Gruyitch and A. Kökösy, "High Quality Bounded Robust Robot Stabilizing Control", *Proc. IEEE Conference on Decision and Control*, San Diego, pp. 797-802, December 1997.

[251] Ly. T. Gruyitch and A. Kökösy, "Robot control for robust stability with finite reachability time in the whole", *Journal of Robotic Systems*, Vol. 16, No. 5, pp. 263-283, 1999.

[252] R. W. Gunderson, "On stability over a finite interval", *IEEE Transactions on Automatic Control*, Vol. AC-12, No. 5, 634-635, 1967.

[253] S. Guo, and L. Han, *Stability and Control of Nonlinear Time-varying Systems*, Berlin: Springer, 2018.

[254] B.-Z. Guo and Z.-L. Zhao, *Active Disturbance Rejection Control for Nonlinear Systems: An Introduction*, New Jersey: John Wiley & Sons, 2016.

[255] P. Habets and K. Peiffer, "Attractivity concepts and vector Lyapunov functions", *Nonlinear Vibration Problems (Zagadnienia Drgan Neiliniowych)*, pp. 35-52, 1975.

[256] W. M. Haddad and V. S. Chellaboina, *Nonlinear Dynamical Systems and Control: A Lyapunov-Based Approach*, New Jersey: Princeton University Press, 2008.

[257] W. Hahn, *Stability of Motion*, New York: Springer-Verlag, 1967.

[258] O. Hájek, "Compactness and asymptotic stability", *Math. Systems Theory*, Vol. 4, No. 2, pp. 154-159, 1970.

[259] O. Hájek, "Ordinary and asymptotic stability of noncompact sets", *J. Differential Equations*, Vol. 11, No. 1, pp. 49-65, 1972.

[260] T. G. Hallam and V. Komkov, "Application of Liapunov's functions to finite time stability", *Rev. Roumaine de Math. Pure et Appliqués*, Vol. 14, No. 4, pp. 495-501, 1969.

[261] A. Halanay, "Absolute stability of certain non-linear regulating systems with time-lags", *Avtomatika i Telemehanika* (in Russian), Vol. XXV, No. 3, pp. 290-301, 1964.

[262] A. Halanay, *Differential Equations*, New York: Academic Press, 1966.

[263] J. K. Hale, "Dynamical systems and stability", *J. Math. Anal. Appl.*, Vol. 26, pp. 39-59, 1969.

[264] J. K. Hale, *Ordinary Differential Equations*, New York: Wiley-Interscience, 1969.

[265] C. C. Hang and J. A. Chang, "An algorithm for constructing Lyapunov functions based on the variable gradient method", *IEEE Transactions on Automatic Control*, Vol. AC-15, No. 4, 510 - 512, 1970.

[266] K. M. Hangos, J. Bokor, and G. Szederkényi, *Analysis and Control of Nonlinear Process Systems*, Berlin: Springer, 2004.

[267] P. Hartmann, *Ordinary Differential Equations*, New York: John Wiley and Sons, 1964 (Russian edition published in Moscow: Mir, 1970).

[268] L. Hatvani, "On application of differential inequalities to the theory of stability", *Proc. Amer. Math. Soc.*, Vol. 79, No. 1, pp. 59-62, 1975.

[269] J. A. Heinen and S. H. Wu, "Est stability of differential equations", *Int. J. Systems Sci.*, Vol. 1, No. 3, pp. 269-277, 1971.

[270] J. R. Hewit and C. Storey, "Comparison of numerical methods in stability analysis", *Int. J. Control*, Vol. 10, No. 6, pp. 687-701, 1969.

[271] J. R. Hewit and C. Storey, "Numerical application of Szegö's method for constructing Lyapunov functions", *IEEE Transactions on Automatic Control*, Vol, AC-14, No. 1, pp. 106-108, 1969.

[272] J. R. Hewit and C. Storey, "Optimization of the Zubov and Ingwerson methods for constructing Lyapunov functions", *Electron. Letters*, Vol. 3, No. 5, pp. 211-212, 1967.

[273] J. G. Hocking and G. S. Young, *Topology*, Reading: Addison-Wesley Publ. Comp. Inc., 1961.

[274] A. Huaux, "On the construction of Lyapunov functions", *IEEE Transactions on Automatic Control*, Vol, AC-12, No. 4, pp. 465-466, 1967.

[275] D. G. Hull, *Optimal Control Theory for Applications*, New York: Springer Verlag, 2003.

[276] E. S. Ibrahim and Z, V. Rekasius, "A stability criterion for nonlinear feedback systems", *IEEE Transactions on Automatic Control*, Vol, AC-9, No. 2, pp. 151-159, 1964.

[277] M. Ikeda and D. D. Šiljak, "Decentralized stabilization of linear time-varying systems", *IEEE Transactions on Automatic Control*, Vol, AC-25, No. 1, pp. 106-107, 1980.

[278] D. R. Ingwerson, "A modified Lyapunov method for nonlinear stability analysis", *IEEE Transactions on Automatic Control*, Vol, AC 6, pp. 199-210, 1961.

[279] A. Isidori, *Nonlinear Control Systems*, London: Springer-Verlag, 1995.

[280] S. Jafarpour and A. D. Lewis, "Time-Varying Vector Fields and Their Flows", Springer, 2014

[281] H. Jianxun, "On practical stability of discontinuous systems", *Depart. of Computer Sci.*, Xiamen University, P. R. China, Vol. 8, pp. 1-34, 1982.

[282] Lj. Jocić, "Planar regions of attraction", *IEEE Transactions on Automatic Control*, Vol, AC-27, No. 3, pp. 708-710, 1982.

[283] P. M. Julich, "On estimating the region of asymptotic stability using Lyapunov's method", *IEEE Transactions on Automatic Control*, Vol, AC-12, No. 6, pp. 787-788, 1967.

[284] E. I. Jury and V. V. Lee, "Absolute stability of systems with many nonlinearities" (in Russian), *Avtomatika i Telemehanika*, Vol. 26, No. 6, pp. 945-965, 1965.

[285] E. Kamke, "Zur Theorie der Systeme Gewöhnlicher Differential-Gleichungen II", *Acta Mathematica*, Vol. 58, pp. 57-85, 1932.

[286] R. E. Kalman, "Liapunov functions for the problem of Lur'e in automatic control", *Proc. Nat. Acad. Sci.* U.S.A., Vol. 49, No. 2, pp. 201-205, 1963.

[287] R. E. Kalman and J. E. Bertram, "Control system analysis and design via the 'second method' of Lyapunov", part I, *Trans. of ASME: J. Basic Eng.*, Vol. 82, pp. 371-393, 1960.

[288] R. E. Kalman, P. L. Falb and M. A. Arbib, *Topics in Mathematical System Theory*, Mc Graw-Hill Book Company, New York, 1969.

[289] K. A. Karatcharov and A. G. Pilyutik, *Introduction to the technical theory of the motion stability* (in Russian), Fizmatgiz, Moscow, 1962.

[290] A. A. Kayade, "A theorem on contractive stability", *SIAM J. Applic. Math.*, Vol.21, No. 4, pp. 601-604, 1971.

[291] A. A. Kayande and V. Lakshmikantham, "Conditionally invariant sets and vector Lyapunov functions", *J. Math. Anal. and Applications*, Vol. 14, pp. 285-293.

[292] A. A. Kayande and J. S. W. Wong, "Finite time stability and comparison principles", *Proc. Camb. Phil. Soc.*, Vol. 64, pp. 749-756, 1968.

[293] H. Khalil, *Nonlinear Systems*, Upper Saddle River, New Jersey: Prentice Hall, 2002.

[294] H. Khalil, *Nonlinear Control*, Upper Saddle River, New Jersey: Prentice Hall, 2014.

[295] M. M. Khapayev, "On a Liapunov like theorem" (in Russian), *Dokl. AN SSSR*, Vol. 176, No. 6, pp. 1262-1265, 1967.

[296] E. Kinnen and C. S. Chen, "Lyapunov functions derived from auxiliary exact differential equations", *Automatica*, Vol. 4, pp. 195-204, 1968.

[297] D. E. Kirk, *Optimal Control Theory: An Introduction.* Englewood Cliffs, N. J.: Prentice-Hall 1970, Dover republication 2004.

[298] N. E. Kirin, R. A. Nelepin and V. N. Baidaev, "Construction of the attraction region by Zubov's method" (in Russian), *Diff. Uravn.*, Vol. 17, No. 8, pp. 1347-1361, 1981. (English translation: pp. 871-880, 1982).

[299] P. E. Kloeden, "Aggregation-decomposition and equi-ultimate boundedness", *J. Australian Math. Soc.*, Vol. XIX, Series B, Part 2, pp. 249-258, 1975.

[300] H. W. Knobloch and F. Kappel, *Gewöhnliche Differentialgleichungen*, B. G. Teubner, Stuttgart, 1974.

[301] D. E. Koditschek, "Application of a new Lyapunov function to global adaptive attitude tracking," Proc. 27th IEEE Conference on Decision and Control, pp. 63-68, 1988.

[302] A. Kökösy and Lj. T. Grujic, "Control synthesis for robot motion robust exponential stability. Part II: application and simulations", *IFAC-IFIP-IMACS Conference on Control of Industrial Systems,* Belfort, France, Vol. 3, pp. 710-715, May 20-22, 1997.

[303] A. Kökösy and Lj. T. Grujic, "Control synthesis for robot motion robust exponential stability. Part II: application and simulations", *IFAC-IFIP-IMACS Conference on Control of Industrial Systems*, Vol. 3, pp. 710-715, May 20-22, 1997.

[304] A. Kökösy and Lj. T. Grujic, "Robust stabilizing robot control. Part II: application and simulations", *1996 IMACS-IEEE Conference on Computational Engineering in Systems Applications: Symposium on Modelling, Analysis and Simulation*, Lille, France, Vol. 2, pp. 1089-1094.

[305] A. Kökösy and Ly. T. Gruyitch, "Robot control for robust global asymptotic stability of a motion", *Proc. World Congress on Automation*, TSE Press, Albuquerque, ISIAC 051.1-ISIAC 051.6, 1998.

[306] A. Kökösy and Ly. T. Gruyitch, "Robust robot control for global attraction with finite reachability time", *Proc. of the European Control Conference*, pp. 1-6, 1997.

[307] S. Köksal S. and V. Lakshmikantham, "Higher derivatives of Lyapunov functions and cone-valued Lyapunov functions," *Nonlinear Analysis, Theory, Methods & Applications*, Vol. 26, No. 9, pp. 1555-1564, 1996.

[308] J. Komarnik and C. C. Li, "Decision surface estimate of nonlinear system stability domain by Lie series method", *IEEE Transactions on Automatic Control*, Vol, AC-18, No. 56, pp. 666-669, 1972.

[309] V. I. Kosolapov, "To complex systems stability" (in Russian), *Prikl. Mehanika*, Vol. 25, No. 7, pp. 133-136, 1979.

[310] N. N. Krasovskii, *Some Problems of the Theory of Stability of Motion*, in Russian, Moscow: FIZMATGIZ, 1959.

[311] N. N. Krasovskii, *Stability of Motion*, Stanford: Stanford University Press, 1963.

[312] Y. H. Ku and N. N. Puri, "On Liapunov functions of high order non-linear systems". *J. Franklin Inst.*, Vol. 276, No. 5, pp. 349-364, 1963.

[313] B. C. Kuo, *Automatic Control Systems*, Englewood Cliffs: Prentice-Hall, Inc., 1967.

[314] K. Kuratowskii, *Topology*, Vol. 1, London: Academic Press, 1966.

[315] G. S. Ladde, V. Lakshmikantham, and S. Leela, "Conditionally asymptotically invariant sets and perturbed systems", *Annali di Matematica Pura ed Applicata*, Ser. IV, XCIV, pp. 33-40, 1972.

[316] G. S. Ladde and S. Leela, "Analysis of invariant sets", *Annali di Matematica Pura ed Applicata*, Ser. IV, XCIV, pp. 283-289, 1972.

[317] G. S. Ladde and S. Leela, "Global results and asymptotically self-invariant sets", *Rendiconti Della Classe di Scienze Fisiche, Matematiche e Naturali*, Academia Nazionale dei Lincei, Ser. VIII LIV, No. 3, pp. 321-327, 1973.

[318] V. Lakshmikantham, "Vector Lyapunov functions and conditional stability", *J. Mathematical Analysis and Applications*, pp. 368-377, 1975.

[319] V. Lakshmikantham and S. Leela, "Asymptotically self-invariant sets and conditional stability", in: *Differential Equations and Dynamical Systems*, eds. J. K. Hale, J. P. LaSalle, New York: Academic Press, pp. 363-369, 1967.

[320] V. Lakshmikantham and S. Leela, *Differential and Integral Inequalities*, New York: Academic Press, 1969.

[321] V. Lakshmikantham, S. Leela and A. A. Martynyuk, *Stability Analysis of Nonlinear Systems*, Basel: Birkhauser, 2015.

[322] J. P. LaSalle, *The Stability of Dynamical Systems*, Philadelphia, Society for Industrial and Applied Mathematics, 1976.

[323] J. P. LaSalle and S. Lefschetz, *Stability by Liapunov's Direct Method*, New York: Academic Press, 1961.

[324] E. B. Lee and L. Markus, *Foundations of Optimal Control Theory*, London: John Wiley & Sons, 1967.

[325] S. Lefschetz, *Stability of Nonlinear Control Systems*, New York: Academic Press, 1965.

[326] W. Leighton, "On the construction of Lyapunov functions for certain autonomous nonlinear differential equations", in *Contributions to Differential Equations*, ed.: J. P. LaSalle, New York: Wiley, Vol. II, pp. 367-383, 1963.

[327] J. Levine, *Analysis and Control of Nonlinear Systems, A Flatness-based Approach*, Berlin: Springer, 2009.

[328] X. Liao and P. Yu, *Absolute Stability of Nonlinear Control Systems*, Berlin: Springer, 2008.

[329] S. Lipschutz, *Set Theory and Related Topics*, New York: McGraw Hill, 1998.

[330] K. A. Loparo and G. L. Blankenship, "Estimating the domain of attraction of nonlinear feedback systems", *IEEE Transactions on Automatic Control*, Vol, AC-23, No. 4, pp. 602-608, 1978.

[331] A. I. Lur'e, *Certain Non-linear Problems of the Theory of Automatic Control* (in Russian), Moscow: Gostehizdat, 1951.

[332] A. I. Lur'e and V. N. Postnikov, "On the stability theory of control systems" (in Russian), *Prikladnaya Matematika i Mehanika*, Vol. VIII, No. 3, pp. 188-1991, 1954.

[333] A. I. Lurie and E. W. Rozenwasser, "On methods for generating Lyapunov functions in the theory of nonlinear regulating systems", *Proc. 1st IFAC Congress: Theory of Continuous Systems-Special Mathematical Problems*, Moscow: Academy of Sciences USSR, pp. 709-717, 1961.

[334] A. C. J. Luo, *Discontinuous Dynamical Systems on Time-Varying Domains*, Springer, Berlin, 2009.

[335] A. M. Lyapunov, *The General Problem of Stability of Motion* (in Russian), Kharkov Mathematical Society, Kharkov, 1892; in Academician A. M. Lyapunov: "Collected Papers", U.S.S.R. Academy of Science, Moscow, II, pp. 5-263, 1956. French translation: "Problème général de la stabilité du mouvement", *Ann. Fac. Toulouse*, 9, pp. 203-474; also in: *Annals of Mathematics Study*, No. 17, Princeton University Press, 1949. English translation: *Intern. J. Control*, 55, pp. 531-773, 1992; also the book, Taylor and Francis, London, 1992.

[336] S. M. Madani-Esfahani, S. Hui and S. H. Żak, "On the estimation of sliding domains and stability regions of variable structure control systems with bounded controllers", Tampa: *Proc. 28th Conference on Decision and Control*, pp. 793-794, 1989.

[337] A. A. Mailybaev and A. P. Seiranyan, "The stability domains of Hamiltonian systems", *Journal of Applied Mathematics and Mechanics*, Volume 63, No. 4, pp. 545-555, 1999.

[338] I. G. Malkin, *Motion Stability Theory* (in Russian), Moscow: Nauka, 1968.

[339] I. G. Malkin, "On the question of the reciprocal Liapunov's theorem on asymptotic stability of control systems" (in Russian), *Avtom. i Telemeh.*, No. 1, pp. 188-191, 1954.

[340] S. G. Margolis and W. G. Vogt, "Control engineering applications of V. I. Zubov's construction procedure for Liapunov functions", *IEEE Transactions on Automatic Control*, Vol, AC-8, No. 2, pp. 104-113, 1963.

[341] R. Marino and S. Nicosia, "Hamiltonian-type Lyapunov functions", *IEEE Transactions on Automatic Control*, Vol, AC-28, No. 11, pp. 1055-1057, 1983.

[342] A. A. Martynyuk, "Decomposition and aggregation in the systems analysis" (in Russian), *Teoret. i Prikl. Mehanika*, No. 1, pp. 87-93, 1975.

[343] A. A. Martynyuk, "Lyapunov matrix-function and stability theory", *Proc. IMACS-IFAC Symp.*, IDN, Villeneuve d'Ascq, France, pp. 261-265, June 3-6, 1986.

[344] A. A. Martynyuk, "On a Liapunov like theorem on stability of a multidimensional system" (in Russian), *Ukr. Mat. Sb.*, Vol. 24, No. 4, pp. 532-537, 1972.

[345] A. A. Martynyuk, "On development of the Liapunov functions in the theory of stability of complex systems" (in Russian), *Prikl. Mehanika*, Vol. 15, No. 10, pp. 3-23, 1979.

[346] A. A. Martynyuk, "On instability of the equilibrium position of a multidimensional system composed of 'neutrally stable subsystems' ", *Prikl. Mehanika*, Vol. 8, No. 6, pp. 77-82, 1972.

[347] A. A. Martynyuk, "On matrix Lyapunov functions and stability of motion" (in Russian), *Proc. Acad. Sci. USSR*, Vol. 280, No. 5, pp. 1062-1066, 1985.

[348] A. A. Martynyuk, "Practical stability and stabilization of control processes" (in Russian), *Applied Mechanics*, Vol. XVII, No. 10, pp. 3-19, 1981.

[349] A. A. Martynyuk, *Stability of Motion of Complex Systems*, Kiev: Naukova dumka, 1975.

[350] A. A. Martynyuk, "The Lyapunov matrix function", *Nonlinear Analysis, Theory and Applications*, Vol. 8, No. 10, pp. 1223-1226, 1984.

[351] A. A. Martynyuk, *Qualitative Analysis of Set-Valued Differential Equations*, Birkhäuser,2019

[352] A. A. Martynyuk and R. Gutowski, *Integral Inequalities and Stability of Motion* (in Russian), Kiev: Naukova dumka, 1979.

[353] A. A. Martynyuk and V. I. Kosolapov, "The comparison principle and the averaging method in the problem of stability of not-asymptotically stable motions under persistent action of disturbances" (in Russian), Preprint: *AN Ukr. SSR*, Inst. Matem., No. 33, pp. 1-24, 1978.

[354] A. A. Martynyuk and N. V. Nikitina, "On nonlinear systems of comparison in the problem of stability of large-scale systems" (in Russian), *Prikl. Mehanika*, Vol. 17, No. 12, pp. 97-102, 1981.

[355] J. L. Massera, "On Liapunov's conditions of stability", *Ann. of Math.*, Vol. 50, pp. 705-721, 1949.

[356] J. L. Massera, "Contributions to stability theory", *Ann. of Math.*, Vol. 64, pp. 182-206, 1956.

[357] V. M. Matrosov, "Comparison Principle and vector Lyapunov functions I" (in Russian), *Diff. Uravn.*, Vol. IV, No. 8, pp. 1374-1386, 1968.

[358] V. M. Matrosov, "Comparison Principle and vector Lyapunov functions II" (in Russian), *Diff. Uravn.*, Vol. IV, No. 10, pp. 1739-1752, 1968.

[359] V. M. Matrosov, "Comparison Principle and vector Lyapunov functions I,II", *Diff. Eqs.*, Academic Press, New York, Vol. 6, pp. 209-242, 1968.

[360] V. M. Matrosov, "On the theory of stability of motion" (in Russian), *Prikl. Matem. i Mehanika*, Vol. 26, pp. 92-100, 1962.

[361] V. M. Matrosov, "On the theory of stability of motion, II" (in Russian), *Trudi Kazan. Aviac. Inst.*, Vol. 80, pp. 22-33, 1963.

[362] V. M. Matrosov, "The method of vector Lyapunov functions in feedback systems" (in Russian), *Automatika i Telemehanika*, No. 9, pp. 63-75, 1972.

[363] V. M. Matrosov, "To the theory of stability of motion" (in Russian), *Prikl. Math. Mekh.*, Vol. 26, No. 5, pp. 885-895, 1962.

[364] V. M. Matrosov, "Vector Lyapunov functions in the analysis of nonlinear interconnected systems", *Proc. Symp. Mathematica*, Bologna, Vol. 6, pp. 209-242, 1971.

[365] V. M. Matrosov, *Vector Lyapunov Function Method: Analysis of Dynamical Properties of Nonlinear Systems* (in Russian), Moscow: FIZMATLIT, 2001.

[366] V. I. Matyukhin, "Motional stability of manipulator robots in a decomposition mode" (in Russian), *Avtomatika i Telemekhanika*, No. 3, pp. 33-44, March 1989.

[367] V. I. Matyukhin, "Stability of the motions of manipulators under persistent perturbations" (in Russian), *Avtomatika i Telemekhanika*, No. 11, pp. 124-134, November 1993.

[368] V. I. Matyukhin, "Strong stability of motions of mechanical systems" (in Russian), *Avtomatika i Telemekhanika*, Vol. 57, No. 1, pp. 28-44, January 1996.

[369] E. J. McShane, *Integration*, Princeton: Princeton University Press, 1944.

[370] K. R. Meyer, "Liapunov functions for the problem of Lur'e", *Proc. Nat. Acad. Sci.*, U.S.A., Vol. 53, pp. 501-503, 1965.

[371] A. N. Michel, "On the bounds of the trajectories of differential systems", *Int. J. Control*, Vol. 10, No. 5, pp. 593-600, 1969.

[372] A. N. Michel, "Quantitative analysis of simple and interconnected systems: stability, boundedness, and trajectory behavior", *IEEE Trans. on Circuit Thoery*, Vol. CT-17, No. 3, pp. 292-301, 1970.

[373] A. N. Michel, "Stability analysis of interconnected systems", *Berichte der Mathematisch - Statistischen Sektion*, Graz, No. 4, pp. 1-107, 1973.

[374] A. N. Michel, "Stability analysis of interconnected systems", *J. SIAM Control*, Vol. 12, No. 3, pp. 554-579, 1974.

[375] A. N. Michel, "Stability, transient behavior and trajectory bounds of interconnected systems", *Int. J. Control*, Vol. 11, No. 4, pp. 703-715, 1970.

[376] A. N. Michel and Lj. T. Grujic, "Mathematical modeling of continuous-time neural networks under pure structural variations", *Proc. 13th World Congress on Computation and Applied Mathematics*, IMACS' 91, Dublin, Ireland, Vol. 3, pp. 1254-1255, 1991.

[377] A. N. Michel and Lj. T. Grujic, "Qualitative analysis of continuous-time neural networks under pure structural variations" Ibid, pp. 1260-1261, 1991.

[378] A. N. Michel and C. J. Herget, *Algebra and Analysis for Engineers and Scientists*, Boston: Birkhäuser, 2007.

[379] A. N. Michel, L. Hou and D. Liu, *Stability of Dynamical Systems. Continuous, Discontinuous, and Discrete Systems*, Boston: Birkhäuser, 2008.

[380] A. N. Michel, L. Hou and D. Liu, *Stability of Dynamical Systems. On the Role of Monotonic and Non-Monotonic Lyapunov Functions*, Boston: Birkhäuser, 2015.

[381] A. N. Michel and R. K. Miller, *Qualitative Analysis of Large-Scale Dynamical Systems*, New York: Academic Press, 1977.

[382] A. N. Michel, R. K. Miller, and B. H. Nam, "Stability analysis of interconnected systems using computer generated Lyapunov functions", *IEEE Transactions on Circuits and Systems*, Vol, CAS-29, No. 7, pp. 431-440, 1982.

[383] A. N. Michel, B. H. Nam, and V. Vittal, "Computer generated Lyapunov functions for interconnected systems: improved results with application to power systems", *IEEE Transactions on Circuits and Systems*, Vol, CAS-31, No. 2, pp. 189-198, 1984.

[384] A. N. Michel and D. W. Porter, "Practical stability and finite-time stability of discontinuous systems", *IEEE Transactions on Circuits and Systems*, Vol, CT-19, No. 2, pp. 123-129, 1972.

[385] A. N. Michel and D. W. Porter, "Stability analysis of composites systems", *IEEE Trans. Aut. Control*, Vol, AC-17, No. 2, pp. 222-226, 1970.

[386] A. N. Michel, N. R. Sarabudla and R. K. Miller, "Stability analysis of complex dynamical systems", *Circuits, Systems, Signal Processing*, Vol. 1, No. 2, pp. 171-202, 1982.

[387] A. N. Michel, K. Wang and B. Hu, *Qualitative Theory of Dynamical Systems: The Role of Stability Preserving Mappings*", Second Edition, New York: Marcel Dekker, 2001.

[388] R. K. Miller and A. N. Michel, *Ordinary Differential Equations*, New York: Academic Press, 1982; Mineola, New York: Dover Publications, Inc., 2007.

[389] B. R. Milojković and Lj. T. Grujić, *Automatic Control*, the textbook in Serbo-Croatian, Belgrade: Faculty of Mechanical Engineering, University of Belgrade, 1977.

[390] B. R. Milojković and Lj. T. Grujić, *Automatic Control*, the textbook in Serbo-Croatian, Belgrade: Faculty of Mechanical Engineering, University of Belgrade, 1981.

[391] D. Mitra and H. C. So, "Existence conditions for L_1 Lyapunov functions for a class of nonautonomous systems", *IEEE Transactions on Circuit Theory*, Vol. CT-19, No. 6, pp. 594-598, 1972.

[392] H. Miyagi and T. Taniguchi, "Construction of Lyapunov function for power systems", *Proc. IEEE*, Vol. 124, No. 12, pp. 1197-1202, 1977.

[393] N. D. Moiseyev, *Essays on Development of the Theory of Stability* (in Russian), Moscow-Leningrad: GITTL, 1949.

[394] J. B. Moore and B. D. O. Anderson, "A generalization of the Popov criterion", *J. Franklin Institute*, Vol. 285, No. 6, pp. 488-492, 1968.

[395] M. Morari, G. Stephanopoulos and R. Aris, "Finite stability regions for large-scale systems with stable and unstable subsystems", *Int. J. Control*, Vol. 28, No. 5, pp. 805-815, 1977.

[396] T. Nagaraja and V. V. Chalam, "Generation of Lyapunov function: a new approach", *Int. J. Control*, Vol. 19, No. 4, pp. 781-787, 1974.

[397] K. S. Narendra and J. H. Taylor, *Frequency Domain Criteria for Absolute Stability*, New York: Academic Press, 1973.

[398] H. Nijmeijer and Arjan van der Schaft, *Nonlinear Dynamical Control Systems*, New York: Springer, 2016.

[399] R. M. Nasyirov, "Stability over a finite time interval in case of two zero roots" (in Russian), *Izv. Vyssh. Uchebn. Zav. (Matematika)*, Vol. 57, No. 2, pp. 48-56, 1967.

[400] V. N. Naumov and Y. Z. Tsypkin, "Frequency criteria for absolute stability of processes in nonlinear automatic control systems" (in Russian), *Avtomatika i Telemehanika*, Vol. XXV, No. 6, pp. 852-867, 1964.

[401] R. A. Nelepin, "On the problem of exact boundary of the region of absolute stability of control systems" (in Russian), *Avtomatika i Telemehanika*, No. 4, pp. 30-37, 1967.

[402] V. V. Nemytskii and V. V. Stepanov, *Qualitative Theory of Differential Equations*, Princeton: Princeton University Press, 1960.

[403] I. Newton, *Mathematical Principles of Natural Philosophy: Book I. The Motion of Bodies*, William Benton, Publisher, ENCYCLOPAEDIA BRITANNICA, Inc., Chicago (first publication: 1687) 1952.

[404] H. Nijmeijer and A. van der Scahft, *Nonlinear Dynamical Control Systems*, New York: Springer-Verlag, 1991.

[405] H. Nijmeijer and A. J. van der Schaft, *Nonlinear Dynamical Control Systems*, 3rd edition, New York: Springer Verlag, 1996.

[406] E. Noldus, A. Galle and L. Josson, "The computation of stability regions for systems with many singular points", *Int. J. Control.*, Vol. 17, No. 3, pp. 644-652, 1973.

[407] K. Ogata, *State Space Analysis of Control Systems*, Englewood Cliffs: Prentice Hall, 1967.

[408] K. Ogata, *Modern Control Engineering*, Englewood Cliffs: Prentice Hall, 1970.

[409] A. S. Oziraner and V. V. Rumyantsev, "Method of Liapunov functions in the problem of stability of motion concerning a part of variables" (in Russian), *Matematika i Mehanika*, Vol. 36, No. 2, pp. 364-384, 1972.

[410] M. A. Pai and C. L. Narayana, "Finite regions of attraction of multilinear systems and its application to the power system stability problem", *IEEE tranc. Automat. Contr.*, Vol. 21, No. 5, pp. 716-721, 1976.

[411] P. C. Parks and A. J. Pritchard, "On the construction and use of Lyapunov functionals", *Proc. 4th IFAC Congress*, pp. 59-76, 1969.

[412] S. Partovi and N. E. Nahi, "Absolute stability of dynamic systems containing non-linear functions of several state variables", *Automatica*, Vol. 5, No. 4, pp. 465-473, 1969.

[413] K. Peiffer and N. Rouche, "Liapunov's second method applied to partial stability", *J. de Mécanique*, Vol. 8, No. 2, pp. 323-334, 1969.

[414] K. P. Persidskii, "On stability of motion at first approximation" (in Russian), *Matematicheski Sbornik*, Vol. 40, pp. 284-293, 1933.

[415] S. K. Persidskii, "On the problem of absolute stability" (in Russian), *Avtomatika i Telemehanika*, No. 12, pp. 5-11, 1969.

[416] E. S. Piatnitskii, "Absolute stability of nonstationary nonlinear systems" (in Russian), *Avtomatika i Telemehanika*, No. 1, pp. 5-15, 1970.

[417] E. S. Piatnitskii, "New investigations on absolute stability of automatic control systems" (in Russian), Survey, *Avtomatika i Telemehanika*, No. 6, pp. 5-36, 1968.

[418] E. S. Piatnitskii, "On existence of absolutely stable systems which do not obey the criterion by Popov" (in Russian), *Avtomatika i Telemehanika*, No. 1, pp. 30-37, 1973.

[419] W. Peruquetti, *Sur la stabilité et l'estimation des comportements non linéaires non stationnaires perturbés*, Doctoral Dissertation, University of Lille 1, France, No. 1286, 25 February 1994.

[420] W. Peruquetti and J.-P. Richard, "Connecting Wazewski's conditions with M-matrices: application to constrained stabilization", *Dynamic Systems and Applications*, Vol. 5, No. 1, pp. 81-96, 1996.

[421] W. Peruquetti, J.-P. Richard and P. Borne, "Vector Lyapunov functions: recent developments for stability, robustness, practical stability and constrained control", *Nonlinear Times and Digest*, Vol. 2, pp. 227-258, 1995.

[422] W. Peruquetti, J.-P. Richard, Lj. T. Grujić and P. Borne, "On practical stability with the settling time via vector norms", *Int. J. of Control*, Vol. 62, No. 1, pp. 173-189, 1995.

[423] A. A. Piontkovskii and L. D. Rutkovkaya, "Investigation of certain stability theory problems by vector Lyapunov function method" (in Russian), *Avtomatika i Telemehanika*, No. 10, pp. 23-31, 1967.

[424] H. Poincaré, "Sur le courbes définies par une équation différentielle, *Journal de Mathématiques*, série 3, No. 7, pp. 375-422, 1881-1882.

[425] J. P. Ponzo, "On the stability of certain nonlinear differential equations", *IEEE tranc. Automat. Contr.*, Vol. 10, No. 5, pp. 470-472, 1965.

[426] L. S. Pontryagin, *Ordinary Differential Equations* (in Russian), Moscow: Nauka, 1970.

[427] V. M. Popov, "Dichotomy and stability by frequency domain method", *Proceedings of IEEE*, Vol. 62, No. 5, pp. 548-562, 1974.

[428] V. M. Popov, "Criterii fr stabilitate pentru sistemele neliniare de reglare automata, bazate pe utilizarea transformatei Laplace", *Studii si cerretari de energetica*, Acad. R. P. R., Vol. IX, No. 4, pp. 647-680, 1959.

[429] V. M. Popov, *Hyperstability of Control Systems*, New York: Springer Verlag, 1973.

[430] V. M. Popov, "Noi criterii de stabilitate petru sistemele automate neliniare", *Studii si cerretari de energetica*, Acad. R. P. R., Vol. X, No. 3, pp. 73-88, 1960.

[431] V. M. Popov, "On absolute stability of nonlinear automatic control systems" (in Russian), *Avtomatika i Telemehanika*, Vol. XXII, No. 8, 961-979, 1961.

[432] V. M. Popov, "On a critical case of the absolute stability" (in Russian), *Avtomatika i Telemehanika,* Vol. XXIII, No. 1, 3-24, 1962.

[433] V. M. Popov, "On an absolute stability theory problem of control systems" (in Russian), *Avtomatika i Telemehanika*, Vol. XXV, No. 9, 1257-1262, 1964.

[434] V. M. Popov, "Solution of a new stability problem of control systems" (in Russian), *Avtomatika i Telemehanika*, Vol. XXIV, No. 1, 7-28, 1964.

[435] D. W. Porter and A. N. Michel, "Stability of composite systems", *Proc. of 4th Asilomar Conf. on Circuits, Systems and Computers*, pp. 634-638, 1970.

[436] D. W. Porter and A. N. Michel, "Stability of composite systems with nonlinear interconnections", *Proc. of 1971 Midwest Symp.*, pp. 6.6-1-6.6-10, 1971.

[437] E. S. Pyatnitsky, "To problem of control black box of mechanical nature," *Problems of Nonlinear Analysis in Engineering Systems*, Kazan, Russia, Vol. 2, No. 4, pp. 1-10, 1996.ons Method and Applications, J. C. Baltzer AG, Sc. Pub. Co., Vol. 8, Sect, IV, pp. 81-88, 1990.

[438] R. Reiss and G. Geiss, "The construction of Lyapunov functions", *IEEE Transactions on Automatic Control*, Vol. AC-8, No. 4, pp. 382-383, May 1963.

[439] M. Rezaiee-Pajand and B. Moghaddasie, " Determination of stability domains for nonlinear dynamical systems using the weighted residuals method", *Civil Engineering Infrastructures Journal*, 46(1), pp. 27–50, June 2013.

[440] M. Ribbens-Pavella, "Critical survey of transient stability studies of multimachine power systems by Lyapunov's direct method", *Proc. 9th Allerton Conf. Circ. and Syst. Theory*, pp. 151-167, 1971.

[441] J. P. Richard and P. Borne, "State-space modelling and transformations", *Systems & Control Encyclopedia*, Ed. M. G. Singh, Pergamon Press, pp. 4540-4547, 1987.

[442] J. P. Richard, P. Borne and J. C. Gentina, "Estimation of stability domains by use of vector norms", *Information and Decision Technologies*, North-Holland, Vol. 14, pp. 241-251, 1988.

[443] F. Robert, "Normes vectorielles de vecteurs et de matrices", *R.F.T.I. - Chiffres*, Vol. 17, No. 4, pp. 261-299, 1964.

[444] J. J. Rodden, "Numerical applications of Lyapunov stability theory", *Poc. JACC*, Session IX, Paper 2, pp. 261-268, 1972.

[445] H. H. Rosenbrock, "A Lyapunov function with applications to some nonlinear physical systems", *Automatica*, Vol. 1, pp. 31-53, 1962.

[446] H. H. Rosenbrock, "A Lyapunov function for some naturally-occurring linear homogeneous time-dependent equations", *Automatica*, Vol. 2, pp. 97-109, 1963.

[447] H. H. Rosenbrock, *State-space and Multivariable Theory*, London: Thomas Nelson and Sons Ltd., 1970.

[448] N. Rouche and J. Mawhin, *Equations Différentielles Ordinaires*, Masson Cie Editeurs, Paris, 1973.

[449] N. Rouche, P. Habets and M. Laloy, *Stability Theory by Liapunov's Direct Method*, New York: Springer-Verlag, 1977.

[450] V. P. Rudakov, "Solution estimate and stability of pseudolinear systems over a finite time interval" (in Russian), *Differential Equations*, NAU, Vol. 5, No. 8, pp. 1385-1389, 1969.

[451] V. V. Rumyantsev, "To stability of motion with respect to a part of variables" (in Russian), *Vestn. Mosk.. un-ta, Ser. Matem i Mehanika*, No. 4, pp. 9-16, 1957.

[452] M. Saeki, M. Araki and B. Kondo, "Local stability of composite systems-frequency-domain condition and estimate of the domain of attraction", *IEEE Trans. Autom. Control*, Vol. AC-25, No. 5, pp. 936-940, 1980.

[453] N. R. Sandell, P. Varaiya, M. Athans and M. G. Safonov, "Survey of decentralized control methods for large scale systems", *IEEE Trans. Autom. Control*, Vol. AC-23, No. 2, pp. 108-128, 1978.

[454] E. Sarti, "Approximate determination of the stability domain for nonlinear systems", Warszawa: *Proc. 4th IFAC World Congress*, Technical Session 20, pp. 43-58, June 16-21, 1969.

[455] S. Sastry, *Nonlinear Systems. Analysis, Stability, and Control*, New York: Springer-Verlag, 1999.

[456] D. G. Schultz, "The generation of Lyapunov functions, in *Advances in Control Systems*, ed. C. Leondes, New York: Academic Press, Vol. 2, pp. 1-64, 1965.

[457] D. G. Schultz and J. E. Gibson, "The variable method for generating Lyapunov functions", *IEEE Trans. Appl. Ind.*, Vol. 81, pp. 203-210, 1962.

[458] D. G. Shields and C. Storey, "The behavior of optimal Lyapunov functions", *Int. J. Control*, Vol. 21, No. 4, pp. 561-573, 1975.

[459] D. D. Šiljak, "Competitive economic systems: stability, decomposition and aggregation", *IEEE Conf. Dec. and Control*, pp. 265-275, 1973.

[460] D. D. Šiljak, *Large-Scale Dynamic Systems: Stability and Structure*, New York: North Holland, 1978.

[461] D. D. Šiljak, *Nonlinear Systems*, New York: John Wiley and Sons, 1969.

[462] D. D. Šiljak, "On large-scale systems stability", *Proc. 9th Ann. Allerton Conf on Circ. and Syst. Theory*, University of Illinois, pp. 731-741, 1971.

[463] D. D. Šiljak, "On pure structure of dynamic systems", *Nonlinear Analysis; Theory, Methods and Applications*, Vol 1, pp. 397-413, 1977.

[464] D. D. Šiljak, "On Stability of the arms race", Report No. NGR: 05-017-010-7508, University of Santa Clara, *NSF Conf. on Control Theory in Intern. Relations Research*, Indiana University, pp. 1-44, 1975.

[465] D. D. Šiljak, "On Stability of large-scale systems under structural perturbations", *IEEE Trans. SMC*, Vol. SMC-3, pp. 415-417, 1973.

[466] D. D. Šiljak, "On the stability of the arms race", Chapter 9 in: *Mathematical Systems in International Relations Research*, Eds. J. V. Gillespie and D. A. Zinnes, New York: Praeger Publ., 1977.

[467] D. D. Šiljak, *Stability of Control Systems* (in Serbo-Croatian), Belgrade: Faculty of Electrical Engineering, 1974.

[468] D. D. Šiljak, "Stability of large-scale systems", *Proc. 1972 IFAC 5th World Congress*, ISA, Pittsburgh, Paper C-32, pp. 1-11, 1972.

[469] D. D. Šiljak, "Stability of large-scale systems under structural perturbations", *IEEE Trans. SMC*, Vol. SMC-2, No. 2, pp. 657-663, 1972.

[470] D. D. Šiljak, "When is a complex ecosystem stable?" *Mathematical Bioscience*, Vol. 25, pp. 25-50, 1975.

[471] D. D. Šiljak and K. C. Sun, "On exponential absolute stability", *Int. J. Control*, Vol. 16, 1003-1018, 1972.

[472] D. D. Šiljak and M. B. Vukčević, "On hierarchic stabilization of large-scale linear systems", *8th Asilomar Conf. on Circuits, Systems and Computers*, Pacific Grove, pp. 503-507, 1974.

[473] D. D. Šiljak and S. Weisenberger, "Regions of exponential stability for the problem of Lur'e", *Regelungstechnik*, Vol. 17, No. 1, pp. 27-29, 1969.

[474] D. D. Šiljak and S. Weisenberger, "Regions of exponential stability for the problem of Lur'e", *Regelungstechnik und Prozess Datenverarbeitung*, Vol. 18, No. 2, pp. 69-71, 1970.

[475] A. S. C. Sinha, "Liapunov functions for a class of large-scale systems", *IEEE Trans. Aut. Control*, Vol. AC-25, No. 3, pp. 558-560, 1980.

[476] J.-J. Slotine and W. Li, *Applied Nonlinear Control*, New York: Springer, 1999.

[477] Y.- D. Song, *Control of Nonlinear Systems via PI, PD and PID: Stability and Performance*, Boca Raton, FL: CRC Press, 2018.

[478] M. Spivak, *Calculus on Manifolds*, New York: W. A. Benjamin, Inc. , 1965.

[479] Y. Stepanenko and J. Yuan, "Robust adaptive control of a class of nonlinear mechanical systems with unbounded and fast-varying uncertainties," *Automatica*, Vol. 28, No. 2, pp. 265-276, 1992.

[480] D. F. Stewart and E. T. Wall, "A matrix transformation for the direct generation of Liapunov state equations", *ActaTechnica ČSAV*, No. 6, pp. 667-682, 1969.

[481] G. P. Szegö, "New methods for constructing Lyapunov functions for time invariant control systems", *Proc. 2nd IFAC World Congress*, London: Butterworths, pp. 584-589, 1964.

[482] G. P. Szegö, "On a new partial differential equation for the stability analysis of time invariant control systems", *J. SIAM Control*, Ser. A, Vol. 1, No. 1, pp. 63-75, 1962.

[483] G. P. Szegö and G. R. Geiss, "A remark on 'A new partial differential equation for the stability analysis of time invariant control systems'", *J. SIAM Control*, Ser. A, Vol. 1, No. 3, pp. 369-376, 1963.

[484] C. Sy-In, "On stability over a finite time interval" (in Russian), *Applied Mathematics and Mechanics*, Vol. XXIII, pp. 230-238, 1959.

[485] S. Tarbouriech and C. Burgat, "Positively invariant sets for constrained continuous-time systems with cone properties", *IEEE Trans. Autom. Control*, Vol. 39, No. 2, pp. 401-405, 1994.

[486] N. Tesla, "On neutrons speed and the light speed". Citation on internet: http://images.search.yahoo.com/search/images?_adv_prop=image&fr=yfp-t-900-s&va=tesla+about+einstein#index=frame

[487] J. Texter, "Numerical algorithm for implementing Zubov's construction in two-dimensional systems", *IEEE Trans. Autom. Control*, Vol. AC-20, No. 1, pp. 62-63, 1974.

[488] W. E. Thompson, "Exponential stability of interconnected systems", *IEEE Transactions on Automatic Control*, Vol. AC-17, No. 2, pp. 222-226, 1972.

[489] W. E. Thompson and H. E. Koening, "Stability of a class of interconnected systems", *Int. J. Control*, Vol. 15, No. 4, pp. 751-763, 1972.

[490] H. Tokumaru and N. Saito, "On the absolute stability of automatic control systems with many nonlinear characteristics", *Memoire of the Faculty of Engineering*, Kyoto University, Vol. 27, No. 3, pp. 347-379, 1965.

[491] I. Troch, "The evaluation of the domain of attraction of nonlinear control systems with hybrid computing systems", Paris: *Proc. 5th IFAC World Congress*, Paper No. 32.1, pp. 1-5, 1972.

[492] J. Tsiniaas and J. Karafyllis, "ISS property for time-varying systems and application to partial-static feedback stabilization and asymptotic tracking," *IEEE Trans. Automatic Control*, Vol. 44, No. 11, pp. 2179-2184, November 1999.

[493] S. Vaidyanathan and C. Volos, Editors, *Advances and Applications in Nonlinear Control Systems*, Berlin: Springer, 2016.

[494] G. S. Vakhonina, A. S. Zemlyakov and V. M. Matrosov, "On techniques for construction of Lyapunov vector functions for linear systems" (in Russian), *Avtomatika i Telemehanika*, No. 2, pp. 5-16, 1973.

[495] A. Vanelli and M. Vidyasagar, "Maximal Lyapunov functions and domains of attraction for autonomous nonlinear systems", *Automatica*, Vol. 21, No. 1, pp. 69-80, 1985.

[496] M. Vidyasagar, "Decomposition techniques for large-scale systems with nonadditive interactions: stability and stabilizability", *IEEE Transactions on Automatic Control*, Vol. AC-25, No. 4, pp. 773-778, 1980.

[497] M. Vidyasagar, *Nonlinear Systems Analysis*, Englewood Cliffs, New Jersey: Prentice-Hall, 1978.

[498] E. T. Wall, "A synthesis of Lyapunov functions for non-linear time-varying control systems", *Int.J. Sys. Science*, Vol. 4., No. 4, pp. 565-575, 1973.

[499] E. T. Wall, "A topological approach to the generation of Lyapunov functions", *Acta Technica SCAV*, Vol. 2, pp. 159-177, 1968.

[500] E. T. Wall, "The generation of Lyapunov functions in control theory by an energy matric algorithm", *Proc. 1968 J.A.C.C.*, pp. 172-179, 1968.

[501] E. T. Wall and M. L. Moe, "An energy metric algorithm for the generation of Lyapunov functions", *IEEE Transactions on Automatic Control*, Vol. AC-13, pp. 121-122, 1968.

[502] E. T. Wall and M. L. Moe, "Generation of Lyapunov functions for time-varying nonlinear systems", *IEEE Transactions on Automatic Control*, Vol. AC-14, p. 211, 1969.

[503] J. A. Walker and H. H. McClamroch, "Finite regions of attraction for the problem of Lur'e", *Int. J. Control*, Vol. 6, No. 4, pp. 331-336, 1967.

[504] J. W. Watson and A. R. Sturberud, "Stability of systems operating in a finite time interval", *IEEE Transactions on Automatic Control*, Vol. AC-12, No. 1 p. 116, 1967 (see also: "Correction to 'Stability of Systems Operating in a Finite Time Interval'", *IEEE Transactions on Automatic Control*, Vol. AC-12, No. 5 p. 635, 1967.

[505] T. Ważewski, "Systèmes des équations et des inégalités différentielles ordinaires aux seconds membres monotones et leurs applications", *Ann. Soc. Polon. Math.*, Vol. 23, pp. 112-166, 1950.

[506] L. Weiss, "Converse theorems for finite time stability", *Proc. 1st Asilomar Conf. on Circuits and Systems*, pp. 1006-1041, 1967.

[507] L. Weiss and E. F. Infante, "Finite time stability under perturbing forces and on product spaaces", *IEEE Transactions on Automatic Control*, Vol. AC-12, No. 1 pp. 54-59, 1967.

[508] L. Weiss and E. F. Infante, "On the stability of systems defined over a finite time interval", *Proc. Nat. Ac. Sci. USA: Mathematics*, Vol. 54, No. 1, pp. 44-48, 1965.

[509] S. Weissenberger, "Application of results from the absolute stability problem to the computation of finite stability domains", *IEEE Transactions on Automatic Control*, Vol. AC-13, No. 1 pp. 124-125, 1968.

[510] S. Weissenberger, "Comments on 'Application of results from the absolute stability problem to the computation of finite stability domains'", *IEEE Transactions on Automatic Control*, Vol. AC-14, No. 1 p. 109, 1969.

[511] S. Weissenberger, "Control synthesis for stability region improvement", *Int. J. Control*, Vol. 12, No. 3, pp. 489-496, 1970.

[512] S. Weissenberger, "Design for stability region maximization", *IEEE Transactions on Automatic Control*, Vol. AC-14, No. 3 pp. 314-315, 1969.

[513] S. Weissenberger, "Piecewise-quadratic and piecewise-linear Lyapunov functions for discontinuous systems", *Int. J. Control*, Vol. 10, No. 2, pp. 171-180, 1969.

[514] S. Weissenberger, "Stability-boundary approximations for relay-control systems via steepest-ascent construction of Lyapunov functions", *Trans. of ASME – Journal of Basic Eng.*, Vol. 88, No. 2, pp. 419-428, June 1966.

[515] S. Weissenberger, "Stability regions of large-scale systems", *Automatica*, Vol. 9, No. 5, pp. 653-663, 1973.

[516] S. Wiggins, *Introduction to Applied Nonlinear Dynamical Systems and Chaos*, New York: Springer, 2003, 2008, 2018.

[517] J. L. Willems, "The computation of finite stability regions by means of open Liapunov surfaces", *Int. J. Control*, Vol. 10, No. 5, pp. 537-544, 1969.

[518] N. van de Wouw and E. Lefeber, *Nonlinear Systems: Techniques for Dynamical Analysis and Control*, Berlin: Springer, 2016.

[519] G. Wright, "Drawing stability domains in MATLAB", 31st October 2018, presented in public at: https://math.boisestate.edu/~wright/courses/m567/StabilityDomains/html/Stability Domains.html

[520] J. T. Xing, *Energy Flow Theory of Nonlinear Dynamical Systems with Applications (Emergence, Complexity and Computation)*, New York: Springer, 2015.

[521] V. A. Yakubovich, "Absolute instability of non-linear control systems - I: General frequency criteria" (in Russian), *Avtomatika i Telemehanika*, Vol. XXIV, No. 3, pp. 5-14, 1970.

[522] V. A. Yakubovich, "Absolute stability of non-linear regulating systems in critical cases - I" (in Russian), *Avtomatika i Telemehanika*, Vol. XXXI, No. 12, pp. 293-303, 1963.

[523] V. A. Yakubovich, "Absolute stability of non-linear regulating systems in critical cases - II" (in Russian), *Avtomatika i Telemehanika*, Vol. XXIV, No. 6, pp. 717-731, 1963.

[524] V. A. Yakubovich, "Absolute stability of non-linear regulating systems in critical cases - III" (in Russian), *Avtomatika i Telemehan*ika, Vol. XXV, No. 5, pp. 601-612, 1964.

[525] V. A. Yakubovich, "Frequency conditions for absolute stability of regulating systems with hysteresis nonlinearities" (in Russian), *Dokl. A. N. SSSR*, Vol. 149, No. 26, pp. 288-291, 1963.

[526] V. A. Yakubovich, "Frequency conditions for absolute stability of control systems with with several non-linear or linear non-stationary blocks" (in Russian), *Avtomatika i Telemehanika*, Vol. XXVIII, No. 6, pp. 5-30, 1967.

[527] V. A. Yakubovich, "Frequency conditions for oscillations in non-linear regulating systems with with single one-one or hysteresis nonlinearity" (in Russian), *Avtomatika i Telemehanika*, Vol. XXXVI, pp. 51-644, 1975.

[528] V. A. Yakubovich, "Method of matrix inequalities in the stability theory of nonlinear regulating systems - I" (in Russian), *Avtomatika i Telemehani*ka, Vol. XXV, No. 7, pp. 1017-1029, 1964.

[529] V. A. Yakubovich, "Method of matrix inequalities in the stability theory of nonlinear regulating systems - III" (in Russian), *Avtomatika i Telemehanika*, Vol. XXVI, No. 4, pp. 577-590, 1965.

[530] V. A. Yakubovich, "Resolution of certain matrix inequalities appearing in the theory of automatic control" (in Russian), *Dokl. A. N. SSSR*, Vol. 143, No. 6, pp. 1304-1307, 1962.

[531] T. Yoshizawa, "Asymptotic behaviour of solutions of ordinary differential equations near sets", *Proc. Int. Symp. on Nonlinear Oscillations*, Kiev, Ukraine, Vol. 1, pp. 213-225, 1963.

[532] T. Yoshizawa, "Eventual properties and quasi-asymptotic stability of a non-compact set", *Funkcialaj Ekvacioj*, Vol. 8, No. 2, pp. 25-90, 1966.

[533] T. Yoshizawa, "Some notes on stability of sets and perturbed system", *Funkcialaj Ekvacio*j, Vol. 6, No. 1, pp. 1-11, 1964.

[534] T. Yoshizawa, *Stability Theory by Lyapunov's Second Method*, Tokyo: The Mathematical Society of Japan, 1966.

[535] V. F. Zadorozhny, "Asymptotic stability analysis of many-dimensional systems by means of the method of vector Liapunov functions" (in Russian), *Kibernetika i Vychislitel'naya Tehnika*, No. 1, pp. 92-97, 1969.

[536] V. F. Zadorozhny and A. A. Martynyuk, "Estimation of influence of connections between sub-systems on stability of a linear non-stationary system" (in Russian), *Prikl. Mehanika*, Vol. 18, No. 9, pp. 65-71, 1972.

[537] V. F. Zadorozhny and A. A. Martynyuk, "On a general problem of aggregation solution as the problem of moments" (in Russian), *Matematicheskaya Fizika*, No. 14, pp. 49-54, 1973.

[538] V. F. Zadorozhny and A. A. Martynyuk, "On solution of the general problem of aggregation solution as the problem of moments" (in Russian), *Matematicheskaya Fizika*, No. 14, pp. 54-55, 1973.

[539] V. F. Zadorozhny and A. A. Martynyuk, "The direct Liapunov method and L-problem of moments in the problems on stability of many-dimensional systems" (in Russian), in: *Problems of Analytical Mechanics, Theories of Stability and Control*, Moscow: Nauka, pp. 151-154, 1975.

[540] A. S. Zemlyakov, "On a problem of the comparison system construction" (in Russian), *T. Kazan. Aviats. in-ta*, No. 144, pp. 46-54, 1972.

[541] V. I. Zubov, *Methods of A. M. Liapunov and Their Applications* (in Russian), Leningrad: Leningrad Gos. University, (English translation: Groningen: P. Noordhoff Ltd., 1964).

Part IX

INDEX

Subject Index